Nonlinear Dynamic in Engineering by Akbari-Ganji's Method

Nonlinear Dynamic in Engineering by Akbari-Ganji's Method

Alireza Ahmadi
Mohammadreza Akbari
Davood Domairry Ganji

Library of Congress Control Number:		2015913937
ISBN:	Hardcover	978-1-5144-0169-9
	Softcover	978-1-5144-0170-5
	eBook	978-1-5144-0171-2

Print information available on the last page.

Rev. date: 11/06/2015

To order additional copies of this book, contact:
Xlibris
1-888-795-4274
www.Xlibris.com
Orders@Xlibris.com
706724

Contents

Preface

Being nonlinear is the challenge of most users dealing with differential equations in mathematical and engineering sciences. In fact, so far, there has not been any especial method to solve various forms of nonlinear differential equations by an easy approach in different fields of study. It is necessary to mention that in recent years, researchers have proposed some semianalytical methods for solving a few groups of nonlinear differential equations, but none of them can be useful for a broad range of nonlinear equations. To understand more, it is better to indicate that some of the existing methods for solving nonlinear equations are HAM, HPM, VIM, DTM, ADM, and so on—in which each method is well applied only for a small category of nonlinear differential equations. Furthermore, in some cases, application of every method has really made significant errors in calculations. Therefore, the aforementioned methods cannot apply for the whole nonlinear problems, and each of them can just answer certain problems. To explain more, the long and complex process is the other weak point of these methods.

In this book, a convenient approach is introduced for all the different kinds of nonlinear differential equations and also for various sets of nonlinear equations, which can analytically solve all the differential equations very easily and obtain the final solution of every differential equation as an algebraic function. This method is named Akbari-Ganji's method (AGM) and is aimed to solve the whole nonlinear differential equations algebraically by choosing an answer function for a differential equation with constant coefficients, which can be computed by applying initial or boundary conditions in a particular manner.

Eventually, the authors are grateful to the ancient Persian mathematician, astronomer, and geographer Muḥammad ibn Mūsā al-Khwārizmī, who—in his *Compendious Book on Calculation by Completion and Balancing*—was the one that presented the first systematic solution of linear and quadratic equations. Furthermore, he has been considered as the original inventor of algebra, and Europeans have derived the term "algebra" from his book. And also, the expression "algorithm" has been taken from his name (the Latin form of his name).

Chapter 1

Nonlinear Differential Equations

1.1. A Brief Explanation to Nonlinear Differential Equations

This book has been published in order to solve nonlinear differential equations governing the procedures of engineering and basic sciences. Definitely, solving nonlinear differential equations governing physical phenomena can be considered as the root of most researches and studies in the applied sciences. And the obtained results have been utilized in order to design, implement, and apply in industry. It is necessary to mention that the behavior of most phenomena in engineering and basic sciences is revealed as nonlinearity,[1, 2] so to comprehend the relationships among the components of each phenomenon, having sufficient mathematical knowledge is very essential. Up to here, a big step has been taken in understanding the relations among the elements of physical phenomena. In the next step, it is important to solve nonlinear differential equations. Since every differential equation consists of a series of independent and dependent variables and also has some parameters with respect to the physical properties of the system, it is better to solve differential equations analytically in terms of their parameters.[3] It is clear that researchers can easily solve these kinds of equations numerically,[4, 5] but there are some weak points in the numerical methods. For example, it is obligatory in a numerical solution to determine the values of existing parameters in differential equations at the beginning of the solution procedure, but the answer of differential equations in analytic methods is always achieved with the existing parameters in their own differential equation. As a result, different answers can be gained by choosing various amounts for these parameters.

1.2. The Application of Nonlinear Differential Equations in Different Sciences

Obviously, the existent analytic methods have not been comprehensive up to now, and they are not suitable for all the nonlinear differential equations,[6–13] and they do not give us the exact answers while comparing them with numerical methods. The results show that the entities of the differential equations governing the systems are different from each other, and to remove this difficulty, we divide all the differential equations governing the system (including engineering and basic sciences) into the following three cases and then solve them by AGM. These analytic methods are useful for a special class of differential equations in a certain domain and just give reliable solutions in this domain.

Analytic solution of nonlinear differential equations by AGM is divided into the following three classes in terms of whether the reaction of the system is harmonic or nonharmonic:

1. Solving equations and set of second- and higher-order nonlinear differential equations for nonharmonic systems (systems whose answers are not a function of time)

These systems are mainly applied in mechanical engineering (heat and fluids), chemistry, civil engineering, and nanomaterials.[14–20]

2. Solving equations and set of first-order nonlinear equations for harmonic and nonharmonic systems (systems whose answers are a function of time)

It is necessary to mention that the basic application of these systems is in electrical engineering (electrical circuits and control), chemistry (kinetics of chemical reactions), and gas dynamics.

3. Solving equations and set of second-order nonlinear differential equations for harmonic systems well-known to vibrational equations.[21–30]

These systems are considerably used when their answers are alternative and vibrational, such as the existent problems of solid mechanics, civil engineering (earthquake), and so on.

1.3. Analytic Methods

1.3.1. The First-Order Nonlinear Differential Equations

The first-order nonlinear differential equations can be considered as the following general form:

$$f(t) = f(u'(t), g(u(t))) = 0. \tag{1.1}$$

And the relevant initial condition is expressed as

$$u(0) = u_0. \tag{1.2}$$

In this step, a compound series consisting of exponential and trigonometric terms has been selected as a suitable answer for this kind of nonlinear differential equation as follows:

$$u(t) = \sum_{k=0}^{n} \{a_k e^{-b_k t} + d_k \cos(\Omega t + \varphi_k)\} = \{a_0 e^{-b_0 t} + a_1 e^{-b_1 t} + \ldots + d_0 \cos(\Omega t + \varphi_0) + d_1 \cos(\Omega t + \varphi_1) + \ldots\}. \tag{1.3}$$

It is citable that equation 1.3 is applicable for solving all the first-order nonlinear differential equations.

Moreover, for analyzing electric circuits in electrical engineering, we can utilize the above relation.

To understand more, it is better to indicate that to facilitate the computational operations in AGM, if the physical system is not vibrational, such as differential equations governing the heat transfer systems (radiation), chemical reactions, or nonvibrational differential equations, then it is logical to omit the trigonometric term of equation 1.3. Therefore, equation 1.3 can be rewritten as follows:

$$u(t) = \sum_{k=0}^{n} \{a_k e^{-b_k t}\} = a_0 e^{-b_0 t} + a_1 e^{-b_1 t} + \ldots \tag{1.4}$$

As an example, some kinds of the first-order nonlinear differential equations governing the chemical, heating, and electric systems have been presented as follows:

$$\frac{d\theta}{d\tau} + (\theta - \theta_a) + \varepsilon(\theta^4 - \theta_s^4) = 0 \quad \rightarrow \text{ for heating systems} \tag{1.5}$$

$$\frac{dc}{dt} = k_1 c_A^2 - k_2 c_B + k_1 k_2 c_A c_B \quad \rightarrow \text{ for chemical reactions} \tag{1.6}$$

$$(R+k\,i)\frac{di}{dt}+\frac{di}{dt}+\frac{i}{c}=v_0\,\omega_0\cos(\omega_0\,t) \qquad \rightarrow \qquad \text{for electrical circuits} \tag{1.7}$$

In the aforementioned equations, equation 1.3 can be a suitable choice for the answer function. Based on the given explanations, the constant coefficients of equation 1.3, $\{a_0, b_0, d_0, \varphi_0$ to $a_n, b_n, d_n, \varphi_n\}$, can be obtained by applying the given initial condition.

1.3.2. The Second- and Higher-Order Differential Equations (Nonvibrational)

Boundary conditions and initial conditions are required for analytic methods of each linear and nonlinear differential equation according to the physic of the problem. Therefore, we can solve every differential equation with any degrees. In order to comprehend the given method, reading the following lines is recommended.

In accordance with the boundary conditions, the general manner of a differential equation is as follows:

The nonlinear differential equation p (which is a function of u), the parameter u (which is a function of x), and their derivatives are considered as follows:

$$p_k : f\left(u, u', u'', \ldots, u^{(m)}\right)=0 \quad ; \quad u=u\left(x\right). \tag{1.8}$$

Boundary conditions :

$$\begin{cases} u\left(0\right)=u_0\ , u'\left(0\right)=u_1\ , \ldots, u^{(m-1)}\left(0\right)=u_{m-1} \\ u\left(L\right)=u_{L_0}\ , u'\left(L\right)=u_{L_1}\ , \ldots, u^{(m-1)}\left(L\right)=u_{L_{m-1}} \end{cases} \tag{1.9}$$

To solve the first differential equation, with respect to the boundary conditions in $x=L$ in equation 1.9, the series of letters in the nth order with constant coefficients, which is the answer of the first differential equation, is considered as follows:

$$u(x)=\lim_{n\to\infty}\sum_{i=0}^{n}a_i x^i=\lim_{n\to\infty}(a_0+a_1x^1+a_2x^2+...+a_nx^n). \qquad (1.10)$$

The more precise the answer of equation 1.8, the more choices of series sentences from equation 1.10. It means by increasing the number of series sentences in equation 1.10, the obtained solution is the real answer, or the exact solution of the problem. In applied problems, approximately five or six sentences from the series are enough to solve nonlinear differential equations. In the answer of differential equation 1.10 regarding the series from the degree (n), there are ($n+1$) unknown coefficients that need ($n+1$) equations to be specified. The boundary conditions of equation 1.9 are used to solve a set of equations that consists of ($n+1$) ones.

The boundary conditions are applied to the functions as follows:

a) The application of the boundary conditions for the answer of differential equation 1.10 is in the form of

If $x=0$

$$\begin{cases} u(0)=a_0=u_0 \\ u'(0)=a_1=u_1 \\ u''(0)=a_2=u_2 \\ \vdots \quad \vdots \quad \vdots \end{cases} \qquad (1.11)$$

And when $x=\mathrm{L}$

$$\begin{cases} u(L)=a_0+a_1L+a_2L^2+...+a_nL^n=u_{L_0} \\ u'(L)=a_1+2a_2L+3a_3L^2+...+na_nL^{n-1}=u_{L_1} \\ u''(L)=2a_2+6a_3L+12a_4L^2+...+n(n-1)a_nL^{n-2}=u_{L_{m-1}} \\ \vdots \quad \vdots \quad \vdots \quad \vdots \quad \vdots \quad \vdots \end{cases} \qquad (1.12)$$

b) After substituting equation 1.12 into equation 1.8, the application of the boundary conditions on differential equation 1.8 is done according to the following procedure:

$$\begin{aligned}
&p_0 : f\left(u(0), u'(0), u''(0), \ldots, u^{(m)}(0)\right)\\
&p_1 : f\left(u(L), u'(L), u''(L), \ldots, u^{(m)}(L)\right). \qquad (1.13)\\
&\vdots \quad \vdots \quad \vdots \quad \vdots \quad \vdots
\end{aligned}$$

With regard to the choice of n; ($n<m$) sentences from equation 1.10 and in order to make a set of equations that consist of *(n+1)* equations and *(n+1)* unknowns, we confront with a number of additional unknowns that are indeed the same coefficients of equation 1.10. Therefore, to remove this problem, we should derive m times from equation 1.8, in accordance with the additional unknowns in the aforementioned set of differential equations, and then this is the time to apply the boundary conditions of equation 1.9 on them.

$$\begin{aligned}
&p'_k : f'\left(u', u'', u''', \ldots, u^{(m+1)}\right)\\
&p''_k : f''\left(u'', u''', u^{(IV)}, \ldots, u^{(m+2)}\right) \qquad (1.14)\\
&\vdots \quad \vdots \quad \vdots \quad \vdots \quad \vdots
\end{aligned}$$

c) Application of the boundary conditions on the derivatives of the differential equation P_k in equation 1.14 is done in the form of

$$p'_k : \begin{cases} f'\left(u'(0), u''(0), u'''(0), \ldots, u^{(m+1)}(0)\right) \\ f'\left(u'(L), u''(L), u'''(L), \ldots, u^{(m+1)}(L)\right) \end{cases} \qquad (1.15)$$

$$p''_k : \begin{cases} f''\left(u''(0), u'''(0), \ldots, u^{(m+2)}(0)\right) \\ f''\left(u''(L), u'''(L), \ldots, u^{(m+2)}(L)\right) \end{cases} \qquad (1.16)$$

The $(n+1)$ equations can be made from equation 1.11 to equation 1.16 so that $(n+1)$ unknown coefficients of equation 1.10 for example, $a_0, a_1, a_2 \ldots a_n$ can be computed. The answer of the nonlinear differential equation 1.8 will be gained by determining coefficients of equation 1.10.

1.3.3. Vibrational Differential Equations

In general, vibrational equations and their initial conditions are defined for different systems as follows:

$$f\left(\ddot{u},\dot{u},u,F_0\sin(\omega_0 t)\right)=0 \tag{1.17}$$

$$\{u(0)=A \quad , \quad \dot{u}(0)=0\} \tag{1.18}$$

1.3.3.1. Choosing the Answer Function of Vibrational Differential Equations by AGM

In AGM, a total answer with constant coefficients is required in order to solve differential equations in various fields of study such as vibrations, structures, fluids, and heat transfer. In vibrational systems, with respect to the kind of vibration, it is necessary to choose the mentioned answer in AGM. To clarify here, we divide vibrational systems into two general forms:

1.3.3.1.a. Vibrational Systems without Any External Force

Differential equations governing this kind of vibrational systems are introduced in the following form:

$$f(\ddot{u},\dot{u},u)=0. \tag{1.19}$$

Now, the answer of this kind of vibrational system is chosen as

$$u(t)=e^{-bt}\{A\cos(\omega t)+B\sin(\omega t)\}. \tag{1.20}$$

According to trigonometric relationships, equation 1.20 is rewritten as follows:

$$u(t)=e^{-bt}\{a\cos(\omega t+\varphi)\}. \tag{1.21}$$

It is notable that in the above equation, $a=\sqrt{A^2+B^2}$ and $\varphi=\arctan(\frac{B}{A})$.

Sometimes, for increasing the precision of the considered answer of equation 1.19, we are able to add another term in the form of cosine by inspiration of Fourier cosine series expansion as

$$u(t)=e^{-bt}\{a\cos(\omega t+\varphi)+d\cos(2\omega t+\varphi)\}. \tag{1.22}$$

In the above equation, we are able to omit the term (e^{-bt}) to facilitate the computational operations in AGM if the system is considered without any damping components.

Generally speaking in AGM, equation 1.21 or equation 1.22 is assumed as the answer of the vibrational differential equation 1.19 where constant coefficients—which are a, b, d, ω (angular frequency) and φ (initial vibrational phase)—can easily be obtained by applying the given initial conditions in equation 1.18. And also, the above procedure will completely be explained through the presented example in the subsequent part.

It is noteworthy that if there is no damping component in the vibrational system, the constant coefficient b in equation 1.21 and equation 1.22 will automatically be computed as zero in AGM solution procedure.

On the contrary, the parameter b in equation 1.21 and equation 1.22, for the other kind of vibrational system with damping component, is obtained as a nonzero parameter in AGM.

1.3.3.1.a.1. Computation of Other Damping Vibrational Parameters

The maximum amplitude of vibrational displacement, velocity, and acceleration equations can be computed as follows:

On the basis of the given explanations in the aforementioned part, the answer of the introduced differential equation is considered as

$$u(t)=be^{-at}\{\cos(\omega t+\varphi)\}. \tag{1.23}$$

The curve of locus for the maximum points of vibrational amplitudes can be achieved from equation 1.23 when $\cos(\omega t+\varphi)=\pm 1$; therefore, we will have

$$u_{\max}(t)=b\,e^{-at}. \qquad (1.24)$$

In the above equation, the constant coefficients *a* and *b* can be obtained on the basis of the given explanations. By substituting $t_k=\frac{k\pi}{\omega}$ into equation 1.24, in which $k\in\{1,\ 2,\ 3,...\}$, it is possible to gain the maximum amplitudes such as $u_{1,\max}$, $u_{2,\max}$ and *etc*. According to the given physical values, the maximum amplitudes can be obtained as follows:

$$u_{k,\max}=b.e^{-a(\frac{k\pi}{\omega})}. \qquad (1.25)$$

Therefore, we will have

$$k=1 \rightarrow t_1=\frac{\pi}{\omega} \quad \text{then} \quad u_{1,\max}=b.e^{-a(\frac{\pi}{\omega})} \qquad (1.26)$$

$$k=2 \rightarrow t_2=\frac{2\pi}{\omega} \quad \text{then} \quad u_{1,\max}=b.e^{-a(\frac{2\pi}{\omega})} \qquad (1.27)$$

$$\vdots \qquad \vdots \qquad \vdots$$

1.3.3.1.a.2. Damping Factor

The damping factor in the vibrational solution of $b\{e^{-at}.\cos(\omega \mathrm{t}+\varphi)\}$ is e^{-at}, since in the vibrational differential equation, the term $e^{-\xi\omega t}$ is the main factor for damping. So ξ can be obtained as follows:

$$e^{-at} \equiv e^{-\xi \omega t} \quad \rightarrow \quad \xi = \frac{a}{\omega}. \tag{1.28}$$

In equation 1.28, the parameters a and ω have been gained in accordance with the given physical values, and the initial vibrational phase can be obtained as follows:

$$\varphi = tg^{-1}(\xi). \tag{1.29}$$

1.3.3.1.a.3. Lost Energy per Cycle

Energy that is dissipated in each cycle because of the existence of damping force (F_d) can be obtained in the following procedure.

It is better to mention that the damping force ($F_d = c\dot{u}$) in every vibrational differential equation can be achieved by multiplying the damping factor (c) by the derivative of displacement (the vibrational velocity $\dot{u}$) as follows:

$$F_d = c\dot{u}. \tag{1.30}$$

In order to compute F_d, we should calculate the derivative of the considered answer function, which is equation 1.21, and substitute it instead of $\dot{u}$ into equation 1.30 as follows:

$$F_d = -cbe^{-at}\{\cos(\omega t + \varphi) + \omega \sin(\omega t + \varphi)\}. \tag{1.31}$$

To evaluate the dissipated energy per cycle, it is necessary to rewrite the obtained F_d from equation 1.31 in terms of equation 1.21, which is our considered answer function, and then substitute the periods per cycle ($t = \frac{2k\pi}{\omega}$, $k = 1, 2, \ldots$) instead of t in the following form:

$$F_d = -c\{au \pm \omega\sqrt{b^2 e^{-\frac{4ak\pi}{\omega}} - u^2}\}. \tag{1.32}$$

If the differential equation governing the vibrational system is linear, the obtained chart for F_d in terms of u will usually be circular, and also, it will be elliptical for nonlinear differential equations. It is citable that the enclosed part of the achieved chart is the lost energy in the considered cycle.

Based on the above explanations, the lost energy in each cycle can be acquired as

$$w_d = \oint F_d\, du \;\rightarrow\; w_d = \oint c\dot{u}^2\, dt \;\rightarrow\; w_d = \int_0^{\frac{2k\pi}{\omega}} c\dot{u}^2\, dt \quad , \quad k = 1,2,.... \tag{1.33}$$

1.3.3.1.b. Vibrational Systems with External Force

In this step, it is assumed that the external forces exerting on the vibrational systems are defined as

$$F(t) = F_0 \sin(\omega_0 t). \tag{1.34}$$

As a result, the differential equation governing the vibrational system is expressed like equation 1.17 as follows:

$$f\left(\ddot{u}, \dot{u}, u, F_0 \sin(\omega_0 t)\right) = 0. \tag{1.35}$$

The answer of the above equation is introduced as the sum of the particular solution (u_p) and the harmonic solution (u_h) as follows:

$$u_h(t) = e^{-bt}\{A\cos(\omega t) + B\sin(\omega t)\} \quad ; \quad u_p(t) = M\cos(\omega_0 t) + N\sin(\omega_0 t) \tag{1.36}$$

Then

$$u(t) = u_p + u_h \tag{1.37}$$

By utilizing trigonometric relationships and substituting the yielded equations into equation 1.37, the desired answer will be obtained in the form of

$$u(t)=e^{-bt}\{a\cos(\omega t+\varphi)\}+d\cos(\omega_0 t+\phi). \tag{1.38}$$

In order to increase the precision of the achieved equation, we are able to add another term in the form of cosine by inspiration of Fourier cosine series expansion as follows:

$$u(t)=e^{-bt}\{a.\cos(\omega t+\varphi)+c.\cos(2\omega t+\varphi)\}+d.\cos(\omega_0 t+\phi). \tag{1.39}$$

Finally, the exact solution of all the vibrational differential equations can be obtained in accordance with the following equation:

$$u(t)=e^{-bt}\{\sum_{k=1}^{\infty}a_k\cos(k\omega t+\varphi_k)\}+d\cos(\omega_0 t+\phi). \tag{1.40}$$

To deeply understand the above procedure, reading the following lines is recommended.

Since the constant coefficient (b) in vibrational systems without damping components is always obtained as zero, we can substitute the term $(b.t)$ instead of (e^{-bt}) into equation 1.39 to decrease computational operations in the following form:

$$u(t)=(b.t)+a.\cos(\omega t+\varphi)+d.\cos(2\omega t+\varphi). \tag{1.41}$$

Based on the above explanations, by applying initial conditions on a system without damping component, the value of parameter (b) is always zero for equations 39–41. Therefore, the role of parameter (b) in equations 39–41, each of which can be considered as the answer of the vibrational problems, is individually considered as a catalyst for increasing the precision of the assumed answer. After applying initial conditions on the vibrational system in both states (with external force and without external force) by AGM, the value of parameter (b) is computed as zero because the mentioned system has a free vibration without any damping component.

Again, we mention that in order to decrease computational operations for systems without damping components and since we know that (b) in the term (e^{-bt}) is zero, (e^{-bt}) can be omitted from equation 1.39. Consequently, equation 1.39, which can be considered as the answer of the systems without any damping component, is rewritten as follows:

$$u(t)=a.\cos(\omega t+\varphi)+d.\cos(\omega_0 t+\phi). \tag{1.42}$$

The constant coefficients of equation 1.39, which are $\{a,\ b,\ \omega,\ \varphi,\ c, d,\ \phi\}$, will easily be computed in AGM by applying the initial conditions of equation 1.18.

1.3.3.2. Application of Initial Conditions to Compute Constant Coefficients and Angular Frequency by AGM

In AGM, the application of initial conditions of equation 1.18 is done in the two following forms:

a) Applying the Initial Conditions on the Answer of Differential Equation

In regard to the kind of vibrational system (with external force and without external force), which was completely discussed in the previous part of this case study, a function is chosen as the answer of the differential equation from equation 1.21 or equation 1.22 for the systems without external forces and from equation 1.39 or equation 1.40 for the defined systems with external forces, and then the initial conditions are applied on the selected function as follows:

$$u(t)=u(IC). \tag{1.43}$$

It is notable that IC is the abbreviation of introduced initial conditions of equation 1.18.

b) Applying the Initial Conditions on the Main Differential Equation and Its Derivatives

After choosing a function as the answer of differential equation according to the kind of vibrational system, this is the best time to substitute the mentioned answer into the main differential equation instead of its dependent variable (u).

Assume a general equation governing the vibrational system such as equation 1.17 with the time-independent parameter (t) and dependent function (u) as

$$f\left(\ddot{u}, \dot{u}, u, F_0 \sin(\omega_0 t)\right) = 0. \tag{1.44}$$

Therefore, on the basis of the kind of vibrational system, a function as the answer of the differential equation, such as equation 1.21 or equation 1.22 and equation 1.39 or equation 1.40, is considered as follows:

$$u = g(t) \tag{1.45}$$

In this step, the aforementioned equation is substituted into equation 1.44 instead of (u) in the following form:

$$f(t) = f\left(g''(t), g'(t), g(t), F_0 \sin(\omega_0 t)\right). \tag{1.46}$$

Eventually, the application of initial conditions on equation 1.46 and its derivatives is expressed as

$$f(IC) = f(g''(IC), g'(IC), g(IC), \ldots) \;,\; f'(IC) = f'(g''(IC), g'(IC), g(IC), \ldots) \;,$$
$$f''(IC) = f''(g''(IC), g'(IC), g(IC), \ldots) \;,\; \ldots\ldots \quad . \tag{1.47}$$

To sum up, it is better to say that in AGM, after applying the initial conditions on equations 1.45,1.46 and also on its derivatives from equation

1.47, according to the order of differential equation and utilizing the two given initial conditions of equation 1.18, a set of algebraic equations consisting of n equations with n unknowns is created. Therefore, the constant coefficients (a, b, c, d, angular frequency "ω"and initial phase) are easily achieved. This procedure will thoroughly be explained in the form of an example in the foregoing part of this paper.

It is noteworthy that in equation 1.47, we are able to use the derivatives of $f(t)$ with higher orders until the number of yielded equations will be equal to the number of the mentioned constant coefficients of the assumed answer.

1.4. Applied Commands in AGM by Maple Software

1.4.1. Analyzing the Second- and Higher-Order Differential Equations (Nonvibrational)

In order to solve nonlinear differential equations by AGM, the instructions of Maple software is expressed in the following steps:

They should exactly be written in the medium of the software. It is important to mention that writing the rows of these steps for Maple instructions should be observed.

1.4.1.1. The First Step

The Command Restart

> *restart;*

In order to write any program in the Maple software medium, the previous memory should be erased so that the new program can be started to work. Consider the following nonlinear differential equation as an example for an engineering system. The solution steps of this equation by AGM can be presented as follows:

Solve the following nonlinear differential equation by AGM.

$$\frac{d^2u(x)}{dx^2}+\alpha u^2(x)+\beta u^3(x)+\gamma_1 u(x)=0 \tag{1.48}$$

And the boundary conditions are expressed as

$$u(0)=1 \quad , \quad u(1)=0. \tag{1.49}$$

1.4.1.2. The Second Step

By choosing an answer function as a finite power series with the following constant coefficients in the Maple medium, we have

$$> u := \sum_{i=0}^{5} a_i \cdot x^i ;$$

The outcome of the program is as follows:

$$u := a_0 + a_1 x + a_2 x^2 + a_3 x^3 + a_4 x^4 + a_5 x^5 .$$

It is important to mention that in order to increase the precision of the answer of differential equation, the number of the series sentences of the answer function can be considered more (for example, $\sum_{i=0}^{7}$ or $\sum_{i=0}^{9}$,…). Thus, the computational operations will be increased significantly.

It is notable that if we want the answer to be observed in the Maple operational outcome, a semicolon (;) should be written at the end of any command; and if we do not want the answer to be seen in the Maple operational outcome, a colon (:) should be written at the end of any command.

1.4.1.3. The Third Step

Writing the Differential Equation in the Maple Medium

The parameter u in the given differential equation is presented as a function of X , and for simplicity in the Maple medium, this differential equation is written without X , which means u is written instead of $u(x)$, and also, a letter such as f is considered at the beginning of the differential equation in the Maple medium as follows:

$$> \mathrm{f} := \frac{d^2}{dx^2} u + \alpha \cdot u^2 + \beta \cdot u^3 + \gamma l\, u = 0;$$

In this step, the considered answer function is substituted into the differential equation automatically, and the outcome is gained as follows:

$$f := 2a_2 + 6a_3 x + 12a_4 x^2 + 20a_5 x^3 + \alpha\left(a_0 + a_1 x + a_2 x^2 + a_3 x^3 + a_4 x^4 + a_5 x^5\right)^2 + \beta\left(a_0 + a_1 x + a_2 x^2 + a_3 x^3 + a_4 x^4 + a_5 x^5\right)^3 + \gamma l\left(a_0 + a_1 x + a_2 x^2 + a_3 x^3 + a_4 x^4 + a_5 x^5\right) = 0$$

1.4.1.4. The Fourth Step

Using the command "unapply"

This command converts the considered answer function and the given differential equation into functions of the parameter x or any other independent parameter (i.e., x and f can be obtained as $u(x)$ and $f(x)$, respectively).

This command is used for applying boundary conditions in the answer function and the differential equation.

$$> u := unapply(u, x);\ f := unapply(f, x);$$

If we write colon (:) at the end of the above commands, the outcome of Maple is not observed, but while putting semicolon (;), the results can be shown as follows:

$$u := x \rightarrow a_0 + a_1 x + a_2 x^2 + a_3 x^3 + a_4 x^4 + a_5 x^5$$

$$f := x \rightarrow 2 a_2 + 6 a_3 x + 12 a_4 x^2 + 20 a_5 x^3 + \alpha \left(a_0 + a_1 x + a_2 x^2 + a_3 x^3 + a_4 x^4 + a_5 x^5\right)^2 + \beta \left(a_0 + a_1 x + a_2 x^2 + a_3 x^3 + a_4 x^4 + a_5 x^5\right)^3 + \gamma l \left(a_0 + a_1 x + a_2 x^2 + a_3 x^3 + a_4 x^4 + a_5 x^5\right) = 0$$

As it has been explained, the command "unapply" can change any equation into a function, wherein for this example in the considered answer function, u is changed into $u(x)$, and the differential equation (f) is converted into $f(x)$, whose main usage is in applying the boundary or initial conditions on these two functions as the following general equations:

$$\text{u(BC or IC)} \quad ; \quad \text{f(u(BC or IC))}. \tag{1.50}$$

(BC) and (IC) are the abbreviations of applying boundary and initial conditions, respectively, on the answer function and also on the differential equation.

1.4.1.5. The Fifth Step

Applying the Boundary Conditions

By substituting the answer function u into the differential equation f, which is the result of doing the second and third steps, the boundary conditions are applied on the answer function (the finite series) and also on the differential equation as follows.

With regard to the given boundary conditions shown by $x = 1$ and $x = 0$, they are applied in the following forms. Of course, it is necessary to mention that the left-hand side of applying the boundary conditions is considered as an arbitrary parameter such as S_1, S_2, ..., S in the following form:

1. Applying the boundary conditions on the considered answer function

> $s1 := u(0) = 1;$ $s2 := u(1) = 0;$

$$s1 := a_0 = 1$$

$$s2 := a_0 + a_1 + a_2 + a_3 + a_4 + a_5 = 0$$

2. Applying the boundary conditions on the differential equation

> $s3 := f(0); s4 := f(1);$

$$s3 := 2\,a_2 + \alpha\,a_0^2 + \beta\,a_0^3 + \gamma l\,a_0 = 0$$

$$\begin{aligned} s4 := {} & 2\,a_2 + 6\,a_3 + 12\,a_4 + 20\,a_5 + \alpha\,(a_0 + a_1 \\ & + a_2 + a_3 + a_4 + a_5)^2 + \beta\,(a_0 + a_1 + a_2 \\ & + a_3 + a_4 + a_5)^3 + \gamma l\,(a_0 + a_1 + a_2 + a_3 \\ & + a_4 + a_5) = 0 \end{aligned}$$

3. Applying the boundary conditions on the derivative of differential equation

The given differential equation in this step is automatically derived by the Maple software, and then we apply the boundary conditions as follows:

$> s5 := f'(0);\ s6 := f'(1);$

$$s5 := 6\,a_3 + 2\,\alpha\,a_0\,a_1 + 3\,\beta\,a_0^2\,a_1 + \gamma l\,a_1 = 0$$

$$s6 := 6\,a_3 + 24\,a_4 + 60\,a_5 + 2\,\alpha\,(a_0 + a_1 + a_2 + a_3 + a_4 + a_5)\,(a_1 + 2\,a_2 + 3\,a_3 + 4\,a_4 + 5\,a_5) + 3\,\beta\,(a_0 + a_1 + a_2 + a_3 + a_4 + a_5)^2\,(a_1 + 2\,a_2 + 3\,a_3 + 4\,a_4 + 5\,a_5) + \gamma l\,(a_1 + 2\,a_2 + 3\,a_3 + 4\,a_4 + 5\,a_5) = 0$$

It is citable that applying the boundary conditions on the considered answer function, on the given differential equation, and also on its higher-order derivatives $(s_1\ ,s_2\ ,\ \dots\ ,\ s_n)$ is continued until the number of the resulting algebraic equations is equal to the number of the constant coefficients of the answer function $(a_0\ ,a_1\ ,\ \dots\ ,\ a_{n-1})$. So these constants can easily be computed by solving (n) equations with (n) unknowns by Maple software.

In this example, the answer function is $u = \sum_{i=0}^{5} a_i\, x^i$, which consisted of six sentences with six constant coefficients (a_0 to a_5), in which the same number of algebraic equations are needed to be made from applying the boundary conditions $(s_1\ ,s_2\ ,\ \dots\ ,\ s_n)$. If the more number of sentences from the answer function was chosen, the boundary conditions would be applied on the second- and higher-order derivatives of the given differential equation, so the numbers of the constant coefficients and the resulting equations from applying the boundary conditions would be equaled.

1.4.1.6. The Sixth Step

The Command "solve"

By using this command, we can obtain the constant coefficients of the answer function (a_0 to a_5), which resulted from applying the boundary conditions $(s_1, s_2, \ldots, s_n)$ as a set of algebraic equations composed of six equations with six unknowns as follows:

$$> S := solve\left([s1, s2, s3, s4, s5, s6], \{a_0, a_1, a_2, a_3, a_4, a_5\}\right);$$

$$S := \Big\{ a_0 = 1, a_1 = -\frac{3\left(240 - 84\alpha - 84\beta - 104\gamma l + 3\alpha\gamma l + 3\beta\gamma l + 3\gamma l^2\right)}{720 - 72\gamma l - 72\alpha - 108\beta + \gamma l^2 + 2\alpha\gamma l + 3\beta\gamma l}, a_2 = -\frac{1}{2}\alpha - \frac{1}{2}\beta - \frac{1}{2}\gamma l, a_3 = \frac{1}{2}\Big(\left(240 - 84\alpha - 84\beta - 104\gamma l + 3\alpha\gamma l + 3\beta\gamma l + 3\gamma l^2\right)(2\alpha + 3\beta + \gamma l)\Big) \Big/ \left(720 - 72\gamma l - 72\alpha - 108\beta + \gamma l^2 + 2\alpha\gamma l + 3\beta\gamma l\right), a_4 = -\frac{3}{2}\left(-44\alpha^2 - 110\alpha\beta - 74\alpha\gamma l - 66\beta^2 - 106\beta\gamma l - 20\gamma l^2 + 2\alpha^2\gamma l + 3\alpha\gamma l^2 + 3\beta^2\gamma l + 4\beta\gamma l^2 + 5\alpha\beta\gamma l + \gamma l^3 + 40\alpha + 120\beta\right) \Big/ \left(720 - 72\gamma l - 72\alpha - 108\beta + \gamma l^2 + 2\alpha\gamma l + 3\beta\gamma l\right), a_5 = \frac{1}{2}\left(-36\alpha^2 - 90\alpha\beta - 60\alpha\gamma l - 54\beta^2 - 90\beta\gamma l - 12\gamma l^2 + 2\alpha^2\gamma l + 3\alpha\gamma l^2 + 3\beta^2\gamma l + 4\beta\gamma l^2 + 5\alpha\beta\gamma l + \gamma l^3 + 72\beta\right) \Big/ \left(720 - 72\gamma l - 72\alpha - 108\beta + \gamma l^2 + 2\alpha\gamma l + 3\beta\gamma l\right) \Big\}$$

Where s is an arbitrary parameter.

The outcome of the above command in solving some of the sets of equations may evaluate the constant coefficients of the answer function (a_0 to a_5) as "RootOf" (i.e., a representative for the roots of equations). Therefore, in order to prevent this manner and because of obtaining the constant coefficients of the answer function (a_0 to a_5) in terms of the parameters α , β , γ_1, it is better to write the command "convert(S, radical)" after the command "Solve" in solving all the nonlinear differential equations by AGM.

> $S := (convert(S, radical));$

By utilizing the outcome of the above command, the constant coefficients of the answer function can be computed as follows:

$$S := \left\{ a_0 = 1, a_1 = -\frac{3\left(240 - 84\,\alpha - 84\,\beta - 104\,\gamma l + 3\,\alpha\,\gamma l + 3\,\beta\,\gamma l + 3\,\gamma l^2\right)}{720 - 72\,\gamma l - 72\,\alpha - 108\,\beta + \gamma l^2 + 2\,\alpha\,\gamma l + 3\,\beta\,\gamma l}, a_2 = -\frac{1}{2}\,\alpha - \frac{1}{2}\,\beta - \frac{1}{2}\,\gamma l, a_3 = \frac{1}{2}\left(\left(240 - 84\,\alpha - 84\,\beta - 104\,\gamma l + 3\,\alpha\,\gamma l + 3\,\beta\,\gamma l + 3\,\gamma l^2\right)\left(2\,\alpha + 3\,\beta + \gamma l\right)\right) \Big/ \left(720 - 72\,\gamma l - 72\,\alpha - 108\,\beta + \gamma l^2 + 2\,\alpha\,\gamma l + 3\,\beta\,\gamma l\right), a_4 = -\frac{3}{2}\left(-44\,\alpha^2 - 110\,\alpha\,\beta - 74\,\alpha\,\gamma l - 66\,\beta^2 - 106\,\beta\,\gamma l - 20\,\gamma l^2 + 2\,\alpha^2\,\gamma l + 3\,\alpha\,\gamma l^2 + 3\,\beta^2\,\gamma l + 4\,\beta\,\gamma l^2 + 5\,\alpha\,\beta\,\gamma l + \gamma l^3 + 40\,\alpha + 120\,\beta\right) \Big/ \left(720 - 72\,\gamma l - 72\,\alpha - 108\,\beta + \gamma l^2 + 2\,\alpha\,\gamma l + 3\,\beta\,\gamma l\right), a_5 = \frac{1}{2}\left(-36\,\alpha^2 - 90\,\alpha\,\beta - 60\,\alpha\,\gamma l - 54\,\beta^2 - 90\,\beta\,\gamma l - 12\,\gamma l^2 + 2\,\alpha^2\,\gamma l + 3\,\alpha\,\gamma l^2 + 3\,\beta^2\,\gamma l + 4\,\beta\,\gamma l^2 + 5\,\alpha\,\beta\,\gamma l + \gamma l^3 + 72\,\beta\right) \Big/ \left(720 - 72\,\gamma l - 72\,\alpha - 108\,\beta + \gamma l^2 + 2\,\alpha\,\gamma l + 3\,\beta\,\gamma l\right) \right\}$$

In order to recall each of the constant coefficients (a_0 to a_5), we will have

$> S[1]; S[2]; S[3]; S[4]; S[5];$

$$a_0 = 1$$

$$a_1 = -\left(3\left(240 - 84\alpha - 84\beta - 104\gamma l + 3\alpha\gamma l + 3\beta\gamma l + 3\gamma l^2\right)\right) \Big/ \left(720 - 72\gamma l - 72\alpha - 108\beta + \gamma l^2 + 2\alpha\gamma l + 3\beta\gamma l\right)$$

$$a_2 = -\frac{1}{2}\alpha - \frac{1}{2}\beta - \frac{1}{2}\gamma l$$

$$a_3 = \frac{1}{2}\left(\left(240 - 84\alpha - 84\beta - 104\gamma l + 3\alpha\gamma l + 3\beta\gamma l + 3\gamma l^2\right)(2\alpha + 3\beta + \gamma l)\right) \Big/ \left(720 - 72\gamma l - 72\alpha - 108\beta + \gamma l^2 + 2\alpha\gamma l + 3\beta\gamma l\right)$$

$$a_4 = -\frac{3}{2}\left(-44\alpha^2 - 110\alpha\beta - 74\alpha\gamma l - 66\beta^2 - 106\beta\gamma l - 20\gamma l^2 + 2\alpha^2\gamma l + 3\alpha\gamma l^2 + 3\beta^2\gamma l + 4\beta\gamma l^2 + 5\alpha\beta\gamma l + \gamma l^3 + 40\alpha + 120\beta\right) \Big/ \left(720 - 72\gamma l - 72\alpha - 108\beta + \gamma l^2 + 2\alpha\gamma l + 3\beta\gamma l\right)$$

1.4.1.7. The Seventh Step

The Command "eval"

In this step, the obtained constant coefficients are substituted into the considered answer function so that its result is the solution of the nonlinear differential equation by AGM. In fact, this step is generally the end of the Maple commands to solve nonlinear differential equations by AGM.

$$> u := eval\left(\sum_{i=0}^{5} a_i \cdot x^i , S\right);$$

$$u := 1 - \frac{3\left(240 - 84\,\alpha - 84\,\beta - 104\,\gamma l + 3\,\alpha\,\gamma l + 3\,\beta\,\gamma l + 3\,\gamma l^2\right) x}{720 - 72\,\gamma l - 72\,\alpha - 108\,\beta + \gamma l^2 + 2\,\alpha\,\gamma l + 3\,\beta\,\gamma l} + \Big($$

$$-\frac{1}{2}\,\alpha - \frac{1}{2}\,\beta - \frac{1}{2}\,\gamma l\Big) x^2 + \frac{1}{2}\,\Big(\big(240 - 84\,\alpha - 84\,\beta - 104\,\gamma l$$

$$+ 3\,\alpha\,\gamma l + 3\,\beta\,\gamma l + 3\,\gamma l^2\big)\,(2\,\alpha + 3\,\beta + \gamma l\,)\,x^3\Big) \Big/ \big(720 - 72\,\gamma l - 72\,\alpha$$

$$- 108\,\beta + \gamma l^2 + 2\,\alpha\,\gamma l + 3\,\beta\,\gamma l\big) - \frac{3}{2}\,\Big(\big(-44\,\alpha^2 - 110\,\alpha\,\beta$$

$$- 74\,\alpha\,\gamma l - 66\,\beta^2 - 106\,\beta\,\gamma l - 20\,\gamma l^2 + 2\,\alpha^2\,\gamma l + 3\,\alpha\,\gamma l^2 + 3\,\beta^2\,\gamma l$$

$$+ 4\,\beta\,\gamma l^2 + 5\,\alpha\,\beta\,\gamma l + \gamma l^3 + 40\,\alpha + 120\,\beta\big)\,x^4\Big) \Big/ \big(720 - 72\,\gamma l - 72\,\alpha$$

$$- 108\,\beta + \gamma l^2 + 2\,\alpha\,\gamma l + 3\,\beta\,\gamma l\big) + \frac{1}{2}\,\Big(\big(-36\,\alpha^2 - 90\,\alpha\,\beta$$

$$- 60\,\alpha\,\gamma l - 54\,\beta^2 - 90\,\beta\,\gamma l - 12\,\gamma l^2 + 2\,\alpha^2\,\gamma l + 3\,\alpha\,\gamma l^2 + 3\,\beta^2\,\gamma l$$

$$+ 4\,\beta\,\gamma l^2 + 5\,\alpha\,\beta\,\gamma l + \gamma l^3 + 72\,\beta\big)\,x^5\Big) \Big/ \big(720 - 72\,\gamma l - 72\,\alpha - 108\,\beta$$

$$+ \gamma l^2 + 2\,\alpha\,\gamma l + 3\,\beta\,\gamma l\big)$$

Where the obtained constant coefficients (a_0 to a_5) are substituted into the considered answer function.

1.4.1.8. Beauty in AGM Approach

As it has been observed from the outcome of Maple, the result is too long after substituting the resulting values of the constant coefficients that are in terms of the parameters (α , β , γ_1) into the answer function, and so we will not have a pleasant outcome. Therefore, in order to make the solution smaller and for beautification of the work, some new variables are introduced. These new variables—such as ψ or ξ or Δ ,...—are chosen instead of similar and repetitive sentences of the answer function.

The command "apply rule" is utilized in this part as follows:

<applyrule([the first similar sentence= ψ , the second similar sentence= ξ] , u).

In regard to the answer function, whose constant coefficients are exactly observed in the aforementioned command, there are repetitive sentences in terms of the parameters (α, β, γ_1), wherein new variables can be defined for them.

It is necessary to mention that after performing the command "eval", whose outcome has been observed, these repetitive sentences can be copied and then entered with the command "apply rule", wherein the repetitive sentences here are attributed to the new variables (choosing the names of these new variables such as ψ_1 , ψ_2 is arbitrary).

>

$$\psi_1 =-72\,\gamma l + 2\,\alpha\,\gamma l + 3\,\beta\,\gamma l + \gamma l^2 + 720 - 72\,\alpha - 108\,\beta;$$

$$\psi_2 = 240 - 104\,\gamma l + 3\,\alpha\,\gamma l + 3\,\gamma l^2 + 3\,\beta\,\gamma l - 84\,\alpha - 84\,\beta;$$

$$\xi = 2\,\alpha^2\,\gamma l + 3\,\alpha\,\gamma l^2 + 3\,\beta^2\,\gamma l + 4\,\beta\,\gamma l^2 + 5\,\beta\,\alpha\,\gamma l + \gamma l^3;$$

$$\Delta_1 = 40\,\alpha + 120\,\beta - 44\,\alpha^2 - 66\,\beta^2 - 20\,\gamma l^2 - 110\,\alpha\,\beta - 74\,\alpha\,\gamma l - 106\,\beta\,\gamma l;$$

$$\Delta_2 = 72\,\beta - 36\,\alpha^2 - 54\,\beta^2 - 12\,\gamma l^2 - 90\,\alpha\,\beta - 60\,\alpha\,\gamma l - 90\,\beta\,\gamma l;$$

$$\Omega_1 =-\frac{1}{2}\,\alpha - \frac{1}{2}\,\beta - \frac{1}{2}\,\gamma l;$$

$$\Omega_2 = 2\,\alpha + 3\,\beta + \gamma l;$$

$$\psi_1 = 720 - 72\,\gamma l - 72\,\alpha - 108\,\beta + \gamma l^2 + 2\,\alpha\,\gamma l + 3\,\beta\,\gamma l$$

$$\psi_2 = 240 - 84\,\alpha - 84\,\beta - 104\,\gamma l + 3\,\alpha\,\gamma l + 3\,\beta\,\gamma l + 3\,\gamma l^2$$

$$\xi = 2\,\alpha^2\,\gamma l + 3\,\alpha\,\gamma l^2 + 3\,\beta^2\,\gamma l + 4\,\beta\,\gamma l^2 + 5\,\alpha\,\beta\,\gamma l + \gamma l^3$$

$$\Delta_1 = 40\,\alpha + 120\,\beta - 44\,\alpha^2 - 66\,\beta^2 - 20\,\gamma l^2 - 110\,\alpha\,\beta - 74\,\alpha\,\gamma l - 106\,\beta\,\gamma l$$

$$\Delta_2 = 72\,\beta - 36\,\alpha^2 - 54\,\beta^2 - 12\,\gamma l^2 - 90\,\alpha\beta - 60\,\alpha\,\gamma l - 90\,\beta\,\gamma l$$

$$\Omega_1 = -\frac{1}{2}\,\alpha - \frac{1}{2}\,\beta - \frac{1}{2}\,\gamma l$$

$$\Omega_2 = 2\,\alpha + 3\,\beta + \gamma l$$

The command (=) without colon (:)—i.e., (:=) at the beginning—is not performed in the Maple medium. Instead, it is just for displaying in the outcome, so none of the above equations are applicable, and they only show us how new variables have been chosen in the outcome.

1.4.1.9. The Eighth Step

The Command "applyrule"

As regards the new introduced variables, the above command can be written as follows:

AGM answer;

$$'u(x)'=applyrule\Big(\Big[-72\,\gamma l + 2\,\alpha\,\gamma l + 3\,\beta\,\gamma l + \gamma l^2 + 720 - 72\,\alpha - 108\,\beta$$

$$= \psi_1,\ 240 - 104\,\gamma l + 3\,\alpha\,\gamma l + 3\,\gamma l^2 + 3\,\beta\,\gamma l - 84\,\alpha - 84\,\beta = \psi_2,$$

$$2\,\alpha^2\,\gamma l + 3\,\alpha\,\gamma l^2 + 3\,\beta^2\,\gamma l + 4\,\beta\,\gamma l^2 + 5\,\beta\,\alpha\,\gamma l + \gamma l^3 = \xi,\ 40\,\alpha + 120\,\beta$$

$$-44\,\alpha^2 - 66\,\beta^2 - 20\,\gamma l^2 - 110\,\alpha\beta - 74\,\alpha\,\gamma l - 106\,\beta\,\gamma l = \Delta_1,\ 72\,\beta$$

$$-36\,\alpha^2 - 54\,\beta^2 - 12\,\gamma l^2 - 90\,\alpha\beta - 60\,\alpha\,\gamma l - 90\,\beta\,\gamma l = \Delta_2, -\frac{1}{2}\,\alpha$$

$$-\frac{1}{2}\,\beta - \frac{1}{2}\,\gamma l = \Omega_1,\ 2\,\alpha + 3\,\beta + \gamma l = \Omega_2\Big], u\Big);$$

AGM answer

$$u(x) = 1 - \frac{3\,\psi_2\,x}{\psi_1} + \Omega_1\,x^2 + \frac{1}{2}\,\frac{\psi_2\,\Omega_2\,x^3}{\psi_1}$$

$$-\frac{3}{2}\,\frac{(\Delta_1 + \xi)\,x^4}{\psi_1} + \frac{1}{2}\,\frac{(\Delta_2 + \xi)\,x^5}{\psi_1}$$

This is the outcome of the above command, which is an appropriate display for the solution of the differential equation with the new variables.

However, the outcome of command "applyrule" cannot be performed because of the existence of the sign = instead of :=, but it can be an appropriate display for the solution of the differential equation with the new introduced variables.

The main equation that is suitable for the next works, such as drawing charts, etc., is the outcome of command "eval", wherein due to its big size, the command "applyrule" is introduced. Therefore, in order to reduce the size of outcomes in Maple software, it is better to close the outcome of the command "eval", which means colon (:) is substituted instead of semicolon (;) at the end. It is citable that the function $u(x)$ at the beginning of the command "applyrule" should get a quotation shown by '$u(x)$'. And if we do not put the quotation, the software shows error.

1.4.1.10. The Ninth Step

Drawing Charts by the Command "plot"

To depict charts of the obtained solution of the differential equation and its derivatives, numerical values are attributed to the physical parameters $(\alpha\ ,\ \beta\ ,\ \gamma_1)$ as follows:

```
> α := 0.7; β := 0.1; γ1 := 0.9;
```

$$\alpha := 0.7$$

$$\beta := 0.1$$

$$\gamma 1 := 0.9$$

```
> AGM; u := evalf(u); du := diff(u, x);
```

$$AGM$$

$$u := 1. - 0.4215212799\,x - 0.8500000000\,x^2 + 0.1826592212\,x^3 + 0.1466495623\,x^4 - 0.05778750375\,x^5$$

$$du := -0.4215212799 - 1.700000000\,x + 0.5479776636\,x^2 + 0.5865982492\,x^3 - 0.2889375188\,x^4$$

The two aforementioned commands show the values of the constant coefficients of the answer function decimally and also the derivative of the function u in terms of x, respectively.

Now the command "plot" is expressed in the following form:

```
> Fig, AGM;
  plot([u], x=0..1, axes=boxed, labels=["x",
     "u(x) AGM"], thickness=2);
  plot([diff(u, x)], x=0..1, axes=boxed, labels
     =["x", "du(x) AGM"], thickness=2);
```

Fig, AGM

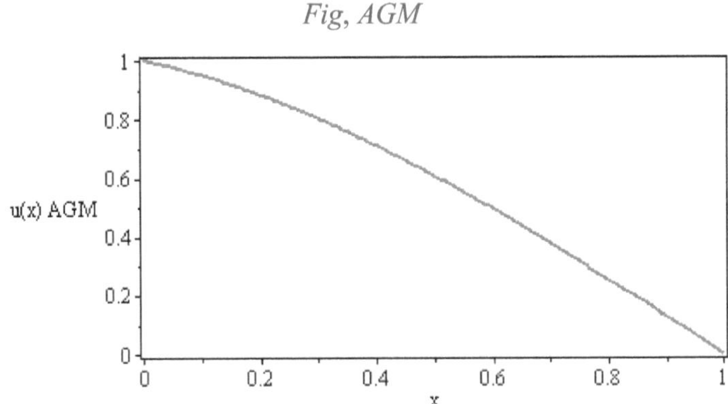

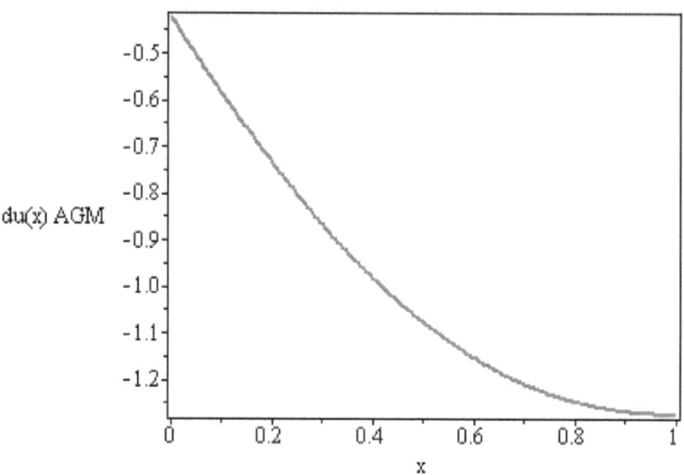

1.4.1.11. The Tenth Step

Solving the Differential Equation by Numerical Method

In this step, the given differential equation can be solved numerically (Runge-Kutta fourth) by choosing physical values in the domain $x \in \{0\ ,\ 1\}$ by helping Maple commands step-by-step as follows:

The Command "restart"

> *restart;*

1.4.1.12. The Eleventh Step

The definition of the physical values for the parameters $(\alpha\ ,\ \beta\ ,\ \gamma_1)$ is done as follows:

> $\alpha := 0.7; \beta := 0.1; \gamma 1 := 0.9;$

$$\alpha := 0.7$$

$$\beta := 0.1$$

$$\gamma 1 := 0.9$$

1.4.1.13. The Twelfth Step

In this step, the given differential equation is written as a function in the software medium in the following form:

$$> f := \frac{d^2}{dx^2} u(x) + \alpha \cdot u(x)^2 + \beta \cdot u(x)^3 + \gamma 1\, u(x) = 0;$$

$$f := \frac{d^2}{dx^2} u(x) + 0.7\, u(x)^2 + 0.1\, u(x)^3 + 0.9\, u(x) = 0$$

1.4.1.14. The Thirteenth Step

The boundary conditions are written as follows:

$$> ic := u(0) = 1,\ u(1) = 0;$$

$$ic := u(0) = 1,\ u(1) = 0$$

It is important to mention that if there is a derivative in the boundary conditions, we will do the following actions:

$$u'(a) = b \xrightarrow{\text{Convert to}} (D\ @@1)(u)(a) = b\ ;$$
$$u''(a) = b \xrightarrow{\text{Convert to}} (D\ @@2)(u)(a) = b\ ;$$
$$u^{(n)}(a) = b \xrightarrow{\text{Convert to}} (D\ @@\mathrm{n})(u)(a) = b\ ;. \tag{1.51}$$

1.4.1.15. The Fourteenth Step

The Command "dsolve"

$$> f1 := dsolve([f, ic], numeric, output = listprocedure);$$

$$f1 := \left[x = \mathrm{proc}(x)\ \ldots\ \mathrm{end\ proc},\ u(x) = \mathrm{proc}(x)\ \ldots\ \mathrm{end\ proc},\ \frac{d}{dx} u(x) = \mathrm{proc}(x)\ \ldots\ \mathrm{end\ proc} \right]$$

It is citable that the letters f and f_1 in steps 12 and 14 are arbitrary, and the user can utilize any other letters. Although the outcomes of the commands "ic" and "dsolve" do not show certain things, they should be written.

1.4.1.16. The Fifteenth Step

The Repetition Statement (For . . . While . . . Do) to Show Numerical Solution in the Domain x∈{0, 1}

In this step, from the domain x ∈ {0 , 1}, eleven numerical values with the step 0.1 are obtained in the outcome of "for loop" in the Maple medium as follows:

```
> for i from 0 by 0.1 while i ≤ 1 do([x=i,f1(i)[2],f1(i)[3]]); end do;
```

$$\left[x=0, u(x)(0)=1., \frac{d}{dx}u(x)(0)=-0.431114957451192\right]$$

$$\left[x=0.1, u(x)(0.1)=0.948591553731791, \frac{d}{dx}u(x)(0.1)=-0.594867797625046\right]$$

$$\left[x=0.2, u(x)(0.2)=0.881520249971018, \frac{d}{dx}u(x)(0.2)=-0.743836540881419\right]$$

$$\left[x=0.3, u(x)(0.3)=0.800409536649850, \frac{d}{dx}u(x)(0.3)=-0.875279643025517\right]$$

$$\left[x=0.4, u(x)(0.4)=0.707108271310672, \frac{d}{dx}u(x)(0.4)=-0.987437160370510\right]$$

$$\left[x=0.5, u(x)(0.5)=0.603592040154117, \frac{d}{dx}u(x)(0.5)=-1.07952671799694\right]$$

$$\left[x=0.6, u(x)(0.6)=0.491868921024819, \frac{d}{dx}u(x)(0.6)=-1.15166189916028\right]$$

$$\left[x=0.7, u(x)(0.7)=0.373896387265424, \frac{d}{dx}u(x)(0.7)=-1.20471498598757\right]$$

$$\left[x=0.8, u(x)(0.8)=0.251513697628966, \frac{d}{dx}u(x)(0.8)=-1.24014859786030\right]$$

$$\left[x=0.9, u(x)(0.9)=0.126391710901670, \frac{d}{dx}u(x)(0.9)=-1.25983903511744\right]$$

$$\left[x=1.0, u(x)(1.0)=0., \frac{d}{dx}u(x)(1.0)=-1.26590946600082\right]$$

The expression $x = i$ in the outcome of Maple is the values of x in the different steps ($x = 0$, 0.1 , 0.2 , ...), and the expression $f_1(\mathrm{i})[2]$ is the numerical value of $u(x)$; and also, $f_1(\mathrm{i})[3]$ is the numerical value of the derivative of the answer function, or $\frac{du(x)}{dx}$ in the Maple outcome.

1.4.1.17. The Sixteenth Step

Drawing the Charts of Numerical Solution in the Domain x∈{0, 1}

The chart of the answer function shown by $u(x)$ and the chart of the derivative of the answer function of the differential equation shown by $u'(x)$ are depicted in the outcomes of the following commands:

```
> Fig, Numerical;
  plots[odeplot](f1, [x, u(x) ], axes = boxed , labels
     = ["x", "u(x)"], thickness = 2);
  plots[odeplot](f1, [x, diff(u(x), x) ], axes
     = boxed , labels = ["x", "du(x)"], thickness = 2)
     ;
```

Fig, Numerical

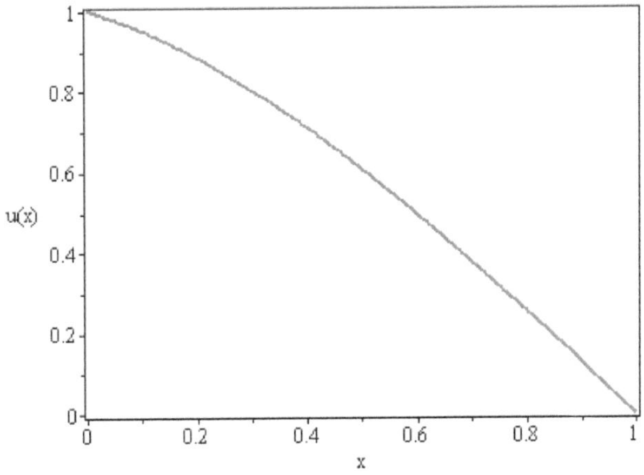

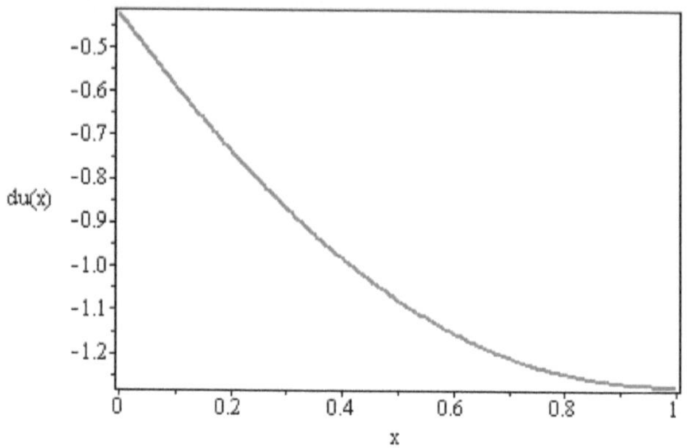

1.4.1.18. The Seventeenth Step

Computing the Absolute Value of the Difference between the Obtained Solution by AGM and Numerical Method

The command "subs" in this manner is utilized for the function $u(x)$ and its derivative $\frac{du(x)}{dx}$ in the following form:

$$> u := subs(f1, u(x)); du := subs\left(f1, \frac{d}{dx}\, u(x)\right);$$

$$u := \textbf{proc}(x) \ \ldots \ \textbf{end proc}$$
$$du := \textbf{proc}(x) \ \ldots \ \textbf{end proc}$$

In this level, we use the obtained answer function and its derivative by AGM from step 9, and then we rewrite them in the Maple medium by attributing arbitrary letters such as U and dU in the following form:

>

$$U := 1. - 0.42152127999x - 0.8500000000x^2 + 0.1826592212x^3 + 0.1466495623x^4 - 0.05778750375x^5;$$

$$dU := -0.4215212799 - 1.700000000x + 0.5479776636x^2 + 0.5865982492x^3 - 0.2889375188x^4;$$

$U := 1. - 0.4215212799x - 0.8500000000x^2 + 0.1826592212x^3 + 0.1466495623x^4 - 0.05778750375x$

$dU := -0.4215212799 - 1.700000000x + 0.5479776636x^2 + 0.5865982492x^3 - 0.2889375188x^4$

In this step, the difference percentage between the achieved solutions by numerical method and AGM, shown by $\left|\frac{u_{NUM} - u_{AGM}}{u_{NUM}}\right|$, $\left|\frac{du_{NUM} - du_{AGM}}{du_{NUM}}\right|$ in the domain $x \in \{0\ ,\ 1\}$ with an arbitrary step (in this example, 0.2), can be obtained as follows:

```
> for x from 0 by 0.1 to 1 do;
Δu = abs((U − u(x))/u(x) · 100), Δdu
   = abs((dU − du(x))/du(x) · 100);
end do;
```

$\Delta u = 0., \Delta du = 2.22531772219427$

$\Delta u = 0.210195412872582, \Delta du = 1.13799914580534$

$\Delta u = 0.443481121146430, \Delta du = 0.378534859788695$

$\Delta u = 0.647515982214374, \Delta du = 0.289017275135007$

$\Delta u = 0.764432032592512, \Delta du = 0.699713014607722$

$\Delta u = HFloat(\infty), \Delta du = 0.787846229689985$

Therefore, the outcomes of the above command that were the difference percent of the answer function resulting from solving by numerical method and AGM and also its derivative are obtained.

The charts of the absolute value of the achieved solution by AGM and numerical method with the command "plot" are depicted as follows:

```
> plot( (abs( (U − u(x) ) ) ), x = 0 ..1, thickness = 2,
    axes = boxed, labels = ["x", "Δu(x)"]);
  plot( abs( dU − du(x) ), x = 0 ..1, thickness = 2,
    axes = boxed, labels = ["x", "Δdu(x)"]);
```

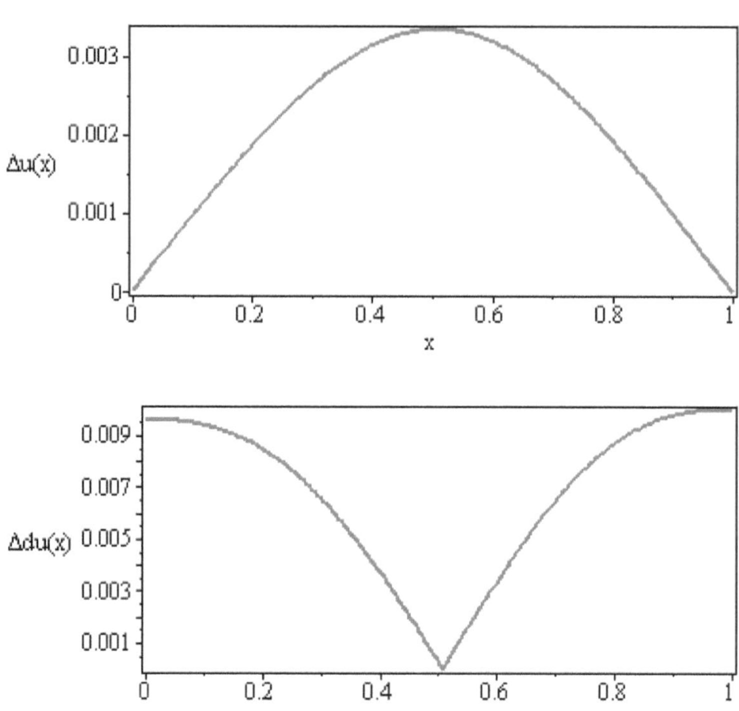

1.4.1.19. Some Important Tips that Must Be Observed while Writing Codes by Maple Software in Order to Solve Differential Equations

1. Chosen arbitrary letters like the letter f from step 3 and S_1 , S_2 , … , S_6 from step 5 should not be similar to the parameters of the differential equation such as u and x so that errors will not occur.

2. Semicolon $(;)$ is used at the end of any command for showing outcomes, and colon $(:)$ should be written at the end for not displaying the outcomes.

3. In order to run the above computer program only by pushing one Enter at the end of the written algorithm and to beautify and reduce the size of operations, we should utilize the keys ↑ Shift + ↵ Enter at the end of each step to go to the next line for writing the next commands and continue it up to the end of the work.

1.4.2. Applied Instructions for Vibrational Equations without Any External Force

To understand more, consider the following vibrational example with damping component as

$$\left(1+\mu u^{m}(t)\right).\frac{d^{2}u(t)}{dt^{2}}+\alpha u^{n}(t).\frac{du(t)}{dt}+\eta\,\frac{du(t)}{dt}+\lambda\sin\left(u(t)\right)=0. \tag{1.52}$$

The parameter η in this differential equation that is the coefficient of $\frac{d\,u(t)}{dt}$ is defined as damping factor in vibrational differential equations.
And the initial conditions are introduced as follows:

$$u(0)=\mathrm{A} \quad , \quad \dot{u}(0)=0. \tag{1.53}$$

In AGM, an answer function as a trigonometric and power series is considered in the following form:

$$u=e^{-at}\{b\cos(\omega t+\varphi)\}. \tag{1.54}$$

Where a , b, ω and φ can be obtained by applying the initial conditions.

1.4.2.1. The First Step

> *restart;*

This command cleans previous memory and makes the program ready to work.

1.4.2.2. The Second Step

Selecting an answer function as a trigonometric and power series from equation 1.3:

$$> u := e^{-a t} \cdot (b \cdot \cos(\omega t + \varphi));$$

$$u := e^{-a t} b \cos(\omega t + \varphi).$$

1.4.2.3. The Third Step

Here, equation 1.52 is written, while the term u is substituted instead of the expression $u(t)$ in order to put the considered answer function into it as follows:

$$> f := (1 + \mu u^m) \frac{d^2}{dt^2} u + \alpha u^n \cdot \left(\frac{d}{dt} u\right) + \eta \left(\frac{d}{dt} u\right) + \lambda \cdot \sin(u) = 0;$$

$$f := \left(1 + \mu \left(e^{-a t} b \cos(\omega t + \varphi)\right)^m\right) \left(a^2 e^{-a t} b \cos(\omega t + \varphi) + 2 a e^{-a t} b \sin(\omega t + \varphi) \omega - e^{-a t} b \cos(\omega t + \varphi) \omega^2\right) + \alpha \left(e^{-a t} b \cos(\omega t + \varphi)\right)^n \left(-a e^{-a t} b \cos(\omega t + \varphi) - e^{-a t} b \sin(\omega t + \varphi) \omega\right) + \eta \left(-a e^{-a t} b \cos(\omega t + \varphi) - e^{-a t} b \sin(\omega t + \varphi) \omega\right) + \lambda \sin\left(e^{-a t} b \cos(\omega t + \varphi)\right) = 0$$

It is notable that selecting a name like f for the obtained equation is necessary to utilize in the foregoing parts.

1.4.2.4. The Fourth Step

Converting the resulting equations from the second and third steps as the functions of $u(t)$ and $f(t)$ in order to apply the initial conditions on them as

```
> u := unapply(u, t); f := unapply(f, t);
```

$$u := t \to e^{-at} b \cos(\omega t + \varphi)$$

$$f := t \to \left(1 + \mu \left(e^{-at} b \cos(\omega t + \varphi)\right)^m\right) \left(a^2 e^{-at} b \cos(\omega t + \varphi) + 2 a e^{-at} b \sin(\omega t + \varphi) \omega - e^{-at} b \cos(\omega t + \varphi) \omega^2\right) + \alpha \left(e^{-at} b \cos(\omega t + \varphi)\right)^n \left(-a e^{-at} b \cos(\omega t + \varphi) - e^{-at} b \sin(\omega t + \varphi) \omega\right) + \eta \left(-a e^{-at} b \cos(\omega t + \varphi) - e^{-at} b \sin(\omega t + \varphi) \omega\right) + \lambda \sin\left(e^{-at} b \cos(\omega t + \varphi)\right) = 0$$

1.4.2.5. The Fifth Step

Applying the initial conditions on the answer function and its derivative as trigonometric and power series from the second step is written in the following form:

```
> s1 := u(0) = A;   s2 := u'(0) = 0;
```

$$s1 := b \cos(\varphi) = A$$

$$s2 := -a b \cos(\varphi) - b \sin(\varphi) \omega = 0$$

It is notable that the initial conditions can be applied by optional parameters like s_1, s_2 ,... up to the sixth step, in which these parameters can be utilized for the command "solve" in the seventh step.

1.4.2.6. The Sixth Step

```
> s3 := f(0);
```

$$s3 := \left(1 + \mu \left(b \cos(\varphi)\right)^m\right) \left(a^2 b \cos(\varphi) + 2 a b \sin(\varphi) \omega - b \cos(\varphi) \omega^2\right) + \alpha \left(b \cos(\varphi)\right)^n \left(-a b \cos(\varphi) - b \sin(\varphi) \omega\right) + \eta \left(-a b \cos(\varphi) - b \sin(\varphi) \omega\right) + \lambda \sin\left(b \cos(\varphi)\right) = 0$$

Applying the initial conditions on the considered answer function of the differential equation from the fourth step is done by this command.

1.4.2.7. The Seventh Step

Apply the initial conditions on the derivative of the obtained equation from step 6 as

> $s4 := f'(0);$

$s4 :=$

$$\frac{1}{b\cos(\varphi)}\left(\mu\,(b\cos(\varphi))^m m\,(-a b\cos(\varphi) - b\sin(\varphi)\,\omega)\,(a^2 b\cos(\varphi) + 2ab\sin(\varphi)\,\omega - b\cos(\varphi)\,\omega^2)\right) + (1+\mu\,(b\cos(\varphi))^m)\,(-a^3 b\cos(\varphi) - 3a^2 b\sin(\varphi)\,\omega + 3ab\cos(\varphi)\,\omega^2 + b\sin(\varphi)\,\omega^3) + \frac{\alpha\,(b\cos(\varphi))^n n\,(-ab\cos(\varphi) - b\sin(\varphi)\,\omega)^2}{b\cos(\varphi)} + \alpha\,(b\cos(\varphi))^n\,(a^2 b\cos(\varphi) + 2ab\sin(\varphi)\,\omega - b\cos(\varphi)\,\omega^2) + \eta\,(a^2 b\cos(\varphi) + 2ab\sin(\varphi)\,\omega - b\cos(\varphi)\,\omega^2) + \lambda\cos(b\cos(\varphi))\,(-ab\cos(\varphi) - b\sin(\varphi)\,\omega) = 0$$

In accordance with step 4, which means the achieved differential equation is written in terms of t, this command ($f'(0)$) computes the derivative of the mentioned equation automatically.

1.4.2.8. The Eighth Step

> $S := solve([s1, s2, s3, s4], \{b, a, \omega, \varphi\});$

$$S := \left\{a = \frac{1}{2}\,\frac{\alpha e^{\ln(A)\,n} + \eta}{1 + \mu e^{\ln(A)\,m}},\ b = A\Big/\left(RootOf\left((4\sin(A)\,\lambda + 4\sin(A)\,\lambda\mu e^{\ln(A)\,m})\,\_Z^2 - 4\sin(A)\,\lambda - 4\sin(A)\,\lambda\mu e^{\ln(A)\,m} + A\alpha^2\,(e^{\ln(A)\,n})^2 + 2A\alpha e^{\ln(A)\,n}\eta + A\eta^2\right)\right),\ \omega = -\frac{1}{A}\left(2\,RootOf\left((4\sin(A)\,\lambda + 4\sin(A)\,\lambda\mu A^m)\,\_Z^2 + A\alpha^2\,(A^n)^2 + 2A\alpha A^n\eta - 4\sin(A)\,\lambda - 4\sin(A)\,\lambda\mu A^m + A\eta^2\right) RootOf\left((4\sin(A)\,\lambda + 4\sin(A)\,\lambda\mu A^m)\,\_Z^2 - A\right)\sin(A)\,\lambda\right),\ \varphi = \arctan\left(RootOf\left((4\sin(A)\,\lambda + 4\sin(A)\,\lambda\mu A^m)\,\_Z^2 - A\right)(\alpha A^n + \eta),\ RootOf\left((4\sin(A)\,\lambda + 4\sin(A)\,\lambda\mu A^m)\,\_Z^2 + A\alpha^2\,(A^n)^2 + 2A\alpha A^n\eta - 4\sin(A)\,\lambda - 4\sin(A)\,\lambda\mu A^m + A\eta^2\right)\right)\right\}$$

This command can be utilized by choosing a letter like S, which solves the equations of steps 5 to 7. Since in the outcome we can see the coefficients of the answer function of equation 1.54, a, b, ω and φ, are obtained in the form of "RootOf", it is better to reform this problem or compute the desired roots by going to step 9.

1.4.2.9. The Ninth Step

> $S := convert(S, radical);$

$$S := \left\{ a = \frac{1}{2}\frac{\alpha e^{\ln(A)\,n} + \eta}{1 + \mu e^{\ln(A)\,m}},\ b = \frac{A}{\sqrt{-\frac{-4\sin(A)\lambda - 4\sin(A)\lambda\mu e^{\ln(A)\,m} + A\alpha^2 (e^{\ln(A)\,n})^2 + 2A\alpha e^{\ln(A)\,n}\eta + A\eta^2}{4\sin(A)\lambda + 4\sin(A)\lambda\mu e^{\ln(A)\,m}}}},\ \omega = -\frac{1}{A}\left(2\sqrt{-\frac{A\alpha^2 (A^n)^2 + 2A\alpha A^n\eta - 4\sin(A)\lambda - 4\sin(A)\lambda\mu A^m + A\eta^2}{4\sin(A)\lambda + 4\sin(A)\lambda\mu A^m}}\sqrt{\frac{A}{4\sin(A)\lambda + 4\sin(A)\lambda\mu A^m}}\sin(A)\lambda\right),\ \varphi = \arctan\left(\sqrt{\frac{A}{4\sin(A)\lambda + 4\sin(A)\lambda\mu A^m}}(\alpha A^n + \eta), \sqrt{-\frac{A\alpha^2 (A^n)^2 + 2A\alpha A^n\eta - 4\sin(A)\lambda - 4\sin(A)\lambda\mu A^m + A\eta^2}{4\sin(A)\lambda + 4\sin(A)\lambda\mu A^m}}\right) \right\}$$

This command is utilized in order to remove the term "RootOf". Thus, it is observed that the values a, b, ω and φ have been obtained clearly.

1.4.2.10. The Tenth Step

Extract constant values of the answer function from {}:

```
> a := rhs(S[1]);
  b := rhs(S[2]);
  ω := rhs(S[3]);
  φ := rhs(S[4]);
```

$$a := \frac{1}{2}\,\frac{\alpha e^{\ln(A)\,n} + \eta}{1 + \mu e^{\ln(A)\,m}}$$

$$b := \frac{A}{\sqrt{-\frac{-4\sin(A)\,\lambda - 4\sin(A)\,\lambda\mu e^{\ln(A)\,m} + A\,\alpha^2\,\left(e^{\ln(A)\,n}\right)^2 + 2A\,\alpha e^{\ln(A)\,n}\,\eta + A\,\eta^2}{4\sin(A)\,\lambda + 4\sin(A)\,\lambda\mu e^{\ln(A)\,m}}}}$$

$$\omega := -\frac{1}{A}\left(2\sqrt{-\frac{A\,\alpha^2\,(A^n)^2 + 2A\,\alpha A^n\,\eta - 4\sin(A)\,\lambda - 4\sin(A)\,\lambda\mu A^m + A\,\eta^2}{4\sin(A)\,\lambda + 4\sin(A)\,\lambda\mu A^m}}\,\sqrt{\frac{A}{4\sin(A)\,\lambda + 4\sin(A)\,\lambda\mu A^m}}\,\sin(A)\,\lambda\right)$$

$$\varphi := \arctan\left(\sqrt{\frac{A}{4\sin(A)\,\lambda + 4\sin(A)\,\lambda\mu A^m}}\,(\alpha A^n + \eta),\ \sqrt{-\frac{A\,\alpha^2\,(A^n)^2 + 2A\,\alpha A^n\,\eta - 4\sin(A)\,\lambda - 4\sin(A)\,\lambda\mu A^m + A\,\eta^2}{4\sin(A)\,\lambda + 4\sin(A)\,\lambda\mu A^m}}\right)$$

In this command, S is an optional parameter from step 8, and K belongs to the natural numbers.

For K = 1 the value a, which is the first parameter in {}, can be obtained; and also for K = 2, the value b, which is the second parameter in {}, is achieved, and so on.

The term *rhs* at the beginning of the command *S*[K] shows only the right-hand side of the equation *S*[K], and writing this command is not compulsory in step 10. If it is required to utilize the values a, b, ω and φ, we can write it.

1.4.2.11. The Eleventh Step

$$> u := eval\left(e^{-a\,t}\cdot(b\cdot\cos(\omega\,t+\varphi)), S\right);$$

$$u := \left(e^{-\frac{1}{2}\frac{(\alpha\,e^{\ln(A)\,n}+\eta)\,t}{1+\mu\,e^{\ln(A)\,m}}}\,A\cos\left(-\frac{1}{A}\left(2\sqrt{-\frac{A\,\alpha^2\,(A^n)^2+2\,A\,\alpha\,A^n\,\eta-4\sin(A)\,\lambda-4\sin(A)\,\lambda\,\mu\,A^m+A\,\eta^2}{4\sin(A)\,\lambda+4\sin(A)\,\lambda\,\mu\,A^m}}\sqrt{\frac{A}{4\sin(A)\,\lambda+4\sin(A)\,\lambda\,\mu\,A^m}}\sin(A)\,\lambda\,t\right)+\arctan\left(\sqrt{\frac{A}{4\sin(A)\,\lambda+4\sin(A)\,\lambda\,\mu\,A^m}}\,(\alpha\,A^n+\eta), \sqrt{-\frac{A\,\alpha^2\,(A^n)^2+2\,A\,\alpha\,A^n\,\eta-4\sin(A)\,\lambda-4\sin(A)\,\lambda\,\mu\,A^m+A\,\eta^2}{4\sin(A)\,\lambda+4\sin(A)\,\lambda\,\mu\,A^m}}\right)\right)\right)\sqrt{-\frac{-4\sin(A)\,\lambda-4\sin(A)\,\lambda\,\mu\,e^{\ln(A)\,m}+A\,\alpha^2\,(e^{\ln(A)\,n})^2+2\,A\,\alpha\,e^{\ln(A)\,n}\,\eta+A\,\eta^2}{4\sin(A)\,\lambda+4\sin(A)\,\lambda\,\mu\,e^{\ln(A)\,m}}}$$

Substituting the constant values of step 9—which are a , b, ω and φ —into the answer function of equation 1.54 is utilized from the command "eval".

1.4.2.12. The Twelfth Step

Based on the outcomes of step 11, which is the answer of the differential equation, it is obvious that the solution is very big and massive. Therefore, to contract the obtained solution and make it suitable, using the following command is recommended. Before writing this command, we should recognize the similar sentences of the obtained solution from step 11, and then by introducing a new variable like ψ in the form of =, which shows the mentioned command is not applicable, we will have

$\psi_1 = 4\sin(A)\,\lambda + 4\sin(A)\,\lambda\,\mu\,A^m;$

$\psi_2 = A\,\alpha^2\,(A^n)^2 + 2\,A\,\alpha\,A^n\,\eta + A\eta^2;$

> $\Delta_1 = \alpha\,e^{\ln(A)\,n} + \eta;$

$\Delta_2 = 1 + \mu\,e^{\ln(A)\,m};$

$'u(t)' = applyrule\Big(\Big[4\sin(A)\,\lambda + 4\sin(A)\,\lambda\,\mu\,A^m = \psi_1, -4\sin(A)\,\lambda - 4\sin(A)\,\lambda\,\mu\,A^m = -\psi_1,$
$A\,\alpha^2\,(A^n)^2 + 2\,A\,\alpha\,A^n\,\eta + A\eta^2 = \psi_2, \alpha\,e^{\ln(A)\,n} + \eta = \Delta_1, 1 + \mu\,e^{\ln(A)\,m} = \Delta_2, 4\sin(A)\,\lambda$
$+ 4\sin(A)\,\lambda\,\mu\,e^{\ln(A)\,m} = \psi_1, -4\sin(A)\,\lambda - 4\sin(A)\,\lambda\,\mu\,e^{\ln(A)\,m} = -\psi_1, A\,\eta^2$
$+ 2\,A\,\eta\,\alpha\,e^{\ln(A)\,n} + A\,\alpha^2\left(e^{\ln(A)\,n}\right)^2 = \psi_2\Big], u\Big);$

$$\psi_1 = 4\sin(A)\,\lambda + 4\sin(A)\,\lambda\mu A^m$$

$$\psi_2 = A\,\alpha^2\,(A^n)^2 + 2\,A\,\alpha A^n\,\eta + A\,\eta^2$$

$$\Delta_1 = \alpha\,e^{\ln(A)\,n} + \eta$$

$$\Delta_2 = 1 + \mu\,e^{\ln(A)\,m}$$

$$u(t) = \frac{1}{\sqrt{-\frac{\psi_2-\psi_1}{\psi_1}}}\left(e^{-\frac{1}{2}\frac{\Delta_1 t}{\Delta_2}}\,A\cos\left(-\frac{2\sqrt{-\frac{\psi_2-\psi_1}{\psi_1}}\sqrt{\frac{A}{\psi_1}}\,\sin(A)\,\lambda t}{A}\right.\right.$$
$$\left.\left.+\arctan\left(\sqrt{\frac{A}{\psi_1}}\,(\alpha A^n+\eta), \sqrt{-\frac{\psi_2-\psi_1}{\psi_1}}\right)\right)\right)$$

The term 'u' = is just for displaying and is not applicable.

1.4.2.13. The Thirteenth Step

> $m := 3;\ n := 4;\ \mu := 0.2;\ \alpha := 1.7;\ \lambda := 0.3;\ A := 0.1;\ \eta$
> $:= 0.2;$

$$m := 3$$

$$n := 4$$

$$\mu := 0.2$$

$$\alpha := 1.7$$

$$\lambda := 0.3$$

$$A := 0.1$$

$$\eta := 0.2$$

In this step, values are attributed to the physical parameters of the system, which are the constant parameters of the vibrational differential equation.

1.4.2.14. The Fourteenth Step

Calling on the outcome of step 11

```
> ω := evalf(ω); u := evalf(u); du := evalf(diff(u, t));
```

$$\omega := -0.5379845352$$

$$u := 0.1017150877\, e^{-0.1000649870\, t} \cos(0.5379845352\, t - 0.1838982552)$$

$$du := -0.01017811893\, e^{-0.1000649870\, t} \cos(0.5379845352\, t - 0.1838982552) - 0.05472114418\, e^{-0.1000649870\, t} \sin(0.5379845352\, t - 0.1838982552)$$

Here, the angular frequency ω, the answer function u, and its derivative (du)are visible after attributing the physical values from step 13.

1.4.2.15. The Fifteenth Step

```
>
  fig, AGM;
  plot(u, t=0 ..40, labels= ["t", "u(t)"], thickness=2);
  drivation;
  plot(du, t=0 ..40,  labels= ["t", "du(t)"], thickness=2);
  Phase Plane, AGM;
  plot([u, du, t=0 ..40], scaling=constrained, thickness
      =2, labels= ["u", "du"]);
```

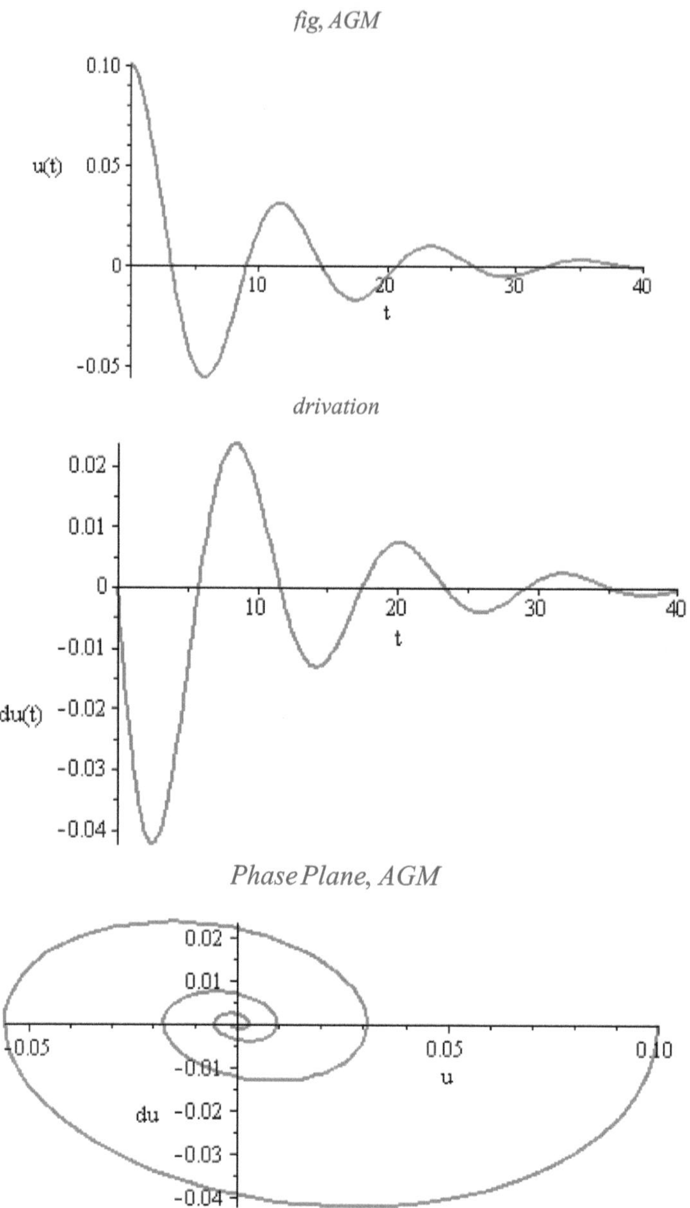

Depicting the charts of the answer function u , its derivative $(du)$and also the related phase plan, which shows the variation of $u(t)$ in terms of $u'(t)$, is carried out in this level.

1.4.2.16. The Sixteenth Step

```
>
fig, max, Domains;
Umax := evalf(b·e^(-a·t));
Um1 := evalf(eval(Umax, t = π/ω));
Um2 := evalf(eval(Umax, t = 2π/ω));
plot([Umax, u, -Umax], t = 0 ..40, axes = boxed, labels
    = ["t", "u(t),umax"], thickness = 2);
```

$$fig, \max, Domains$$

$$Umax := 0.1017150877\,e^{-0.1000649870\,t}$$

$$Um1 := 0.05670374232$$

$$Um2 := 0.03161098779$$

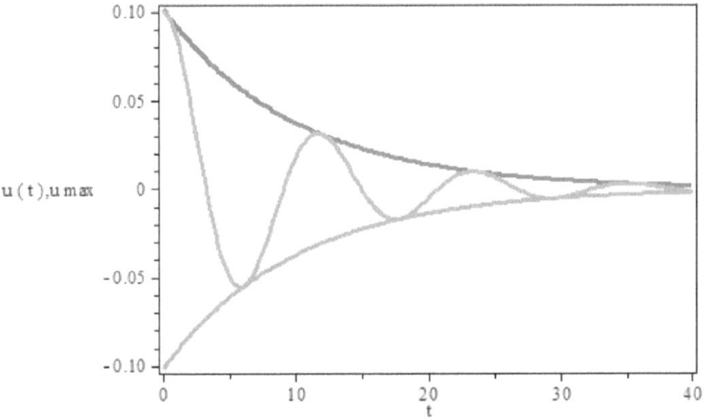

As regards the above command, we are able to substitute $t = \frac{\pi}{\omega}, \frac{2\pi}{\omega}, \ldots$ in the formula of the locus of maximum points obtained by $u_{\max} = b.e^{-at}$ and then simultaneously plot u , $u_{\max}$, $-u_{\max}$. It is necessary to mention that the maximum amplitude of vibration in two successive cycles is shown by u_{m1} and u_{m2}, and the command "evalf" is just used for displaying the results in the form of decimal numbers.

1.4.2.17. The Seventeenth Step

The locus of maximum vibrational velocity and acceleration are obtained as follows:

$$\dot{u} = V_m = b\left(\sqrt{a^2+\omega^2}\right)e^{-at} \quad ; \quad \ddot{u} = a_m = b\left(a^2+\omega^2\right)e^{-at} \qquad (1.55)$$

Based on the given explanations in equation 1.55 and the previous step, we are able to simultaneously plot three charts consisting of vibrational velocity, $+V_m$ and $-V_m$ in $t = \frac{\pi}{\omega}, \frac{2\pi}{\omega}, \ldots$, and the above procedure can be done for the vibrational acceleration too.

```
>
velocity;
Vm := b·sqrt(a^2 + ω^2)·e^(-a t);
Vm1 := evalf(eval(Vm, t = π/(2 ω)));
Vm2 := evalf(eval(Vm, t = 3 π/(2 ω)));
plot([Vm, -Vm, du], t = 0 ..40, axes = boxed, labels = ["t",
    "V(t),Vmax"], thickness = 2);
accelretion;
am := b·(a^2 + ω^2)·e^(-a t);
am1 := evalf(eval(am, t = π/ω));
am2 := evalf(eval(am, t = 2 π/ω));
plot([am, diff(u, t, t), -am], t = 0 ..40, axes = boxed,
    labels = ["t", "a(t),a max"], thickness = 2);
```

velocity

$$Vm := 0.05565965976\, e^{-0.1000649870\, t}$$

$$Vm1 := 0.04155791203$$

$$Vm2 := 0.02316754760$$

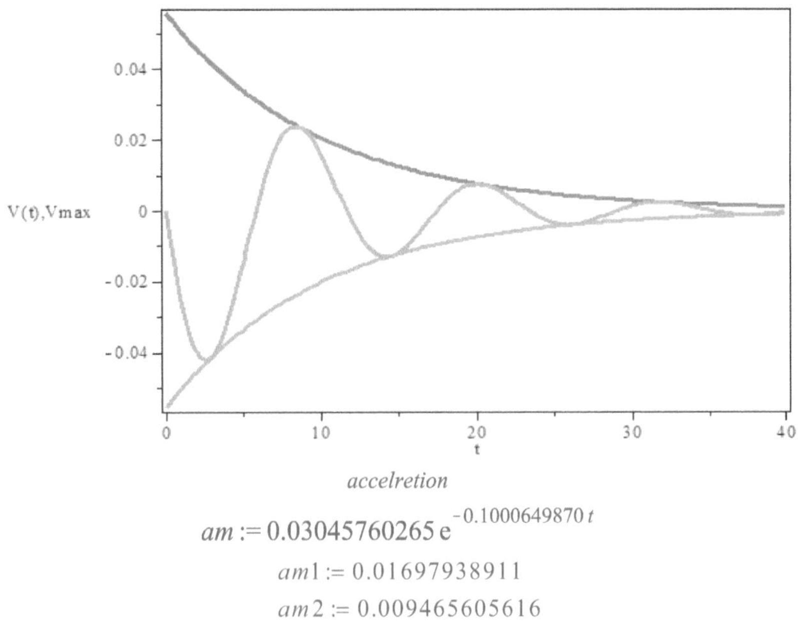

accelretion

$$am := 0.03045760265\,e^{-0.1000649870\,t}$$

$$am1 := 0.01697938911$$

$$am2 := 0.009465605616$$

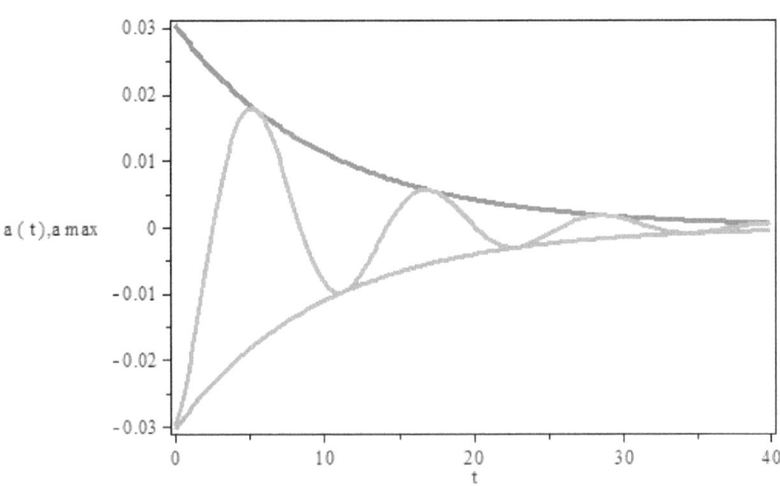

1.4.2.18. The Eighteenth Step

The locus of damping force and lost energy in each vibrational cycle can be obtained as follows:

Here, x is an arbitrary character that is considered instead of a displacement parameter (u).

```
>
Lost energy in each cycle;
k := 1;
```

$$Fd1 := evalf\left(\eta\cdot\left(a\,x+\omega\cdot\sqrt{b^2\cdot e^{-\frac{4\cdot k\cdot a\cdot\pi}{\omega}}-x^2}\right)\right);$$

$$Fd2 := evalf\left(\eta\cdot\left(a\,x-\omega\cdot\sqrt{b^2\cdot e^{-\frac{4\cdot k\cdot a\cdot\pi}{\omega}}-x^2}\right)\right);$$

$$Wd := \int_0^{\frac{2k\pi}{\omega}} \eta\cdot du^2\,dt;$$

```
plot([Fd1, Fd2], x=-0.5..0.5, labels=["u", "Fd"],
    thickness=2);
```

$$k := 1$$

$$Fd1 := 0.02001299740x + 0.1075969070\sqrt{0.0009992545485 - 1.x^2}$$

$$Fd2 := 0.02001299740x - 0.1075969070\sqrt{0.0009992545485 - 1.x^2}$$

$$Wd := 0.001351717567$$

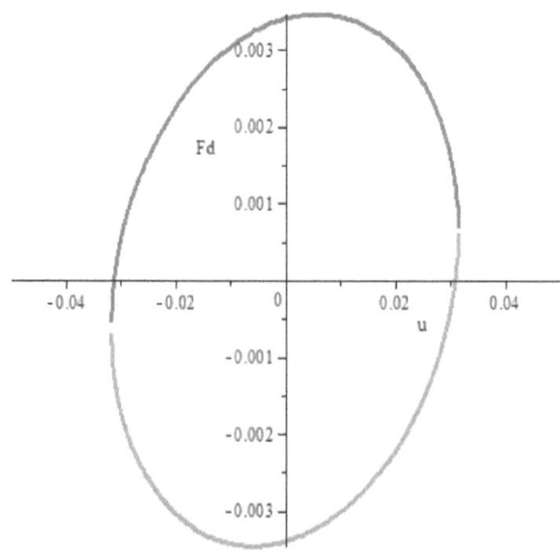

The equation of damping force is introduced as

$$F_d = \eta(ax \pm \omega\sqrt{b^2 e^{\frac{-4k\pi a}{\omega}} - x^2}). \tag{1.56}$$

And also, the equation of damping lost energy is obtained as follows:

$$w_d = \int_0^{\frac{2k\pi}{\omega}} \eta u'(t)^2\, dt. \tag{1.57}$$

It is necessary to mention that the parameter X is the vibrational displacement ($u(t)$) obtained from the solution of differential equation (the reason of choosing $u(t)$ instead of X is just for preventing the error of the program). For the first cycle ($k = 1$), the chart of the damping force (F_d) is visible in the form of an ellipse, and the amount of lost energy (ω_d) in each cycle can be observed by the command "plot".

1.4.2.19. The Nineteenth Step

This step is like step 18, with the exception of considering K = 2 in the second cycle.

```
>
k := 2;
Fd1 := evalf(η·(a x + ω·sqrt(b^2·e^(-(4·k·a·π)/ω) - x^2)));
Fd2 := evalf(η·(a x - ω·sqrt(b^2·e^(-(4·k·a·π)/ω) - x^2)));
Wd := ∫_0^(2kπ/ω) η·du^2 dt;
plot([Fd1, Fd2], x=-1.2..1.2, labels=["x", "Fd"],
    thickness=2);
```

$$k := 2$$

$$Fd1 := 0.02001299740x + 0.1075969070\sqrt{0.00009651204338 - 1.x^2}$$

$$Fd2 := 0.02001299740x - 0.1075969070\sqrt{0.00009651204338 - 1.x^2}$$

$$Wd := 0.001482271913$$

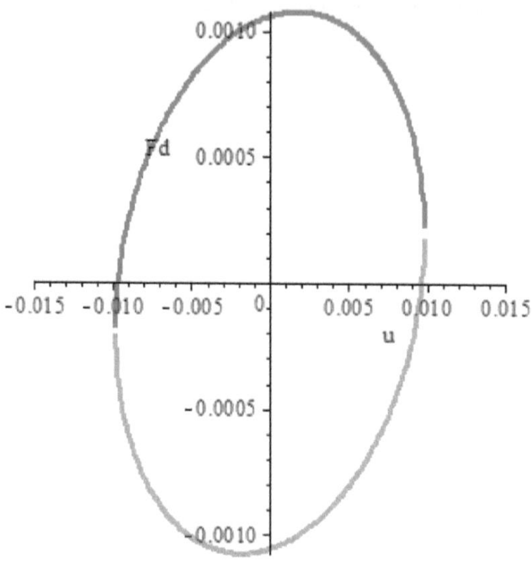

1.4.2.20. The Twentieth Step

This step is exactly like step 19, with the exception of considering K = 3 in the third cycle.

>

$$k := 3;$$

$$Fd1 := evalf\left(\alpha\cdot\left(a\,x+\omega\cdot\sqrt{b^2\cdot e^{-\frac{4\cdot k\cdot a\cdot\pi}{\omega}}-x^2}\right)\right);$$

$$Fd2 := evalf\left(\alpha\cdot\left(a\,x-\omega\cdot\sqrt{b^2\cdot e^{-\frac{4\cdot k\cdot a\cdot\pi}{\omega}}-x^2}\right)\right);$$

$$Wd := \int_0^{\frac{2\,k\,\pi}{\omega}} \eta\cdot du^2\,\mathrm{d}t;$$

$$plot([Fd1, Fd2], x=-3.5\,..3.5,\ labels=["x", "Fd"],\ thickness=2);$$

$$k := 3$$

$Fd1 := 0.1701104779x + 0.9145737098\sqrt{0.000009321523309 - 1.x^2}$

$Fd2 := 0.1701104779x - 0.9145737098\sqrt{0.000009321523309 - 1.x^2}$

$Wd := 0.001494881380$

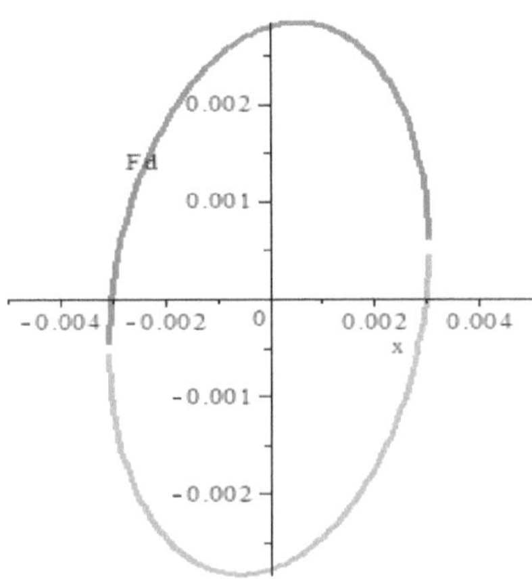

1.4.2.21. The Twenty-First Step

Numerical solution and its comparison with the obtained solution by AGM.

> *restart;*

This command is utilized in order to delete the previous memory.

1.4.2.22. The Twenty-Second Step

> $m := 3; n := 4; \mu := 0.2; \alpha := 1.7; \lambda := 0.3; A := 0.1; \gamma l := 0.2;$

$$m := 3$$

$$n := 4$$

$$\mu := 0.2$$

$$\alpha := 1.7$$

$$\lambda := 0.3$$

$$A := 0.1$$

$$\gamma l := 0.2$$

Identifying the physical parameters of the differential equation like in step 13 is done in this level.

1.4.2.23. The Twenty-Third Step

$$> f := \left(1 + \mu\, u(t)^m\right) \frac{d^2}{dt^2} u(t) + \alpha\, u(t)^n \cdot \left(\frac{d}{dt} u(t)\right) + \gamma l \frac{d}{dt} u(t) + \lambda \cdot \sin(u(t)) = 0;$$

$$f := \left(1 + 0.2\, u(t)^3\right)\left(\frac{d^2}{dt^2} u(t)\right) + 1.7\, u(t)^4 \left(\frac{d}{dt} u(t)\right) + 0.2\left(\frac{d}{dt} u(t)\right) + 0.3 \sin(u(t)) = 0$$

In this part, we write the differential equation in terms of $u(t)$.

1.4.2.24. The Twenty-Fourth Step

$$> ic := u(0) = A, D(u)(0) = 0;$$

$$ic := u(0) = 0.1, D(u)(0) = 0$$

The initial conditions are introduced completely in this step.

1.4.2.25. The Twenty-Fifth Step

> *fl* := *dsolve*([*f*, *ic*], *numeric*, *output* = *listprocedure*);

$$fl := \left[t = \text{proc}(t) \;\ldots\; \text{end proc}, u(t) = \text{proc}(t) \;\ldots\; \text{end proc}, \frac{d}{dt} u(t) = \text{proc}(t) \;\ldots\; \text{end proc} \right]$$

1.4.2.26. The Twenty-Sixth Step

```
> Fig, Numerical;
  plots[odeplot](f1, [t, u(t) ], t=0 ..40, thickness=2);
  plots[odeplot](f1, [t, diff(u(t), t) ],  t=0 ..40, thickness
      =2);
  plots[odeplot](f1, [u(t), diff(u(t), t) ], t=0 ..40, scaling
      = constrained, labels= [ "u(t)", "du(t)" ], thickness=2);
```

Fig, Numerical

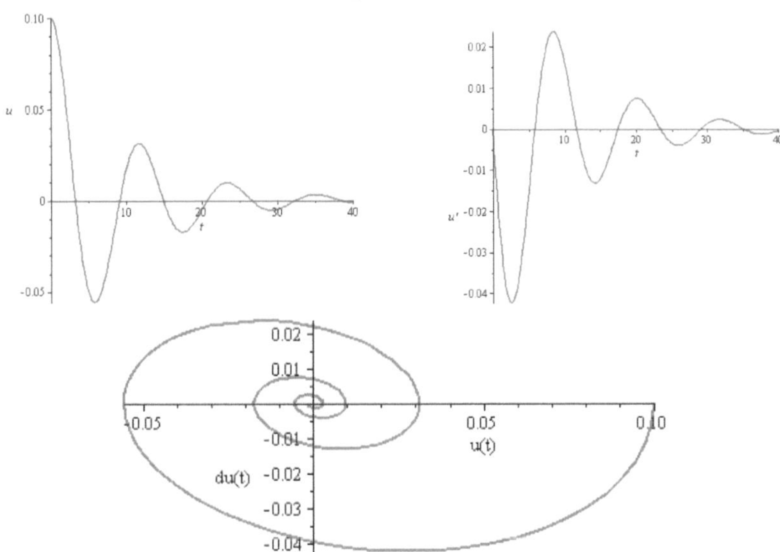

In this step, the obtained results from the differential equation—which are vibrational displacement, velocity, and the related phase plan in the specified domain $t \in \{0, 40\}$—are depicted graphically.

1.4.2.27. The Twenty-Seventh Step

```
> for i from 0  by 4  while  i ≤ 40 do;
  ([x=i, f1(i)[2], f1(i)[3]]);
  end do;
```

$$\left[x=0,\ u(t)(0) = 0.100000000000000,\ \frac{d}{dt}u(t)(0) = 0.\right]$$

$$\left[x=4, u(t)(4) = -0.0264353156866241, \frac{d}{dt}u(t)(4) = -0.0311854128238309\right]$$

$$\left[x=8, u(t)(8) = -0.0254104238026797, \frac{d}{dt}u(t)(8) = 0.0229880771419245\right]$$

$$\left[x=12, u(t)(12) = 0.0306259594553295, \frac{d}{dt}u(t)(12) = -0.00295332161354923\right]$$

$$\left[x=16, u(t)(16) = -0.0111912258934972, \frac{d}{dt}u(t)(16) = -0.00814895180463276\right]$$

$$\left[x=20, u(t)(20) = -0.00549821817420073, \frac{d}{dt}u(t)(20) = 0.00734319117373460\right]$$

$$\left[x=24, u(t)(24) = 0.00908792819077401, \frac{d}{dt}u(t)(24) = -0.00176016142153155\right]$$

$$\left[x=28, u(t)(28) = -0.00423964477560042, \frac{d}{dt}u(t)(28) = -0.00199988943965136\right]$$

$$\left[x=32, u(t)(32) = -0.000953066377886321, \frac{d}{dt}u(t)(32) = 0.00226756261458643\right]$$

$$\left[x=36, u(t)(36) = 0.00260874690018108, \frac{d}{dt}u(t)(36) = -0.000775703770473621\right]$$

$$\left[x=40, u(t)(40) = -0.00149803322730843, \frac{d}{dt}u(t)(40) = -0.000446135199785692\right]$$

This command shows the numerical values of the differential equation consisting of $u(t)$ and $u'(t)$ in the specified domain with the step of four seconds.

1.4.2.28. The Twenty-Eighth Step

>

$$u := subs(f1, u(t));$$
$$du := subs\left(f1, \frac{d}{dt}u(t)\right);$$
$$U := 0.1017150877\, e^{-0.1000649870\, t} \cos(0.5379845352\, t - 0.1838982552);$$
$$dU := -0.01017811893\, e^{-0.1000649870\, t} \cos(0.5379845352\, t - 0.1838982552) - 0.05472114418\, e^{-0.1000649870\, t} \sin(0.5379845352\, t - 0.1838982552);$$

$$u := \mathbf{proc}(t)\ \ldots\ \mathbf{end\ proc}$$
$$du := \mathbf{proc}(t)\ \ldots\ \mathbf{end\ proc}$$

$$U := 0.1017150877\,e^{-0.1000649870\,t}\cos(0.5379845352\,t - 0.1838982552)$$

$$dU := -0.01017811893\,e^{-0.1000649870\,t}\cos(0.5379845352\,t - 0.1838982552) - 0.05472114418\,e^{-0.1000649870\,t}\sin(0.5379845352\,t - 0.1838982552)$$

This command converts the parameters u and du, which are the displacement and velocity of vibration into $u(t)$ and $\frac{du(t)}{dt}$, and then the equations of answer function and its derivative should be written from step 14, and each of them is named as U and dU, respectively.

1.4.2.29. The Twenty-Ninth Step

```
> fig, differ, AGM, Numeric;
  plot( (abs( U − u(t) ) ), t = 0 ..80, thickness = 2, labels
     = [ "t", "|uAGM,uNum,u(t)|" ] );
  plot( abs( dU − du(t) ), t = 0 ..80, thickness = 2, labels
     = [ "t", "|duAGM,duNum,du(t)|" ] );
```

fig, differ, AGM, Numeric

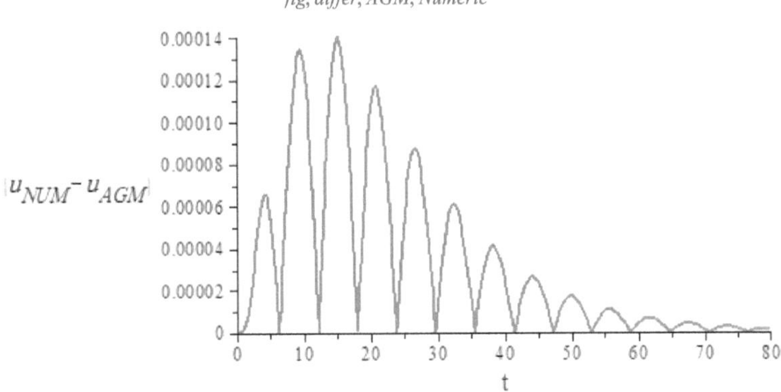

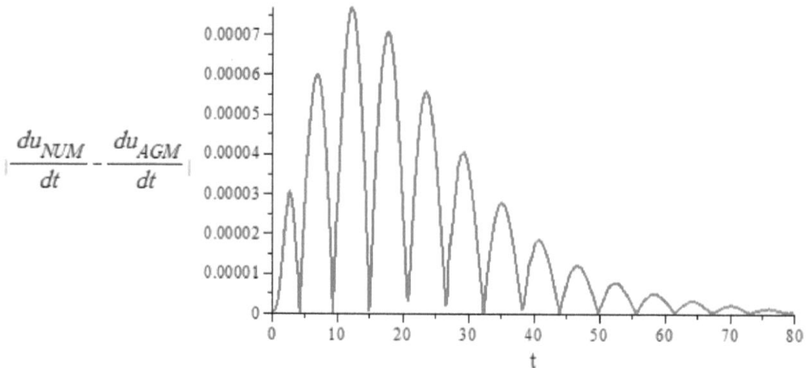

This command is utilized for depicting the absolute value of the difference between the obtained solution by AGM and numerical method.

1.4.2.30. The Thirtieth Step

```
> Numeric, AGM;
  for t from 0 by 4 to 40 do
  't' = t;
  'u(t)' = 0.1017150877 e^(-0.1000649870 t) cos(0.5379845352 t - 0.1838982552) ;
  end do;
```

Numeric, AGM

$t = 0$

$u(t) = 0.1000000001$

$t = 4$

$u(t) = -0.02637113021$

$t = 8$

$u(t) = -0.02550588823$

$t = 12$

$u(t) = 0.03061016995$

$t = 16$

$u(t) = -0.01106812569$

$t = 20$

$u(t) = -0.005603054290$

$t = 24$

$u(t) = 0.009093324296$

$t = 28$

$$u(t) = -0.004174484869$$

$$t = 32$$

$$u(t) = -0.001012219752$$

$$t = 36$$

$$u(t) = 0.002619524740$$

$$t = 40$$

$$u(t) = -0.001472840819$$

By utilizing the outcome of step 14, which is the solution to the differential equation with AGM, we are able to use $u(t)$ in "for loop" in order to obtain the numerical solution. It is notable that for comparing the obtained solution by AGM and numerical method (RKF45), the steps of analytic and numerical solutions should be the same, and the result shows that AGM is very accurate.

1.4.2.31. The Thirty-First Step

Since the damping factor in the considered answer function of vibrational differential equations, $b\{e^{-at}.\cos(\omega t+\varphi)\}$, is e^{-at}—and also since the term $e^{-\xi\omega t}$ is the main factor for damping—we are able to obtain ξ in the following form:

$$e^{-at} \equiv e^{-\xi\omega t} \qquad \rightarrow \qquad \xi = \frac{a}{\omega} \tag{1.58}$$

```
> ξ := abs(a/ω);
```

$$\xi := \frac{1}{4} \left| \left(\left(\eta + \alpha e^{\ln(A)\, n} \right) A \right) \Big/ \left(\left(1 + \mu e^{\ln(A)\, m} \right) \left(- \frac{1}{4 \sin(A) \lambda + 4 \sin(A) \lambda \mu A^m} \left(A \eta^2 + 2 A \eta \alpha A^n - 4 \sin(A) \lambda - 4 \sin(A) \lambda \mu A^m + A \alpha^2 \left(A^n \right)^2 \right) \right)^{1/2} \sqrt{\frac{A}{4 \sin(A) \lambda + 4 \sin(A) \lambda \mu A^m}} \sin(A) \lambda \right) \right|$$

\> 'ξ'= *applyrule*$\Big(\Big[4 \sin(A) \lambda + 4 \sin(A) \lambda \mu A^m = \psi_1,$
$-4 \sin(A) \lambda - 4 \sin(A) \lambda \mu A^m = -\psi_1, A \alpha^2 \left(A^n \right)^2$
$+ 2 A \alpha A^n \eta + A \eta^2 = \psi_2, \alpha e^{\ln(A)\, n} + \eta = \Delta_1, 1$
$+ \mu e^{\ln(A)\, m} = \Delta_2, 4 \sin(A) \lambda + 4 \sin(A) \lambda \mu e^{\ln(A)\, m}$
$= \psi_1, -4 \sin(A) \lambda - 4 \sin(A) \lambda \mu e^{\ln(A)\, m} = -\psi_1, A \eta^2$
$+ 2 A \eta \alpha e^{\ln(A)\, n} + A \alpha^2 \left(e^{\ln(A)\, n} \right)^2 = \psi_2 \Big], \xi \Big);$

$$\xi = \frac{1}{4} \left| \frac{\Delta_1 A}{\Delta_2 \sqrt{- \frac{\psi_2 - \psi_1}{\psi_1}} \sqrt{\frac{A}{\psi_1}} \sin(A) \lambda} \right|.$$

1.4.2.32. The Thirty-Second Step

Overlapping the charts of the obtained solution by AGM and numerical method:

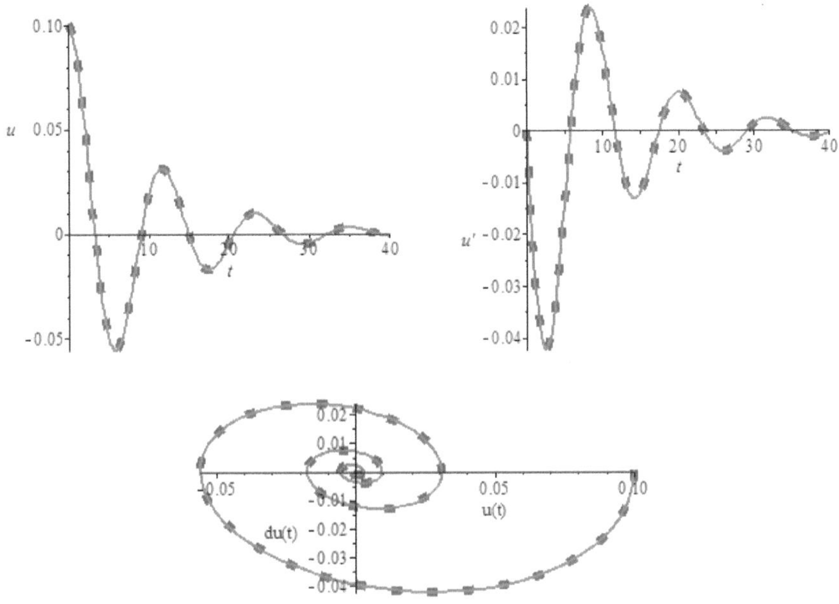

In this level, we are able to graphically compare the obtained results from step 15 and step 26, which are the achieved solution by AGM and numerical method in one chart simultaneously.

To sum up it is better to indicate that we can get the final solution of the problem wholesale by writing every step successively and utilizing the command *Shift* + *Enter* at the end of each step.

1.4.3. Applied Commands for Solving Vibrational Equations with External Force

In this part, one vibrational example with external force and without damping component is considered to be solved.

Consider the following differential equation as

$$\frac{d^2 z(t)}{dt^2}+\alpha z(t)+\beta z^2(t)=\lambda \sin(\Omega t). \tag{1.59}$$

In the above equation, the term $\lambda \sin(\Omega t)$ is the external harmonic force with angular frequency Ω, and λ is defined as the amplitude of external force.

Then, the initial conditions are expressed as

$$z(0) = A \quad , \quad \dot{z}(0)=0. \tag{1.60}$$

In AGM, an answer function for vibrational systems with external harmonic force is considered as follows:

$$z = a\cos(\omega t+\varphi)+b\cos(\Omega t+\phi) \tag{1.61}$$

Where the values a , b , ω , φ and ϕ will be computed by applying the initial conditions.

It is citable that step 1, step 2, …, step 7 are exactly the same as the last example presented as follows:

1.4.3.1. The First Step

> *restart;*

1.4.3.2. The Second Step

$$> z := a\cdot\cos(\omega t+\varphi) + b\cdot\sin(\Omega t+\phi);$$

$$z := a\cos(\omega t+\varphi) + b\sin(\Omega t+\phi)$$

1.4.3.3. The Third Step

$$> f := \frac{d^2}{dt^2} z + \alpha z + \beta z^3 = \lambda\cdot \sin(\Omega t);$$

$$f := -a\cos(\omega t+\varphi)\,\omega^2 - b\sin(\Omega t+\phi)\,\Omega^2 + \alpha\,(a\cos(\omega t+\varphi) + b\sin(\Omega t+\phi)) + \beta\,(a\cos(\omega t+\varphi) + b\sin(\Omega t+\phi))^3 = \lambda\sin(\Omega t)$$

1.4.3.4. The Fourth Step

$> z := \mathit{unapply}(z, t);\ f := \mathit{unapply}(f, t);$

$$z := t \to a\cos(\omega t+\varphi) + b\sin(\Omega t+\phi)$$

$$f := t \to -a\cos(\omega t+\varphi)\,\omega^2 - b\sin(\Omega t+\phi)\,\Omega^2 + \alpha\,(a\cos(\omega t+\varphi) + b\sin(\Omega t+\phi)) + \beta\,(a\cos(\omega t+\varphi) + b\sin(\Omega t+\phi))^3 = \lambda\sin(\Omega t)$$

1.4.3.5. The Fifth Step

$> s1 := z(0) = A;\ s2 := z'(0) = 0;$

$$s1 := a\cos(\varphi) + b\sin(\phi) = A$$

$$s2 := -a\sin(\varphi)\,\omega + b\cos(\phi)\,\Omega = 0$$

1.4.3.6. The Sixth Step

$> s3 := f(0);$

$$s3 := -a\cos(\varphi)\,\omega^2 - b\sin(\phi)\,\Omega^2 + \alpha\,(a\cos(\varphi) + b\sin(\phi)) + \beta\,(a\cos(\varphi) + b\sin(\phi))^3 = 0$$

1.4.3.7. The Seventh Step

$> s4 := f'(0);$

$$s4 := a\sin(\varphi)\,\omega^3 - b\cos(\phi)\,\Omega^3 + \alpha\,(-a\sin(\varphi)\,\omega + b\cos(\phi)\,\Omega) + 3\beta\,(a\cos(\varphi) + b\sin(\phi))^2\,(-a\sin(\varphi)\,\omega + b\cos(\phi)\,\Omega) = \lambda\,\Omega$$

1.4.3.8. The Eighth Step

Apply the initial conditions on the second derivative of the differential equation in terms of t:

$> s5 := f''(0);$

$$s5 := a\cos(\varphi)\,\omega^4 + b\sin(\phi)\,\Omega^4 + \alpha\left(-a\cos(\varphi)\,\omega^2 - b\sin(\phi)\,\Omega^2\right) + 6\beta\left(a\cos(\varphi) + b\sin(\phi)\right)\left(-a\sin(\varphi)\,\omega + b\cos(\phi)\,\Omega\right)^2 + 3\beta\left(a\cos(\varphi) + b\sin(\phi)\right)^2\left(-a\cos(\varphi)\,\omega^2 - b\sin(\phi)\,\Omega^2\right) = 0$$

1.4.3.9. The Ninth Step

$$> S := solve([s1, s2, s3, s4, s5], \{a, b, \omega, \varphi, \phi\});$$

$$S := \Big\{a = (\beta^2 A^4 - 2\beta\Omega^2 A^2 + 2\beta A^2\alpha + \Omega^4 - 2\alpha\Omega^2 + \alpha^2) \Big/ \Big((-2\beta\Omega^2 A^2 + \Omega^4 - 2\alpha\Omega^2 + 3\beta^2 A^4 + 4\beta A^2\alpha + \alpha^2)\, RootOf\big((3\beta^3 A^8 - 4\Omega^2 A^6\beta^2 + 7\beta^2 A^6\alpha - 6\Omega^2 A^4\beta\alpha + A^4\beta\Omega^4 + 5\beta A^4\alpha^2 + A^2\alpha\Omega^4 + A^2\alpha^3 - 2\Omega^2 A^2\alpha^2 + \lambda^2\Omega^2)\,\_Z^2 - 3A^6\beta^3 + 4A^4\beta^2\Omega^2 - 7A^4\beta^2\alpha + 6A^2\beta\alpha\Omega^2 - A^2\beta\Omega^4 - 5\beta A^2\alpha^2 + 2\alpha^2\Omega^2 - \alpha\Omega^4 - \alpha^3\big)\Big), b$$

$$= \frac{1}{-2\beta\Omega^2 A^2 + \Omega^4 - 2\alpha\Omega^2 + 3\beta^2 A^4 + 4\beta A^2\alpha + \alpha^2}\big(RootOf(-2\beta A^2\alpha\lambda^2 + 2\beta A^2\lambda^2\Omega^2 - 4\beta^4 A^{10} - 4\beta^2 A^6\alpha^2 - 8\beta^3 A^8\alpha - \beta^2 A^4\lambda^2 + 2\Omega^2\alpha\lambda^2 - \alpha^2\lambda^2 - \Omega^4\lambda^2 + \_Z^2)\big), \phi$$

$$= \arctan\Big((2A^3\beta(\beta A^2 + \alpha)) \Big/ \big(RootOf(-2\beta A^2\alpha\lambda^2 + 2\beta A^2\lambda^2\Omega^2 - 4\beta^4 A^{10} - 4\beta^2 A^6\alpha^2 - 8\beta^3 A^8\alpha - \beta^2 A^4\lambda^2 + 2\Omega^2\alpha\lambda^2 - \alpha^2\lambda^2 - \Omega^4\lambda^2 + \_Z^2)\big), (\lambda(\beta A^2 - \Omega^2 + \alpha)) \Big/ \big(RootOf(-2\beta A^2\alpha\lambda^2 + 2\beta A^2\lambda^2\Omega^2 - 4\beta^4 A^{10} - 4\beta^2 A^6\alpha^2 - 8\beta^3 A^8\alpha - \beta^2 A^4\lambda^2 + 2\Omega^2\alpha\lambda^2 - \alpha^2\lambda^2 - \Omega^4\lambda^2 + \_Z^2)\big)\Big), \omega = RootOf\big((\beta A^2 - \Omega^2 + \alpha)\,\_Z^2 + \beta\Omega^2 A^2 + \alpha\Omega^2 - 3\beta^2 A^4 - 4\beta A^2\alpha - \alpha^2\big), \varphi = \arctan\Big(\big(\lambda\Omega\, RootOf\big((3\beta^3 A^8 - 4\Omega^2 A^6\beta^2 + 7\beta^2 A^6\alpha - 6\Omega^2 A^4\beta\alpha + A^4\beta\Omega^4 + 5\beta A^4\alpha^2 + A^2\alpha\Omega^4 + A^2\alpha^3 - 2\Omega^2 A^2\alpha^2 + \lambda^2\Omega^2)\,\_Z^2 - 3A^6\beta^3 + 4A^4\beta^2\Omega^2 - 7A^4\beta^2\alpha + 6A^2\beta\alpha\Omega^2 - A^2\beta\Omega^4 - 5\beta A^2\alpha^2 + 2\alpha^2\Omega^2 - \alpha\Omega^4 - \alpha^3\big)\big) \Big/ \big((\beta A^2 - \Omega^2 + \alpha)\, RootOf((\beta A^2 - \Omega^2 + \alpha)\,\_Z^2 + \beta\Omega^2 A^2 + \alpha\Omega^2 - 3\beta^2 A^4 - 4\beta A^2\alpha - \alpha^2)\big), RootOf\big((3\beta^3 A^8 - 4\Omega^2 A^6\beta^2 + 7\beta^2 A^6\alpha - 6\Omega^2 A^4\beta\alpha + A^4\beta\Omega^4 + 5\beta A^4\alpha^2 + A^2\alpha\Omega^4 + A^2\alpha^3 - 2\Omega^2 A^2\alpha^2 + \lambda^2\Omega^2)\,\_Z^2 - 3A^6\beta^3 + 4A^4\beta^2\Omega^2 - 7A^4\beta^2\alpha + 6A^2\beta\alpha\Omega^2 - A^2\beta\Omega^4 - 5\beta A^2\alpha^2 + 2\alpha^2\Omega^2 - \alpha\Omega^4 - \alpha^3\big)\, A\Big)\Big\}$$

This command is used in order to solve the obtained equations from steps 5 to 8.

1.4.3.10. The Tenth Step

> $S := convert(S, radical);$

$$S := \left\{a = \left(A^4\beta^2 - 2A^2\Omega^2\beta + 2A^2\alpha\beta + \Omega^4 - 2\Omega^2\alpha + \alpha^2\right) \Big/ \left(\left(3A^4\beta^2 - 2A^2\Omega^2\beta + 4A^2\alpha\beta + \Omega^4 - 2\Omega^2\alpha + \alpha^2\right)\sqrt{-\frac{-3A^6\beta^3 + 4A^4\Omega^2\beta^2 - 7A^4\alpha\beta^2 - A^2\Omega^4\beta + 6A^2\Omega^2\alpha\beta - 5A^2\alpha^2\beta - \Omega^4\alpha + 2\Omega^2\alpha^2 - \alpha^3}{3A^8\beta^3 - 4A^6\Omega^2\beta^2 + 7A^6\alpha\beta^2 + A^4\Omega^4\beta - 6A^4\Omega^2\alpha\beta + 5A^4\alpha^2\beta + A^2\Omega^4\alpha - 2A^2\Omega^2\alpha^2 + A^2\alpha^3 + \Omega^2\lambda^2}}\right), b\right.$$

$$= \frac{\sqrt{4A^{10}\beta^4 + 8A^8\alpha\beta^3 + 4A^6\alpha^2\beta^2 + A^4\beta^2\lambda^2 - 2A^2\Omega^2\beta\lambda^2 + 2A^2\alpha\beta\lambda^2 + \Omega^4\lambda^2 - 2\Omega^2\alpha\lambda^2 + \alpha^2\lambda^2}}{3A^4\beta^2 - 2A^2\Omega^2\beta + 4A^2\alpha\beta + \Omega^4 - 2\Omega^2\alpha + \alpha^2}, \phi$$

$$= \arctan\left(\frac{2A^3\beta\left(A^2\beta + \alpha\right)}{\sqrt{4A^{10}\beta^4 + 8A^8\alpha\beta^3 + 4A^6\alpha^2\beta^2 + A^4\beta^2\lambda^2 - 2A^2\Omega^2\beta\lambda^2 + 2A^2\alpha\beta\lambda^2 + \Omega^4\lambda^2 - 2\Omega^2\alpha\lambda^2 + \alpha^2\lambda^2}}, \frac{\lambda\left(A^2\beta - \Omega^2 + \alpha\right)}{\sqrt{4A^{10}\beta^4 + 8A^8\alpha\beta^3 + 4A^6\alpha^2\beta^2 + A^4\beta^2\lambda^2 - 2A^2\Omega^2\beta\lambda^2 + 2A^2\alpha\beta\lambda^2 + \Omega^4\lambda^2 - 2\Omega^2\alpha\lambda^2 + \alpha^2\lambda^2}}\right), \omega$$

$$= \sqrt{-\frac{-3A^4\beta^2 + A^2\Omega^2\beta - 4A^2\alpha\beta + \Omega^2\alpha - \alpha^2}{A^2\beta - \Omega^2 + \alpha}}, \varphi$$

$$\left.= \arctan\left(\frac{\lambda\Omega\sqrt{-\frac{-3A^6\beta^3 + 4A^4\Omega^2\beta^2 - 7A^4\alpha\beta^2 - A^2\Omega^4\beta + 6A^2\Omega^2\alpha\beta - 5A^2\alpha^2\beta - \Omega^4\alpha + 2\Omega^2\alpha^2 - \alpha^3}{3A^8\beta^3 - 4A^6\Omega^2\beta^2 + 7A^6\alpha\beta^2 + A^4\Omega^4\beta - 6A^4\Omega^2\alpha\beta + 5A^4\alpha^2\beta + A^2\Omega^4\alpha - 2A^2\Omega^2\alpha^2 + A^2\alpha^3 + \Omega^2\lambda^2}}}{\left(A^2\beta - \Omega^2 + \alpha\right)\sqrt{-\frac{-3A^4\beta^2 + A^2\Omega^2\beta - 4A^2\alpha\beta + \Omega^2\alpha - \alpha^2}{A^2\beta - \Omega^2 + \alpha}}}, \sqrt{-\frac{-3A^6\beta^3 + 4A^4\Omega^2\beta^2 - 7A^4\alpha\beta^2 - A^2\Omega^4\beta + 6A^2\Omega^2\alpha\beta - 5A^2\alpha^2\beta - \Omega^4\alpha + 2\Omega^2\alpha^2 - \alpha^3}{3A^8\beta^3 - 4A^6\Omega^2\beta^2 + 7A^6\alpha\beta^2 + A^4\Omega^4\beta - 6A^4\Omega^2\alpha\beta + 5A^4\alpha^2\beta + A^2\Omega^4\alpha - 2A^2\Omega^2\alpha^2 + A^2\alpha^3 + \Omega^2\lambda^2}}\,A\right)\right\}$$

This step is utilized to remove "Root Of" like step 10 in example 1.

1.4.3.11. The Eleventh Step

Extract the constant values of the answer function of equation 1.61:

```
> a := rhs(S[1]);
  b := rhs(S[2]);
  ϕ := rhs(S[3]);
  ω := rhs(S[4]);
  φ := rhs(S[5]);
```

$$a := \left(\beta^2 A^4 - 2\beta\Omega^2 A^2 + 2\beta A^2\alpha + \Omega^4 - 2\alpha\Omega^2 + \alpha^2\right) \Big/ \Bigg(\left(-2\beta\Omega^2 A^2 + \Omega^4 - 2\alpha\Omega^2 + 3\beta^2 A^4 + 4\beta A^2\alpha + \alpha^2\right)$$

$$\left(-\left(-3A^6\beta^3 + 4A^4\beta^2\Omega^2 - 7A^4\beta^2\alpha + 6A^2\beta\alpha\Omega^2 - A^2\beta\Omega^4 - 5\beta A^2\alpha^2 + 2\alpha^2\Omega^2 - \alpha\Omega^4 - \alpha\right.\right.$$

$$\left.\left. + 7\beta^2 A^6\alpha - 6\Omega^2 A^4\beta\alpha + A^4\beta\Omega^4 + 5\beta A^4\alpha^2 + A^2\alpha\Omega^4 + A^2\alpha^3 - 2\Omega^2 A^2\alpha^2 + \lambda^2\Omega^2\right)\right)^{1/2}\Bigg)$$

$$b := \frac{1}{-2\beta\Omega^2 A^2 + \Omega^4 - 2\alpha\Omega^2 + 3\beta^2 A^4 + 4\beta A^2\alpha + \alpha^2} \left(2\beta A^2\alpha\lambda^2 - 2\beta A^2\lambda^2\Omega^2 + 4\beta^4 A^{10} + 4\beta^2 A^6\alpha^2 + 8\beta^3 A^8\alpha + \beta^2 A^4\lambda^2 - 2\Omega^2\alpha\lambda^2 + \alpha^2\lambda^2 + \Omega^4\lambda^2\right)^{1/2}$$

$$\phi := \arctan\Bigg(\left(2A^3\beta\left(\beta A^2 + \alpha\right)\right) \Big/ \left(2\beta A^2\alpha\lambda^2 - 2\beta A^2\lambda^2\Omega^2 + 4\beta^4 A^{10} + 4\beta^2 A^6\alpha^2 + 8\beta^3 A^8\alpha + \beta^2 A^4\lambda^2 - 2\Omega^2\alpha\lambda^2 + \alpha^2\lambda^2 + \Omega^4\lambda^2\right)^{1/2}, \left(\lambda\left(\beta A^2 - \Omega^2 + \alpha\right)\right) \Big/ \left(2\beta A^2\alpha\lambda^2 - 2\beta A^2\lambda^2\Omega^2 + 4\beta^4 A^{10} + 4\beta^2 A^6\alpha^2 + 8\beta^3 A^8\alpha + \beta^2 A^4\lambda^2 - 2\Omega^2\alpha\lambda^2 + \alpha^2\lambda^2 + \Omega^4\lambda^2\right)^{1/2}\Bigg)$$

$$\omega := \sqrt{-\frac{\beta\Omega^2 A^2 + \alpha\Omega^2 - 3\beta^2 A^4 - 4\beta A^2\alpha - \alpha^2}{\beta A^2 - \Omega^2 + \alpha}}$$

$$\varphi := \arctan\Bigg(\Bigg(\lambda\Omega\left(-\left(-3A^6\beta^3 + 4A^4\beta^2\Omega^2 - 7A^4\beta^2\alpha + 6A^2\beta\alpha\Omega^2 - A^2\beta\Omega^4 - 5\beta A^2\alpha^2 + 2\alpha^2\Omega^2 - \alpha\Omega^4 - \alpha^3\right) \Big/ \left(3\beta^3 A^8 - 4\Omega^2 A^6\beta^2 + 7\beta^2 A^6\alpha - 6\Omega^2 A^4\beta\alpha + A^4\beta\Omega^4 + 5\beta A^4\alpha^2 + A^2\alpha\Omega^4 + A^2\alpha^3 - 2\Omega^2 A^2\alpha^2 + \lambda^2\Omega^2\right)\right)^{1/2}\Bigg) \Big/ \Bigg(\left(\beta A^2 - \Omega^2\right.$$

$$+\alpha)\sqrt{-\frac{\beta\Omega^2 A^2+\alpha\Omega^2-3\beta^2A^4-4\beta A^2\alpha-\alpha^2}{\beta A^2-\Omega^2+\alpha}}\Bigg),$$

$$(-(-3A^6\beta^3+4A^4\beta^2\Omega^2-7A^4\beta^2\alpha+6A^2\beta\alpha\Omega^2-A^2\beta\Omega^4-5\beta A^2\alpha^2$$

$$+2\alpha^2\Omega^2-\alpha\Omega^4-\alpha^3)/(3\beta^3A^8-4\Omega^2A^6\beta^2+7\beta^2A^6\alpha-6\Omega^2A^4\beta\alpha+A^4\beta\Omega^4+5\beta A^4\alpha^2$$

$$+A^2\alpha\Omega^4+A^2\alpha^3-2\Omega^2A^2\alpha^2+\lambda^2\Omega^2))^{1/2}A\Big)$$

1.4.3.12. The Twelfth Step

Substitute the constant values a , b , ω , φ and ϕ from step 10 into the answer function of equation 1.61:

$$> z := (eval((a\cdot\cos(\omega t+\varphi))+b\cdot\sin(\Omega t+\phi),S));$$

$$z := \Bigg((\beta^2A^4-2\beta\Omega^2A^2+2\beta A^2\alpha+\Omega^4-2\alpha\Omega^2$$

$$+\alpha^2)\cos\Bigg(\sqrt{-\frac{\beta\Omega^2A^2+\alpha\Omega^2-3\beta^2A^4-4\beta A^2\alpha-\alpha^2}{\beta A^2-\Omega^2+\alpha}}\,t$$

$$+\arctan\Bigg(\Big(\lambda\Omega(-(-3A^6\beta^3+4A^4\beta^2\Omega^2-7A^4\beta^2\alpha+6A^2\beta\alpha\Omega^2-A^2\beta\Omega^4$$

$$-5\beta A^2\alpha^2+2\alpha^2\Omega^2-\alpha\Omega^4-\alpha^3)/(3\beta^3A^8-4\Omega^2A^6\beta^2+7\beta^2A^6\alpha-6\Omega^2A^4\beta\alpha+A^4\beta\Omega^4$$

$$+5\beta A^4\alpha^2+A^2\alpha\Omega^4+A^2\alpha^3-2\Omega^2A^2\alpha^2+\lambda^2\Omega^2))^{1/2}\Big)\Big/\Big((\beta A^2-\Omega^2$$

$$+\alpha)\sqrt{-\frac{\beta\Omega^2A^2+\alpha\Omega^2-3\beta^2A^4-4\beta A^2\alpha-\alpha^2}{\beta A^2-\Omega^2+\alpha}}\Big),$$

$$(-(-3A^6\beta^3+4A^4\beta^2\Omega^2-7A^4\beta^2\alpha+6A^2\beta\alpha\Omega^2-A^2\beta\Omega^4-5\beta A^2\alpha^2$$

$$+2\alpha^2\Omega^2-\alpha\Omega^4-\alpha^3)/(3\beta^3A^8-4\Omega^2A^6\beta^2+7\beta^2A^6\alpha-6\Omega^2A^4\beta\alpha+A^4\beta\Omega^4+5\beta A^4\alpha^2$$

$$+A^2\alpha\Omega^4+A^2\alpha^3-2\Omega^2A^2\alpha^2+\lambda^2\Omega^2))^{1/2}A\Big)\Bigg)\Bigg)\Big/\Big((-2\beta\Omega^2A^2+\Omega^4-2\alpha\Omega^2$$

$$+3\beta^2A^4+4\beta A^2\alpha+\alpha^2)$$

$$(-(-3A^6\beta^3+4A^4\beta^2\Omega^2-7A^4\beta^2\alpha+6A^2\beta\alpha\Omega^2-A^2\beta\Omega^4-5\beta A^2\alpha^2+2\alpha^2\Omega^2-\alpha\Omega^4-\alpha$$

$$+7\beta^2A^6\alpha-6\Omega^2A^4\beta\alpha+A^4\beta\Omega^4+5\beta A^4\alpha^2+A^2\alpha\Omega^4+A^2\alpha^3-2\Omega^2A^2\alpha^2+\lambda^2\Omega^2))^{1/2}\Big)$$

$$+\frac{1}{-2\beta\Omega^2A^2+\Omega^4-2\alpha\Omega^2+3\beta^2A^4+4\beta A^2\alpha+\alpha^2}\Bigg($$

$$(2\beta A^2\alpha\lambda^2-2\beta A^2\lambda^2\Omega^2+4\beta^4A^{10}+4\beta^2A^6\alpha^2+8\beta^3A^8\alpha+\beta^2A^4\lambda^2-2\Omega^2\alpha\lambda^2$$

$$+\alpha^2\lambda^2+\Omega^4\lambda^2)^{1/2}\sin\Bigg(\Omega t+\arctan\Bigg((2A^3\beta(\beta A^2+\alpha))\Big/$$

$$(2\beta A^2\alpha\lambda^2-2\beta A^2\lambda^2\Omega^2+4\beta^4A^{10}+4\beta^2A^6\alpha^2+8\beta^3A^8\alpha+\beta^2A^4\lambda^2-2\Omega^2\alpha\lambda^2$$

$$+\alpha^2\lambda^2+\Omega^4\lambda^2)^{1/2},(\lambda(\beta A^2-\Omega^2+\alpha))\Big/$$

$$(2\beta A^2\alpha\lambda^2-2\beta A^2\lambda^2\Omega^2+4\beta^4A^{10}+4\beta^2A^6\alpha^2+8\beta^3A^8\alpha+\beta^2A^4\lambda^2-2\Omega^2\alpha\lambda^2$$

$$+\alpha^2\lambda^2+\Omega^4\lambda^2)^{1/2}\Bigg)\Bigg)\Bigg)$$

1.4.3.13. The Thirteenth Step

$$\psi_1=-(\alpha\Omega^2-3\beta^2A^4-4\beta A^2\alpha+\beta\Omega^2A^2-\alpha^2);$$
$$\psi_2=-(-3A^6\beta^3+4A^4\beta^2\Omega^2-7A^4\beta^2\alpha-5\beta A^2\alpha^2-A^2\beta\Omega^4+6A^2\beta\alpha\Omega^2-\alpha^3+2\alpha^2\Omega^2-\alpha\Omega^4);$$
$$\psi_3=3\beta^3A^8-4\Omega^2A^6\beta^2+7\beta^2A^6\alpha+A^4\beta\Omega^4+5\beta A^4\alpha^2-6\Omega^2A^4\beta\alpha+A^2\alpha^3+A^2\alpha\Omega^4$$
$$-2\Omega^2A^2\alpha^2+\lambda^2\Omega^2;$$
$$\psi_4=4\beta^4A^{10}+8\beta^3A^8\alpha+\beta^2A^4\lambda^2+4\beta^2A^6\alpha^2+\alpha^2\lambda^2+\Omega^4\lambda^2-2\Omega^2\alpha\lambda^2+2\beta A^2\alpha\lambda^2$$
$$-2\beta A^2\lambda^2\Omega^2;$$
$$\Delta_1=-2\beta\Omega^2A^2+\Omega^4-2\alpha\Omega^2+3\beta^2A^4+4\beta A^2\alpha+\alpha^2;$$
$$\Delta_2=\beta^2A^4-2\beta\Omega^2A^2+2\beta A^2\alpha+\Omega^4-2\alpha\Omega^2+\alpha^2;$$
$$\Lambda=\alpha+\beta A^2;$$
$$'z(t)\coloneqq applyrule\big([\alpha\Omega^2-3\beta^2A^4-4\beta A^2\alpha+\beta\Omega^2A^2-\alpha^2=-\psi_1,-3A^6\beta^3+4A^4\beta^2\Omega^2-7A^4\beta^2\alpha$$
$$-5\beta A^2\alpha^2-A^2\beta\Omega^4+6A^2\beta\alpha\Omega^2-\alpha^3+2\alpha^2\Omega^2-\alpha\Omega^4=-\psi_2,3\beta^3A^8-4\Omega^2A^6\beta^2$$
$$+7\beta^2A^6\alpha+A^4\beta\Omega^4+5\beta A^4\alpha^2-6\Omega^2A^4\beta\alpha+A^2\alpha^3+A^2\alpha\Omega^4-2\Omega^2A^2\alpha^2+\lambda^2\Omega^2=\psi_3,$$
$$4\beta^4A^{10}+8\beta^3A^8\alpha+\beta^2A^4\lambda^2+4\beta^2A^6\alpha^2+\alpha^2\lambda^2+\Omega^4\lambda^2-2\Omega^2\alpha\lambda^2+2\beta A^2\alpha\lambda^2$$
$$-2\beta A^2\lambda^2\Omega^2=\psi_4,-2\beta\Omega^2A^2+\Omega^4-2\alpha\Omega^2+3\beta^2A^4+4\beta A^2\alpha+\alpha^2=\Delta_1,\beta^2A^4-2\beta\Omega^2A^2$$
$$+2\beta A^2\alpha+\Omega^4-2\alpha\Omega^2+\alpha^2=\Delta_2,\alpha+\beta A^2=\Lambda],z\big);$$
$$'\omega\coloneqq applyrule\big([\alpha\Omega^2-3\beta^2A^4-4\beta A^2\alpha+\beta\Omega^2A^2-\alpha^2=-\psi_1,\alpha+\beta A^2=\Lambda],\omega\big);$$

$$\psi_1 = -\beta\Omega^2 A^2 - \alpha\Omega^2 + 3\beta^2 A^4 + 4\beta A^2\alpha + \alpha^2$$

$$\psi_2 = 3A^6\beta^3 - 4A^4\beta^2\Omega^2 + 7A^4\beta^2\alpha - 6A^2\beta\alpha\Omega^2 + A^2\beta\Omega^4 + 5\beta A^2\alpha^2 + \alpha\Omega^4 + \alpha^3 - 2\alpha^2\Omega^2$$

$$\psi_3 = 3\beta^3 A^8 - 4\Omega^2 A^6\beta^2 + 7\beta^2 A^6\alpha - 6\Omega^2 A^4\beta\alpha + A^4\beta\Omega^4 + 5\beta A^4\alpha^2 + A^2\alpha\Omega^4 + A^2\alpha^3 - 2\Omega^2 A^2\alpha^2 + \lambda^2\Omega^2$$

$$\psi_4 = 2\beta A^2\alpha\lambda^2 - 2\beta A^2\lambda^2\Omega^2 + 4\beta^4 A^{10} + 4\beta^2 A^6\alpha^2 + 8\beta^3 A^8\alpha + \beta^2 A^4\lambda^2 - 2\Omega^2\alpha\lambda^2 + \alpha^2\lambda^2 + \Omega^4\lambda^2$$

$$\Delta_1 = -2\beta\Omega^2 A^2 + \Omega^4 - 2\alpha\Omega^2 + 3\beta^2 A^4 + 4\beta A^2\alpha + \alpha^2$$

$$\Delta_2 = \beta^2 A^4 - 2\beta\Omega^2 A^2 + 2\beta A^2\alpha + \Omega^4 - 2\alpha\Omega^2 + \alpha^2$$

$$\Lambda = \beta A^2 + \alpha$$

$$z(t) = \frac{\Delta_2\cos\left(\sqrt{\frac{\psi_1}{\Lambda-\Omega^2}}\,t + \arctan\left(\frac{\lambda\Omega\sqrt{\frac{\psi_2}{\psi_3}}}{\left(\Lambda-\Omega^2\right)\sqrt{\frac{\psi_1}{\Lambda-\Omega^2}}}, \sqrt{\frac{\psi_2}{\psi_3}}\,A\right)\right)}{\Delta_1\sqrt{\frac{\psi_2}{\psi_3}}} + \frac{\sqrt{\psi_4}\sin\left(\Omega t + \arctan\left(\frac{2A^3\beta\Lambda}{\sqrt{\psi_4}}, \frac{\lambda\left(\Lambda-\Omega^2\right)}{\sqrt{\psi_4}}\right)\right)}{\Delta_1}$$

$$\omega = \sqrt{\frac{\psi_1}{\Lambda-\Omega^2}}$$

In order to beautify the outcome of step 12 and choose the suitable new variable, this command is applied; and also, the angular frequency in this part is written in terms of the new variables.

1.4.3.14. The Fourteenth Step

Select numerical values for the physical parameters α , β , λ , Ω , A in the differential equation and its initial conditions:

```
> λ:= 0.2; α:= 0.6; β:=0.1; A:=0.1; Ω:=2;
```

$$\lambda := 0.2$$
$$\alpha := 0.6$$
$$\beta := 0.1$$
$$A := 0.1$$
$$\Omega := 2.$$

1.4.3.15. The Fifteenth Step

Calling on the outcome of step 12:

```
> ω := ω; z := z; dz := diff(z, t);
```

$$\omega := 0.7750137848$$

$$z := 0.1817963426\cos(0.7750137848\,t - 0.9884214942) + 0.05883471527\sin(2\,t + 3.141415837)$$

$$dz := -0.1408946715\sin(0.7750137848\,t - 0.9884214942) + 0.1176694305\cos(2\,t + 3.141415837)$$

1.4.3.16. The Sixteenth Step

Depict the charts of the answer function, its derivative, and the related phase plan:

```
>
Fig, AGM;
plot(z, t=0 ..50, axes=boxed, labels=["t", "z(t)"], thickness=2);
Derivation;
plot(dz, t=0 ..50, axes=boxed, labels=["t", "dz(t)"], thickness=2);
Phase Plane, AGM;
plot([z, dz, t=0 ..20], scaling=constrained, thickness=2, axes=boxed,
    labels=["z(t)", "dz(t)"]);
```

Fig, AGM

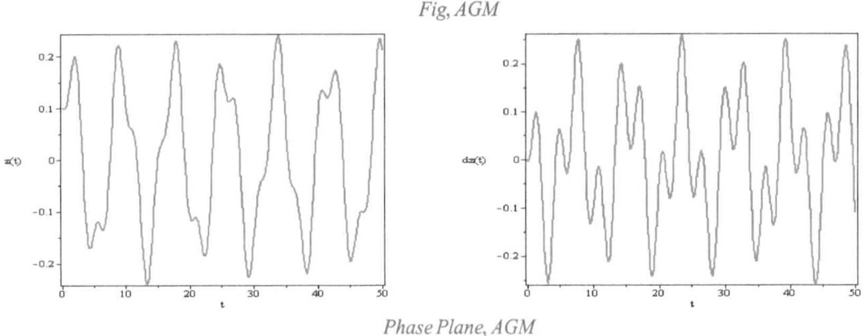

Phase Plane, AGM

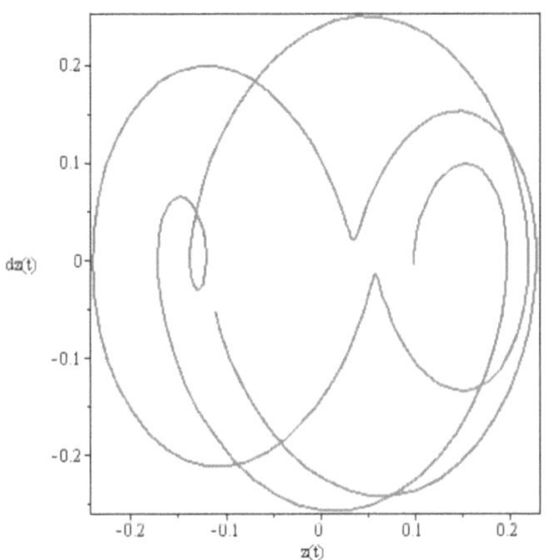

1.4.3.17. The Seventeenth Step

> *restart;*

$$a := \left(\beta^2 A^4 - 2\beta\Omega^2 A^2 + 2\beta A^2\alpha + \Omega^4 - 2\alpha\Omega^2 + \alpha^2\right) \Big/ \Big(\left(-2\beta\Omega^2 A^2 + \Omega^4 - 2\alpha\Omega^2 + 3\beta^2 A^4 + 4\beta A^2\alpha + \alpha^2\right)$$

$$\left(-\left(-3A^6\beta^3 - 7A^4\beta^2\alpha + 4A^4\beta^2\Omega^2 + 6A^2\beta\alpha\Omega^2 - A^2\beta\Omega^4 - 5\beta A^2\alpha^2 + 2\alpha^2\Omega^2 - \alpha\Omega^4 - o + 7\beta^2 A^6\alpha + A^4\beta\Omega^4 - 6\Omega^2 A^4\beta\alpha + 5\beta A^4\alpha^2 + A^2\alpha^3 + A^2\alpha\Omega^4 - 2\Omega^2 A^2\alpha^2 + \lambda^2\Omega^2\right)\right)^{1/2}\Big)$$

;

b

$$:= \frac{1}{-2\beta\Omega^2A^2+\Omega^4-2\alpha\Omega^2+3\beta^2A^4+4\beta A^2\alpha+\alpha^2}$$
$$\left(4\beta^4A^{10}+4\beta^2A^6\alpha^2+2\beta A^2\alpha\lambda^2+\Omega^4\lambda^2+\alpha^2\lambda^2+8\beta^3A^8\alpha-2\beta A^2\lambda^2\Omega^2\right.$$
$$\left.-2\Omega^2\alpha\lambda^2+\beta^2A^4\lambda^2\right)^{1/2};$$

$$\phi := \arctan\Big(\left(2A^3\beta(\alpha+\beta A^2)\right) \Big/$$
$$\left(4\beta^4A^{10}+4\beta^2A^6\alpha^2+2\beta A^2\alpha\lambda^2+\Omega^4\lambda^2+\alpha^2\lambda^2+8\beta^3A^8\alpha-2\beta A^2\lambda^2\Omega^2\right.$$
$$\left.-2\Omega^2\alpha\lambda^2+\beta^2A^4\lambda^2\right)^{1/2}, \left(\lambda(\beta A^2-\Omega^2+\alpha)\right) \Big/$$
$$\left(4\beta^4A^{10}+4\beta^2A^6\alpha^2+2\beta A^2\alpha\lambda^2+\Omega^4\lambda^2+\alpha^2\lambda^2+8\beta^3A^8\alpha-2\beta A^2\lambda^2\Omega^2\right.$$
$$\left.-2\Omega^2\alpha\lambda^2+\beta^2A^4\lambda^2\right)^{1/2}\Big);$$

$$\omega := \sqrt{-\frac{\beta\Omega^2A^2+\alpha\Omega^2-3\beta^2A^4-4\beta A^2\alpha-\alpha^2}{\beta A^2-\Omega^2+\alpha}};$$

$$a := \left(\beta^2A^4-2\beta\Omega^2A^2+2\beta A^2\alpha+\Omega^4-2\alpha\Omega^2+\alpha^2\right) \Big/ \Big(\left(-2\beta\Omega^2A^2+\Omega^4-2\alpha\Omega^2+3\beta^2A^4\right.$$
$$\left.+4\beta A^2\alpha+\alpha^2\right)$$
$$\left(-\left(-3A^6\beta^3+4A^4\beta^2\Omega^2-7A^4\beta^2\alpha+6A^2\beta\alpha\Omega^2-A^2\beta\Omega^4-5\beta A^2\alpha^2+2\alpha^2\Omega^2-\alpha\Omega^4-o\right.\right.$$
$$\left.\left.+7\beta^2A^6\alpha-6\Omega^2A^4\beta\alpha+A^4\beta\Omega^4+5\beta A^4\alpha^2+A^2\alpha\Omega^4+A^2\alpha^3-2\Omega^2A^2\alpha^2+\lambda^2\Omega^2\right)\right)^{1/2}\Big)$$

$$b :=$$
$$\frac{1}{-2\beta\Omega^2A^2+\Omega^4-2\alpha\Omega^2+3\beta^2A^4+4\beta A^2\alpha+\alpha^2}$$
$$\left(2\beta A^2\alpha\lambda^2-2\beta A^2\lambda^2\Omega^2+4\beta^4A^{10}+4\beta^2A^6\alpha^2+8\beta^3A^8\alpha+\beta^2A^4\lambda^2-2\Omega^2\alpha\lambda^2\right.$$
$$\left.+\alpha^2\lambda^2+\Omega^4\lambda^2\right)^{1/2}$$

$$\phi := \arctan\Big(\left(2A^3\beta(A^2\beta+\alpha)\right) \Big/$$
$$\left(2\beta A^2\alpha\lambda^2-2\beta A^2\lambda^2\Omega^2+4\beta^4A^{10}+4\beta^2A^6\alpha^2+8\beta^3A^8\alpha+\beta^2A^4\lambda^2-2\Omega^2\alpha\lambda^2\right.$$
$$\left.+\alpha^2\lambda^2+\Omega^4\lambda^2\right)^{1/2}, \left(\lambda(\alpha-\Omega^2+A^2\beta)\right) \Big/$$
$$\left(2\beta A^2\alpha\lambda^2-2\beta A^2\lambda^2\Omega^2+4\beta^4A^{10}+4\beta^2A^6\alpha^2+8\beta^3A^8\alpha+\beta^2A^4\lambda^2-2\Omega^2\alpha\lambda^2\right.$$
$$\left.+\alpha^2\lambda^2+\Omega^4\lambda^2\right)^{1/2}\Big)$$

$$\omega := \sqrt{-\frac{\beta\Omega^2A^2+\alpha\Omega^2-3\beta^2A^4-4\beta A^2\alpha-\alpha^2}{\alpha-\Omega^2+A^2\beta}}$$

The command "restart" is used to obtain the other vibrational parameters.

The program in this part is started by the command "restart". It is necessary to mention that the response of the external harmonic force while vibrating can be computed by copying the constant values of the answer function—which are a , b , ω , φ and ϕ from step 11.

1.4.3.18. The Eighteenth Step

The response of external harmonic force while vibration is achieved in the following form:

```
>
Fd := b·cos(Ω·t + ϕ);
'Fd(t)' = applyrule([4β^4 A^10 + 8β^3 A^8 α + β^2 A^4 λ^2 + 4β^2 A^6 α^2 + α^2 λ^2 + Ω^4 λ^2 − 2Ω^2 α λ^2
    + 2β A^2 α λ^2 − 2β A^2 λ^2 Ω^2 = ψ4, -2β Ω^2 A^2 + Ω^4 − 2α Ω^2 + 3β^2 A^4 + 4β A^2 α + α^2 = Δ1, α
    + β A^2 = Λ], Fd);
```

$$Fd := \frac{1}{-2\beta\Omega^2A^2+\Omega^4-2\alpha\Omega^2+3\beta^2A^4+4\beta A^2\alpha+\alpha^2}\Big((2\beta A^2\alpha\lambda^2-2\beta A^2\lambda^2\Omega^2+4\beta^4A^{10}+4\beta^2A^6\alpha^2+8\beta^3A^8\alpha+\beta^2A^4\lambda^2-2\Omega^2\alpha\lambda^2$$

$$+\alpha^2\lambda^2+\Omega^4\lambda^2)^{1/2}\cos\Big(\Omega t+\arctan\Big((2A^3\beta(A^2\beta+\alpha))\Big/$$

$$(2\beta A^2\alpha\lambda^2-2\beta A^2\lambda^2\Omega^2+4\beta^4A^{10}+4\beta^2A^6\alpha^2+8\beta^3A^8\alpha+\beta^2A^4\lambda^2-2\Omega^2\alpha\lambda^2$$

$$+\alpha^2\lambda^2+\Omega^4\lambda^2)^{1/2}, (\lambda(\alpha-\Omega^2+A^2\beta))\Big/$$

$$(2\beta A^2\alpha\lambda^2-2\beta A^2\lambda^2\Omega^2+4\beta^4A^{10}+4\beta^2A^6\alpha^2+8\beta^3A^8\alpha+\beta^2A^4\lambda^2-2\Omega^2\alpha\lambda^2$$

$$+\alpha^2\lambda^2+\Omega^4\lambda^2)^{1/2}\Big)\Big)\Big)$$

$$Fd(t)=\frac{\sqrt{\psi_4}\cos\left(\Omega t+\arctan\left(\frac{2A^3\beta\Lambda}{\sqrt{\psi_4}},\frac{\lambda(\Lambda-\Omega^2)}{\sqrt{\psi_4}}\right)\right)}{\Delta_1}$$

By identifying the values of b and ϕ from step 17, we will have

$$x_p = b\cos(\Omega t+\phi). \tag{1.62}$$

And like the previous example, we use the command "applyrule".

1.4.3.19. The Nineteenth Step

The ratio of amplitude before and after the vibration of external harmonic force is computed thus:

>

$$B := \frac{b}{\lambda};$$

$$'B' = applyrule\Big(\Big[4\beta^4 A^{10} + 8\beta^3 A^8\alpha + \beta^2 A^4\lambda^2 + 4\beta^2 A^6\alpha^2 + \alpha^2\lambda^2 + \Omega^4\lambda^2 - 2\Omega^2\alpha\lambda^2 + 2\beta A^2\alpha\lambda^2 - 2\beta A^2\lambda^2\Omega^2 = \psi_4, -2\beta\Omega^2 A^2 + \Omega^4 - 2\alpha\Omega^2 + 3\beta^2 A^4 + 4\beta A^2\alpha + \alpha^2 = \Delta_1, \alpha + \beta A^2 = \Lambda\Big], B\Big);$$

$B :=$

$$\left(2\beta A^2\alpha\lambda^2 - 2\beta A^2\lambda^2\Omega^2 + 4\beta^4 A^{10} + 4\beta^2 A^6\alpha^2 + 8\beta^3 A^8\alpha + \beta^2 A^4\lambda^2 - 2\Omega^2\alpha\lambda^2 + \alpha^2\lambda^2 + \Omega^4\lambda^2\right)^{1/2} / \left(\left(-2\beta\Omega^2 A^2 + \Omega^4 - 2\alpha\Omega^2 + 3\beta^2 A^4 + 4\beta A^2\alpha + \alpha^2\right)\lambda\right)$$

$$B = \frac{\sqrt{\psi_4}}{\Delta_1\lambda}$$

The initial amplitude of external force before vibration in accordance with equation 1.59 is equal to λ while vibration is b. Therefore, the ratio of these terms is introduced as

$$B = \frac{b}{\lambda}. \tag{1.63}$$

The parameter (B) is a suitable base to identify the initial value for the amplitude of vibration of external force, which is λ. It is necessary to mention that the value of parameter λ should not be bigger than a specific limit in order to keep the vibrational system safe.

1.4.3.20. The Twentieth Step

Convert the amplitude ratio (B) in terms of $\frac{\Omega}{\omega}$:

>

$\eta = '\frac{\Omega}{\omega}';$

$\eta = \frac{\Omega}{\omega};$

$\lambda := 0.2;\ \alpha := 0.6;\ \beta := 0.1;\ A := 0.1;$

$M := \frac{\Omega}{\omega} = \eta;$

$\Omega := solve(M, \Omega)[1];$

$B := \frac{b}{\lambda};$

$$\eta = \frac{\Omega}{\omega}$$

$$\eta = \frac{\Omega}{\sqrt{-\frac{\beta\,\Omega^2 A^2 + \alpha\,\Omega^2 - 3\,\beta^2 A^4 - 4\,\beta\,A^2\,\alpha - \alpha^2}{\alpha - \Omega^2 + A^2\,\beta}}}$$

$$\lambda := 0.2$$

$$\alpha := 0.6$$

$$\beta := 0.1$$

$$A := 0.1$$

$$M := \frac{\Omega}{\sqrt{-\frac{0.601\,\Omega^2 - 0.362403}{0.601 - \Omega^2}}} = \eta$$

$$\Omega := 0.7752418977\,\eta \left(\left(-302.5000000 + 300.5000000\,\eta^2 + 0.5000000000\sqrt{3.61201\,10^5\,\eta^4 - 7.27210\,10^5\,\eta^2 + 3.61201\,10^5} \right) \Big/ \left(-300.5000000 + 300.5000000\,\eta^2 + 0.5000000000\sqrt{3.61201\,10^5\,\eta^4 - 7.27210\,10^5\,\eta^2 + 3.61201\,10^5} \right) \right)^{1/2}$$

$B :=$

$$\Bigg(5.000000000\Bigg(0.01444805445-\big(0.02889608000\eta^2\big(-302.5000000+300.5000000\eta^2$$

$$+0.5000000000\sqrt{3.61201\,10^5\,\eta^4-7.27210\,10^5\,\eta^2+3.61201\,10^5}\big)\big)\big/\big(-300.5000000$$

$$+300.5000000\eta^2+0.5000000000\sqrt{3.61201\,10^5\,\eta^4-7.27210\,10^5\,\eta^2+3.61201\,10^5}\big)$$

$$000000000\sqrt{3.61201\,10^5\,\eta^4-7.27210\,10^5\,\eta^2+3.61201\,10^5}\big)^2\Bigg)^{1/2}\Bigg)\Bigg/\Bigg($$

$$-\big(0.7224019998\eta^2\big(-302.5000000+300.5000000\eta^2$$

$$+0.5000000000\sqrt{3.61201\,10^5\,\eta^4-7.27210\,10^5\,\eta^2+3.61201\,10^5}\big)\big)\big/\big(-300.5000000$$

$$+300.5000000\eta^2+0.5000000000\sqrt{3.61201\,10^5\,\eta^4-7.27210\,10^5\,\eta^2+3.61201\,10^5}\big)$$

$$+\big(0.3612009999\eta^4\big(-302.5000000+300.5000000\eta^2+0.5000000000\sqrt{3.61201\,10^5\,\eta^4-7.2}$$

$$+0.5000000000\sqrt{3.61201\,10^5\,\eta^4-7.27210\,10^5\,\eta^2+3.61201\,10^5}\big)^2+0.362403\big)$$

Whenever we consider $\eta=\dfrac{\Omega}{\omega}$, which means the ratio of the angular frequency of external force (Ω) to the natural frequency of the system (ω), equation 1.63 can be obtained in terms of η from step 19. And also, the physical parameters of the vibrational system are introduced in this level.

1.4.3.21. The Twenty-First Step

The command "plot" for depicting the ratio of amplitude of external harmonic force (B) can be acquired as follows:

```
> plot( B, η=0.5 ..3, y=0 ..5 axes=boxed, labels= ["η",
    "B"], thickness=2);
```

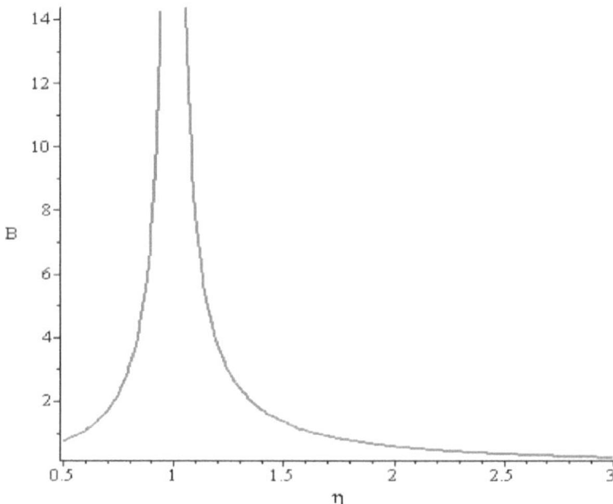

This chart displays some important tips for designing vibrational systems as follows:

a) If $\eta = \dfrac{\Omega}{\omega} = 1$:

In this state, the vibrational system is at the resonance point, and the amplitude of vibration tends to be infinite.

$$\eta \to 1 \quad \xrightarrow{\text{Resonance}} \quad \Omega = \omega \;\Rightarrow\; B \to \infty \tag{1.64}$$

This state is not suitable because of the obtained chart that shows the vibrational system in this state is on the asymptote; and as a result, the system can be harmed.

b) If $\eta = \dfrac{\Omega}{\omega} < 1$:

$$\frac{\Omega}{\omega} < 1 \quad \Rightarrow \quad \omega > \Omega \tag{1.65}$$

In this state, the natural angular frequency system (ω) will be bigger than the angular frequency of the external force. And according to the obtained charts by increasing η , the system tends to the resonance point; or we can say that the system is divergent.

c) If $\eta = \dfrac{\Omega}{\omega} > 1$:

$$\frac{\Omega}{\omega} > 1 \quad \Rightarrow \quad \omega < \Omega \tag{1.66}$$

In this state, the natural frequency of the system (ω) is less than the external force frequency. And in accordance with the obtained chart when $\eta \to \infty$, the system is convergent, which shows the desired state.

It is concluded that in selecting an external force for every vibrational system, the angular frequency (Ω) should be bigger than the natural frequency of the system (ω) in order to keep the system safe.

1.4.3.22. The Twenty-Second Step

The command "plot" for displaying the response of harmonic forces can be applied in the following form:

```
> η := 2;

  plot( Fd, t=0 ..20, axes=boxed, labels= [ "η", "Fd(t)" ],
    thickness=2 );
```

$\eta := 2$

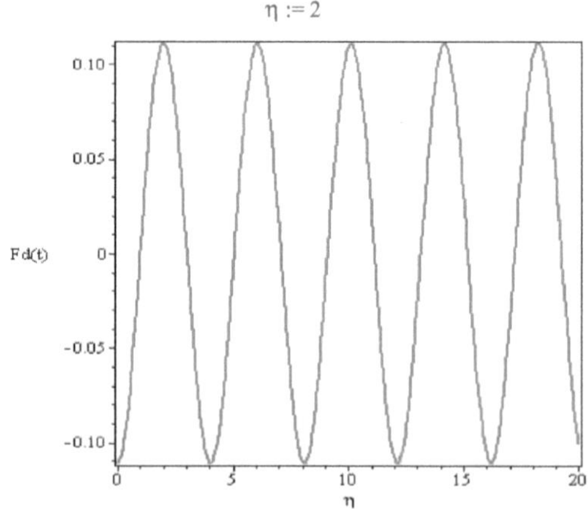

On the basis of the given explanation in step 21, which indicates $\eta > 1$, it is observed by choosing the value $\eta = 2$, the harmonic response vibrates safely without exerting any defect to the system.

1.4.3.23. The Twenty-Third Step

Solving the differential equation by numerical method and its comparison with AGM

```
> restart;   α := 0.6; β := 0.1; A := 0.1; λ := 0.2; Ω := 2;
```

$$\alpha := 0.6$$
$$\beta := 0.1$$
$$A := 0.1$$
$$\lambda := 0.2$$
$$\Omega := 2$$

The command "restart" and choosing physical values for the system are done in this step.

1.4.3.24. The Twenty-Fourth Step

The differential equation, initial condition, and the command "dsolve" are written as follows:

$$> f := \frac{d^2}{dt^2} z(t) + \alpha z(t) + \beta z(t)^3 = \lambda \sin(\Omega t);$$

```
ic := z(0) = A, D(z)(0) = 0;
f1 := dsolve([f, ic], numeric, output = listprocedure);
```

$$f := \frac{d^2}{dt^2} z(t) + 0.6 z(t) + 0.1 z(t)^3 = 0.2 \sin(2t)$$

$$ic := z(0) = 0.1, D(z)(0) = 0$$

$$f1 := \left[t = \text{proc}(t) \;\ldots\; \text{end proc}, z(t) = \text{proc}(t) \;\ldots\; \text{end proc}, \frac{d}{dt} z(t) = \text{proc}(t) \;\ldots\; \text{end proc} \right]$$

1.4.3.25. The Twenty-Fifth Step

The command "for loop" for numerical solution:

```
> for i from 0 by 5 while i ≤ 50 do;
  ([t=i,f1(i)[2],f1(i)[3]]);
  end do;
```

$$\left[t=0, z(t)(0)=0.100000000000000, \frac{d}{dt}z(t)(0)=0.\right]$$

$$\left[t=5, z(t)(5)=-0.144292399602701, \frac{d}{dt}z(t)(5)=0.0640689734232589\right]$$

$$\left[t=10, z(t)(10)=0.106279207461034, \frac{d}{dt}z(t)(10)=-0.114624235944169\right]$$

$$\left[t=15, z(t)(15)=-0.00163219196426219, \frac{d}{dt}z(t)(15)=0.114704070629085\right]$$

$$\left[t=20, z(t)(20)=-0.115558928163186, \frac{d}{dt}z(t)(20)=-0.0511069282115519\right]$$

$$\left[t=25, z(t)(25)=0.181026732188391, \frac{d}{dt}z(t)(25)=-0.0552513917546779\right]$$

$$\left[t=30, z(t)(30)=-0.154584142555597, \frac{d}{dt}z(t)(30)=0.155824240932523\right]$$

$$\left[t=35, z(t)(35)=0.0434371892254350, \frac{d}{dt}z(t)(35)=-0.197242534516762\right]$$

$$\left[t=40, z(t)(40)=0.0995646320675167, \frac{d}{dt}z(t)(40)=0.150461430250177\right]$$

$$\left[t=45, z(t)(45)=-0.202220444926970, \frac{d}{dt}z(t)(45)=-0.0274698302053926\right]$$

$$\left[t=50, z(t)(50)=0.209472439687285, \frac{d}{dt}z(t)(50)=-0.120918153244225\right]$$

1.4.3.26. The Twenty-Sixth Step

Depicting the charts of the obtained solution, its derivative, and the related phase plan:

```
>
  Fig, Numerical;
  plots[odeplot](f1, [t, z(t) ], t=0..50, thickness=2, axes
      =boxed);
  plots[odeplot](f1, [t, diff(z(t), t) ],  t=0..50, thickness
      =2, axes=boxed);
  plots[odeplot](f1, [z(t), diff(z(t), t) ], t=0..20, scaling
      =constrained, axes=boxed, labels=["dz(t)", "z(t)"],
      thickness=2);
```

Fig, Numerical

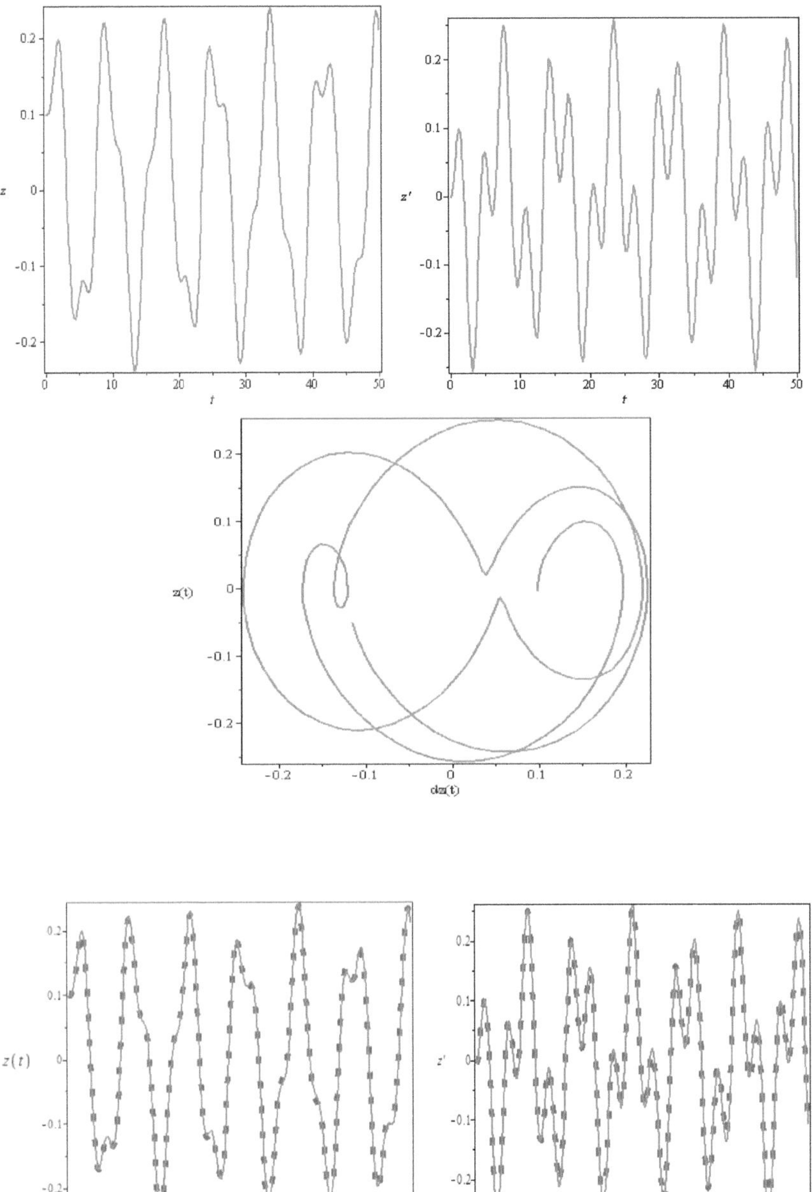

z
t
z'
t
z(t)
dz(t)
z(t)
t
z'
t
Numeric
AGM
Curve 1
Curve 2

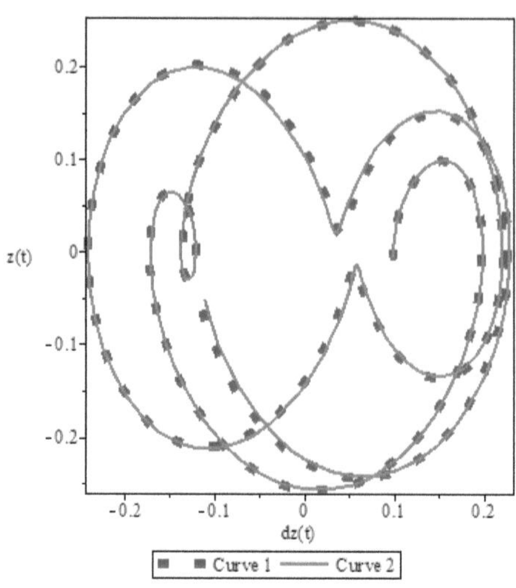

1.4.3.27. The Twenty-Seventh Step

\>

$$z := subs(f1, z(t));$$
$$dz := subs\left(f1, \frac{d}{dt} z(t)\right);$$
$$U := 0.1817963426 \cos(0.7750137848\,t - 0.9884214942) + 0.05883471527 \sin(2\,t + 3.141415837);$$
$$dU := -0.1408946715 \sin(0.7750137848\,t - 0.9884214942) + 0.1176694305 \cos(2\,t + 3.141415837);$$

$$z := \textbf{proc}(t) \;\ldots\; \textbf{end proc}$$
$$dz := \textbf{proc}(t) \;\ldots\; \textbf{end proc}$$

$$U := 0.1817963426 \cos(0.7750137848\,t - 0.9884214942) + 0.05883471527 \sin(2\,t + 3.141415837)$$

$$dU := -0.1408946715 \sin(0.7750137848\,t - 0.9884214942) + 0.1176694305 \cos(2\,t + 3.141415837)$$

1.4.3.28. The Twenty-Eighth Step

```
> fig, differ, AGM, Numeric;
  plot( (abs(U − z(t)) ), t = 0 ..50, thickness = 2, labels
      = [ "t", "|uAGM-uNum,u(t)|" ] );
  plot( abs(dU − dz(t) ), t = 0 ..50, thickness = 2, labels
      = [ "t", "|duAGM-duNum,du(t)|" ] );
```

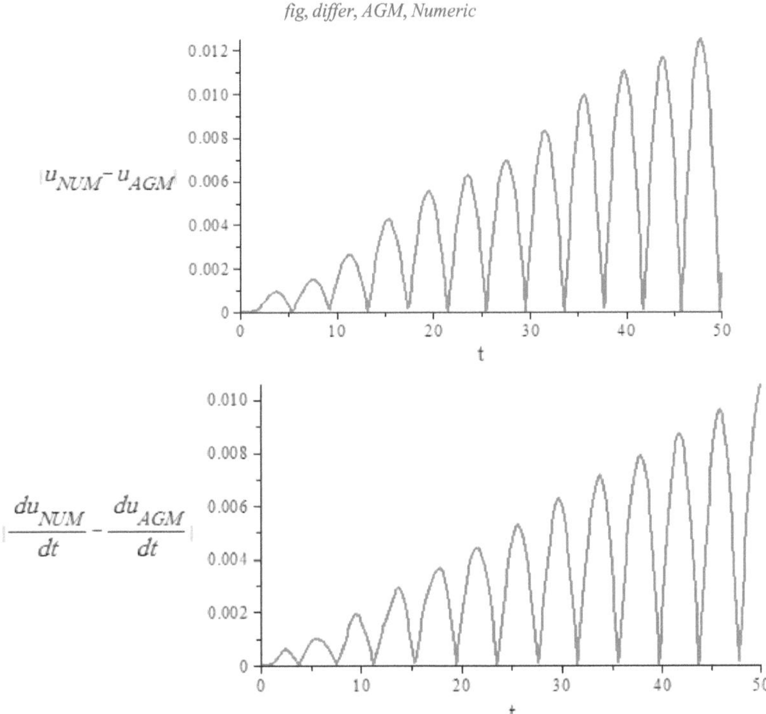

The command "plot" is utilized for depicting the absolute value of the difference between the obtained solution by AGM and numerical method.

Again, we mention that the final solution of the problem can be achieved wholesale by writing every step successively and using the command Shift + Enter at the end of each step.

1.5. References

1. D. W. Jordan and P. Smith, *Nonlinear Ordinary Differential Equations: An Introduction for Scientists and Engineers*, 4th ed. (New York: Oxford University Press, 2007).

2. D. Zwillinger, *Handbook of Differential Equations* (New York: Academic Press, 1992).

3. R. V. Dukkipati, *Mathematical Methods* (New Delhi: New Age International, 2009).

4. R. V. Dukkipati, *Numerical Methods* (New Delhi: New Age International, 2009).

5. R. V. Dukkipati, *Numerical Methods through Solved Problems* (New Delhi: New Age International, 2009).

6. D. D. Ganji, A. Sadighi, and I. Khatami, "Assessment of Two Analytical Approaches in Some Nonlinear Problems Arising in Engineering Sciences," *Physics Letters A*, vol. 372, no. 24 (2008): 4399–4406.

7. D. D. Ganji et al., "Solution of the Epidemic Model by Homotopy Perturbation Method," *Applied Mathematics and Computation* 187 (2007): 1056–1062.

8. D. D. Ganji et al., "Application of He's Homotopy Perturbation Method to Strongly Nonlinear Coupled Systems," *Journal of Physics: Conference Series* 96, no. 012078 (2008): 6.

9. J. H. He, "Some Asymptotic Methods for Strongly Nonlinear Equations," *International Journal of Modern Physics B* 20, no. 10 (2006): 1141–1199.

10. J. H. He, "A Review on Some New Recently Developed Nonlinear Analytical Techniques," *International Journal of Non-Linear Science and Numerical Simulation*, no. 1 (2000): 51–70.

11. S. Monami, *Communications in Nonlinear Science and Numerical Simulation* 12 (2007): 1283–1290.

12. Mark D. Ardema, *Analytical Dynamics: Theory and Applications* (New York: Kluwer Academic / Plenum, 2005).

13. A. M. Wazwaz, "The Tanh and the Sine-Cosine Methods for the Complex Modified K-dV and the Generalized K-dV Equations," *Computers and Mathematics with Applications* 49 (2005): 1101–1112.

14. W. Addo-Asah, H. C. Akpati, and R. E. Mickens, "Investigation of a Generalized Van Der Pol Oscillator Differential Equation," *Journal of Sound and Vibration*, no. 179 (1995): 733–735.

15. Mark D. Ardema, *Newton-Euler Dynamics* (Springer Science: 2005).

16. M. H. Pashaei, D. D. Ganji, and M. Akbarzade, "Application of the Energy Balance Method for Strongly Nonlinear Oscillators," *Progress in Electromagnetics Research M*, vol.2 (2008): 47–56.

17. D. D. Ganji et al., "Determination of the Frequency-Amplitude Relation for Nonlinear Oscillators with Fractional Potential Using He's Energy Balance Method," *Progress in Electromagnetic Research C*, vol. 5 (2008): 21–33.

18. Richard H. Rand, *Lecture Notes on Nonlinear Vibrations* (Cornell University, 2005).

19. Attilio Maccari, "Approximate Solution of a Class of Nonlinear Oscillators in Resonance with a Periodic Excitation," *Nonlinear Dynamics* 15 (1998): 329–343.

20. H. R. Harrison and T. Nettleton, *Advanced Engineering Dynamics* (London: Arnold,1997).

21. SHA. Hashemi Kachapi, Rao V. Dukkipati, and D. D. Ganji, *Nonlinear Vibration: Theory and Application* (New Age International, 2011).

22. SHA. Hashemi Kachapi et al., "Periodic Solution for Strongly Nonlinear Vibration Systems by He's Variational Iteration Method," *Mathematical Methods in the Applied Sciences*, no. 32 (2009): 2339–2349.

23. M. Naghipoour et al., "Analysis of Nonlinear Oscillation Systems Using He's Variational Approach," *Journal of Physics: Conference Series* 96, no. 012077 (2008).

24. SHA. Hashemi Kachapi et al., "Solution of Strongly Nonlinear Oscillation

Systems Using Variational Approach," *Journal of Applied Functional Analysis* 4, no. 3 (2009): 528–535.

25. N. Tolou et al., "Analytical Investigation of Strongly Nonlinear Normal Mode Using Homotopy Perturbation Method and He's Variational Method," *Journal of Applied Functional Analysis* 4, no. 4 (2009): 682–690.

26. Mehdipour, D. D. Ganji, and M. Mozaffari, "Application of the Energy Balance Method to Nonlinear Vibrating Equations," *Current Applied Physics*, vol. 10, no. 1 (2010): 104–112.

27. S. S. Ganji et al., "Application of AFF and HPM to the Systems of Strongly Nonlinear Oscillation," *Current Applied Physic*s, vol. 10, no. 5 (2010): 1317–1325.

28. Z. Z. Ganji, D. D.Ganji, and A. Asgari, "Finding General and Explicit Solutions of High Nonlinear Equations by the Exp-Function Method," *Computers and Mathematics with Application*s, vol. 58, no. 11–12 (December 2009): 2124–2130.

29. A. Kimiaeifar et al., "Analytical Solution for Van Der Pol-Duffing Oscillators," *Chaos, Solitons and Fractal*s, vol. 42, no. 5 (2009): 2660–2666.

30. S. S. Ganji, D. D. Ganji, and S. Karimpour, *He's Energy Balance and He's Variational Methods for Nonlinear Oscillations in Engineering* 3 (2009): 461–471.

Chapter 2

The First-Order Nonlinear Differential Equations

2.1. Investigation of the First-Order Differential Equations

A differential equation is a mathematical equation that connects some function with its derivatives. In reality, the functions usually stand for physical quantities, and the derivatives represent the rate of their changes, and the equation describes a relationship between the two. Differential equations play an important role in many disciplines including engineering, physics, biology, etc. Additionally, a differential equation is considered linear if the unknown function and its derivatives have the power 1, and also, the products of the unknown function and its derivatives are not permitted. But nonlinear differential equations do not follow the aforementioned rules. Inasmuch as the order of a differential equation is equal to the highest derivative in the equation, the highest derivation of first-order nonlinear differential equations is only one.

Linear differential equations often appear as approximations to nonlinear equations. These approximations are valid just under restricted conditions. For instance, a harmonic vibrational equation is an approximation to the nonlinear pendulum equation, which is valid only for a small amplitude of vibration. Eventually, we can indicate that finding an exact solution for a nonlinear differential equation is very vital, because the precise simulations of real-world phenomena result in nonlinear equations.

2.2. Applied Examples

In order to solve this kind of nonlinear differential equations by AGM, an answer function is needed in the following general forms.

The answer function for nonharmonic systems (nonvibrational) is defined as follows:

$$u(t) = \sum_{k=1}^{n} a_k e^{-kt}. \tag{2.1}$$

And for harmonic systems (vibrational systems) and also for nonharmonic systems defined in a specified domain, the relevant answer function is proposed as follows:

$$u(t) = \lim_{n\to\infty}\sum_{k=1}^{n}\{a_k e^{-b_k t} + d_k \cos(\Omega t + \varphi_k)\}. \tag{2.2}$$

This kind of nonlinear differential equations (nonharmonic systems) is applied for heat transfer in radiation, fluids, nanomaterials, chemical engineering, and all the first-order nonlinear equations. After that, we can see the presence of harmonic systems in electric circuits and control systems in electrical and chemical engineering.

To understand more, reading the following examples is recommended.

Example 2.2.1

It is supposed that in designing chemical reactors for a certain reaction, the speed of chemical reaction between reactants and the product can be modeled and simplified as a nonlinear differential equation in terms of time as follows:

Solve the following equation by AGM.

$$A \underset{k_2}{\overset{k_1}{\rightleftarrows}} B \quad ; \quad r_A = \frac{-dc_A}{dt} = -k_1 c_A^4 + k_2 c_A^2 + (k_1 k_2 + k_1) c_A \tag{2.3}$$

And the initial concentration of reactants at the beginning of the reaction (boundary conditions) is as follows:

$$c_A(t) = c_0 \quad at \quad t = 0. \tag{2.4}$$

Consider the values of the physical properties of the system as

$$k_1 = 0.1 \quad , \quad k_2 = 0.01 \quad , \quad c_0 = 0.2. \tag{2.5}$$

2.2.1.1. Solving the Differential Equation by AGM

Equation (2.3) can be rewritten as follows:

$$f(t):\ \frac{dc}{dt}=0.1c_A^4-0.01c_A^2-0.101c_A. \tag{2.6}$$

And the related initial conditions of the differential equation are expressed as

$$c_A(0)=0.2. \tag{2.7}$$

Therefore, an answer function is chosen for solving equation 2.6 as follows:

$$c_A=ae^{-bt}+d\cos(\mathrm{t}+\varphi). \tag{2.8}$$

The constant coefficients a, b, d and φ can easily be computed by applying initial conditions.

2.2.1.2. Applying Initial Conditions by AGM

On the basis of given explanations, initial conditions in AGM are applied in the following two ways:

a. Applying the initial conditions on the answer function as follows:

$$c_A=c_A(IC). \tag{2.9}$$

IC is the abbreviation of applying initial equations. So the initial condition is applied on the given answer function as

$$c_A(0)=0.2\ \rightarrow\qquad a+d\cos(\varphi)=0.2. \tag{2.10}$$

b. Initial conditions are applied on equation 2.6, and its derivatives are shown as $f(\mathrm{t})$ in the following general equations:

$$f(c_A(t)) \quad \rightarrow \quad f(c_A(IC)) = 0 \quad \rightarrow \quad f'(c_A(IC)) = 0 \ , \ \ldots . \tag{2.11}$$

Equation 2.11 means that equation 2.8 is substituted into equation 2.6 instead of the dependent parameter C_A, and then the initial condition is applied to it.

Therefore, applying the initial condition is done as follows:

$$f(c_A(0)):$$

$$-ab - d\sin\varphi = (a + d\cos\varphi)^2[0.1(a + d\cos\varphi)^2 - 0.01] - 0.101(a + d\cos\varphi). \tag{2.12}$$

Then the initial condition is applied on the first derivative of the achieved equation as follows:

$$f'(c_A(0)):$$

$$\begin{aligned} &ab^2 - d\cos\varphi = (a + d\cos\varphi)(ab + d\sin\varphi)[0.02 - 0.4(a + d\cos\varphi)^2] + \\ &0.101(ab + d\sin\varphi) = 0 \end{aligned} \quad . \tag{2.13}$$

And for the second derivative of the obtained equation, we will have

$$f''(c_A(0)):$$

$$\begin{aligned} &-ab^3 + d\sin\varphi = (a + d\cos\varphi)^2\{0.4(a + d\cos\varphi)(ab^2 - d\cos\varphi) - 1.2(ab + \\ &d\sin\varphi)\} - 0.02(a + d\cos\varphi)^2 - (ab^2 - d\cos\varphi)\{0.02(a + d\cos\varphi) + 0.101] = 0. \end{aligned} \tag{2.14}$$

By solving the set of algebraic equations consisting of equation 2.10 to equation 2.14, except equation 2.11, the constant coefficients of the equation 2.8, a_0 to a_4, can easily be computed.

$$a = 0.2 \quad , \quad b = 0.1021 \quad , \quad d = 0.0000141 \quad , \quad \varphi = 1.164323 \tag{2.15}$$

By substituting the achieved constant coefficients from equation 2.15 into equation 2.8, the solution of the nonlinear differential equation 2.6 can be obtained as follows:

$$c_A(t) = 0.2e^{-0.1021t} + 0.0000141\cos(t + 1.164323). \tag{2.16}$$

2.2.1.3. A Comparison between the Obtained Answer by AGM and Numerical Solution

In regard to the time domain $t \in \{1, 50\}$, the nonlinear differential equation, equation 2.6, has been solved by numerical method, and the obtained results have been compared with AGM in the following table:

Table 2.1. Comparing the Obtained Solution by AGM and Numerical Method (RKF45)

t	*0*	*10*	*20*	*30*	*40*	*50*
$C_A(t)$	0.2	0.02115	0.026150	0.009508	0.003461	0.001260
$C_A(t)$ AGM	0.2	0.7202	0.025923	0.0093522	0.003349	0.001220

In order to increase the precision of the given answer function by AGM, we can increase the number of the sentences in equation 2.8, and as a result, the mathematical operation will be increased.

The obtained charts by AGM and numerical solution have been compared below:

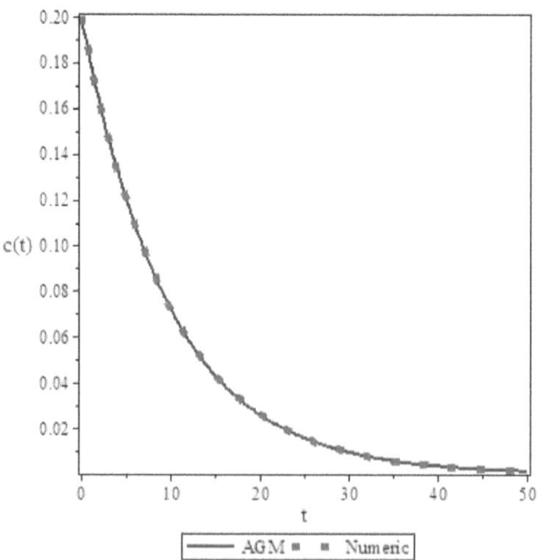

Fig. 2.1. Comparing the achieved solution by AGM and numerical method

As regards the above chart, for all the chemical reactions in reactors in which the differential equation governing their system is nonlinear, this approach is suitable to solve the first-order nonlinear differential equations.

Example 2.2.2

Many processes related to engineering and basic sciences are simulated by nonlinear differential equations with one initial condition such as $u(0) = 0$ (for example, dropping a bullet inside a fluid by considering the fluid friction without initial velocity and also rolling the bullet on a frictional surface). This kind of nonlinear differential equations has many applications like the following example.

Solve the following nonlinear differential equation by AGM.

$$f(t): \quad \frac{du(t)}{dt} + k_1 u(t) + k_2 u^2(t) - k_3 = 0 \tag{2.17}$$

And the related initial condition of this equation is defined as

$$u(0) = 0. \tag{2.18}$$

Consider the constant physical values of the differential equation as follows:

$$k_1 = 0.15 \quad , \quad k_2 = 0.3 \quad , \quad k_3 = 0.05. \tag{2.19}$$

2.2.2.1. The Solution of Differential Equation by AGM

In order to solve the differential equation 2.17, an answer function is needed as a finite power series in the form of

$$u(t) = \sum_{k=1}^{3} a_k e^{-b_k t} = a_1 e^{-b_1 t} + a_2 e^{-b_2 t} + a_3 e^{-b_3 t}. \tag{2.20}$$

2.2.2.2. Applying Initial Conditions by AGM

In AGM, initial conditions are applied to the answer function of equation 2.20 and also on the function of the differential equation as follows:

$$u(0)=0 \quad \rightarrow \quad a_1+a_2+a_3=0. \tag{2.21}$$

The initial condition is applied on the differential equation after substituting the answer function into the differential equation as

$$f(u(0)):$$
$$-a_1b_1-a_2b_2-a_3b_3+0.15(a_1+a_2+a_3)[1+2(a_1+a_2+a_3)]-0.05=0. \tag{2.22}$$

Applying the initial condition on the first derivative of the differential equation,

$$f'(u(0)):$$
$$a_1b_1^2+a_2b_2^2+a_3b_3^2-0.15(a_1b_1+a_2b_2+a_3b_3)[1+0.6(a_1+a_2+a_3)]=0. \tag{2.23}$$

Applying the initial condition on the second derivative of the differential equation,

$$f''(u(0)):$$
$$\begin{aligned}&-a_1b_1^3-a_2b_2^3-a_3 3b_3^3+0.15(a_1b_1^2+a_2b_2^2+a_3b_3^2)[1+4(a_1+a_2+a_3)]+\\&0.6(a_1b_1+a_2b_2+a_3b_3)^2=0\end{aligned} \tag{2.24}$$

And finally, the initial condition is applied on the third and fourth derivatives of the differential equation in the following forms:

$$f'''(u=0))=0 \tag{2.25}$$

$$f''''(u(t=0))=0. \tag{2.26}$$

By solving a set of algebraic equations consisting of six equations with six unknowns from equation 2.21 to equation 2.26, the constant coefficients of the answer function can be computed. And then these values are substituted instead of the constant coefficients of the answer function, and as a result, the nonlinear differential equation 2.17 is easily answered with mathematical simplification as follows:

$$u(t)=-0.054e^{-0.1415t}\cos(0.2182t)+0.161e^{-0.1415t}\sin(0.2182t)+0.054e^{0.133t}. \tag{2.27}$$

2.2.2.3. Numerical Solution of the Differential Equation and Comparing It with AGM

In regard to the given initial condition and the physical values of equation 2.19, the differential equation 2.17 has numerically been solved in the certain domain of $t \in \{0, 5\}$ and compared with AGM in the table below:

Table 2.2. Comparison between the Achieved Solution by AGM and Numerical Method (RKF45)

t	0.0	1	2	3	4	5
$u(t)$ NUM	0.0	0.046126	0.08493	0.116630	0.14208	0.16221
$u(t)$ AGM	0.0	0.04621	0.08493	0.116698	0.14241	0.1633

The obtained solutions are depicted by AGM and numerical method in the following figure:

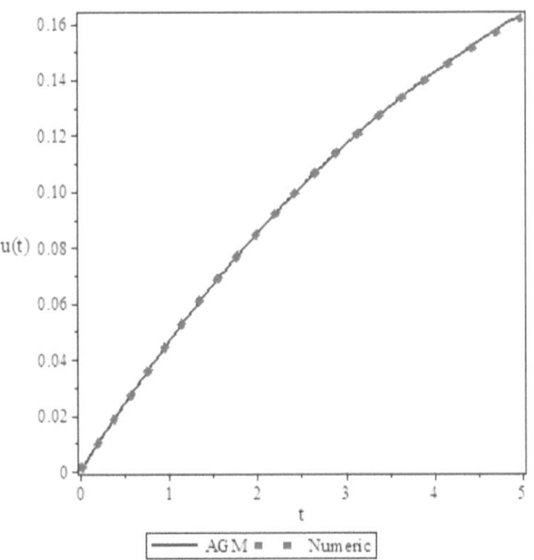

Fig. 2.2. A comparison between the given solution by AGM and numerical method.

Example 2.2.3

Solve the following nonlinear differential equation.

$$f(\tau): \quad \frac{du(\tau)}{d\tau} + u(\tau) + \varepsilon u^4(\tau) = 0 \tag{2.28}$$

And the related initial condition is presented as

$$u(\tau) = 1 \quad at \quad \tau = 0. \tag{2.29}$$

Consider the constant value (ε) as follows:

$$\varepsilon = 0.35. \tag{2.30}$$

2.2.3.1. Solving the Differential Equation by AGM

Like the previous examples, for the nonlinear differential equation 2.28, an answer function is considered as a finite power series as follows:

$$u(\tau) = \sum_{k=1}^{3} a_k e^{-b_k \tau} = a_1 e^{-b_1 \tau} + a_2 e^{-b_2 \tau} + a_3 e^{-b_3 \tau}. \tag{2.31}$$

By applying the initial condition, the constant coefficients a_1 to a_3 and b_1 to b_3 are easily computed.

2.2.3.2. Applying Initial Conditions by AGM

a. As it has been said, the initial condition in AGM is applied on the answer function and also on the main differential equation after substituting the answer function into the differential equation 2.28 as follows:

$$u(0) = 1 \quad \rightarrow \quad a_1 + a_2 + a_3 = 1. \tag{2.32}$$

b. Applying the initial condition on equation 2.28 shown by $f(\mathrm{u}(\tau))$ is done as follows:

$$f(u(0)):$$
$$-a_1b_1-a_2b_2-a_3b_3+(a_1+a_2+a_3)[1+0.35(a_1+a_2+a_3)^3]=0. \quad (2.33)$$

The initial condition is applied on the first derivative of the obtained equation shown by $f'(\mathrm{u}(\tau))$ as

$$f'(u(0)):$$
$$-a_1b_1^3-a_2b_2^3-a_3b_3^3+a_1b_1^2+a_2b_2^2+a_3b_3^2+(a_1+a_2+a_3)^2[4.2(a_1b_1+a_2b_2+ a_3b_3)^2+1.4(a_1+a_2+a_3)(a_1b_1^2+a_2b_2^2+a_3b_3^2)]=0 \quad . (2.34)$$

And also, the initial condition is applied on the second, third, and fourth derivatives shown by $f''(\mathrm{u}(\tau))$, $f'''(\mathrm{u}(\tau))$ and $f''''(\mathrm{u}(\tau))$, respectively, as follows:

$$f''(u(\tau=0))=0 \quad (2.35)$$

$$f'''(u(\tau=0))=0 \quad (2.36)$$

$$f''''(u(\tau=0))=0\,. \quad (2.37)$$

By solving the set of algebraic equations consisting of equation 2.32 to equation 2.37, the constant coefficients of the answer function can easily be computed by using the Maple software as follows:

$$a_1=0.9236 \quad , \quad a_2=0.0743 \quad , \quad a_3=0.00202,$$
$$b_1=1.024 \quad , \quad b_2=5.078 \quad , \quad b_3=13.248 \quad . (2.38)$$

After substituting the obtained constant coefficients into equation 2.31, the solution of the nonlinear differential equation 2.28 can be gained in the following form:

$$u(\tau)=0.9236e^{-1.024\tau}+0.0743e^{-5.078\tau}+0.00202e^{-13.248\tau} \,. \quad (2.39)$$

2.2.3.3. Solving the Differential Equation by Numerical Method and Its Comparison with AGM

As regards $\varepsilon = 0.35$, the nonlinear differential equation 2.28 in $\tau \in \{0, 1\}$ has numerically been solved in the table below and has been compared with AGM.

Table 2.3. Comparing the Achieved Solution with AGM and Numerical Method (RKF45)

τ	*0.0*	*0.2*	*0.4*	*0.6*	*0.8*	*1.0*
$u(\tau)$ NUM	1.0	0.77967	0.62317	0.50387	0.40979	0.3343
$u(\tau)$ AGM	1.0	0.77966	0.62299	0.5032	0.40851	0.3321

The chart obtained by numerical method and AGM has been compared graphically as follows:

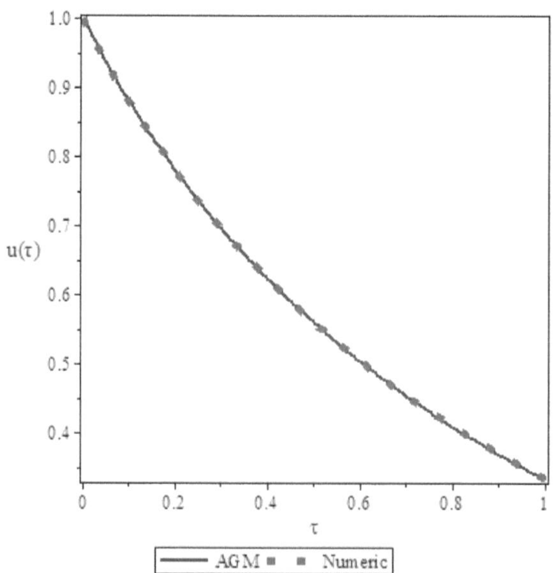

Fig. 2.3. A comparison between the obtained solution by AGM and numerical method

Example 2.2.4

The differential equation governing the combustion systems, along with heat transfer (radiation), is gained like the nonlinear differential equation 2.40 in terms of dimensionless parameters, which is of great importance in engineering of fluids, gases, chemical engineering, and also in other basic sciences. Therefore, consider the following nonlinear differential equation and solve it.

$$f(\tau):\ [1+\varepsilon_1(\theta-\theta_a)]\frac{d\theta}{d\tau}+(\theta-\theta_a)+\varepsilon_2(\theta-\theta_s)^2=0 \tag{2.40}$$

And the related initial condition is expressed as

$$\theta(\tau)=1 \quad at \quad \tau=0. \tag{2.41}$$

And the constant physical values are defined as follows:

$$\varepsilon_1=0.65 \quad , \quad \varepsilon_2=0.3 \quad , \quad \theta_a=0.1 \quad , \quad \theta_s=0.2. \tag{2.42}$$

2.2.4.1. Solving the Differential Equation by AGM

Like the previous examples, consider the following answer function as a finite power series in the form of

$$\theta=\sum_{k=0}^{2}a_k e^{-b_k\tau}=a_1e^{-b_1\tau}+a_2e^{-b_2\tau}. \tag{2.43}$$

In the above equation, the constant coefficients a_1, a_2, b_1, b_2 are easily computed by applying initial conditions.

2.2.4.2. Applying Initial Conditions by AGM

Based on the given explanations in the previous examples, the initial condition is applied in the following general forms:

$$\theta = \theta(IC) \tag{2.44}$$

$$f(\theta(\tau)): \rightarrow f(\theta(\text{IC})) = 0 \quad , \quad f'(\theta(\text{IC})) = 0 \quad , \; .... \tag{2.45}$$

IC is the abbreviation of initial condition. Equation 2.45 means that the answer function is substituted into the differential equation 2.40 instead of the dependent parameter θ.

a. The given initial condition is applied on the answer function as

$$\theta(0) = 1 \quad \rightarrow \quad a_1 + a_2 = 1. \tag{2.46}$$

b. Initial condition is applied on the differential equation with respect to equation 2.45 and also on its derivatives as follows.

$$f(\theta(0)):$$
$$-(0.935 + 0.65a_1 + 0.65a_2)(a_1b_1 + a_2b_2) + a_1 + a_2 = 0.1 + 0.3(a_1 + a_2 - 0.2)^4 = 0 \tag{2.47}$$

On the first derivative,

$$f'(\theta(\tau = 0)) = 0. \tag{2.48}$$

Then for the second derivative, we will have

$$f''(\theta(\tau = 0)) = 0. \tag{2.49}$$

By solving the set of algebraic equations consisting of equation 2.46 to equation 2.49, the constant coefficients of equation 2.43 can be obtained as

$$a_1 = 0.9964 \quad , \quad a_2 = 0.00353 \quad , \quad b_1 = 0.63 \quad , \quad b_2 = 5.09. \tag{2.50}$$

By substituting the achieved constant coefficients into equation 2.43, the solution of the nonlinear differential equation 2.40 is gained in the form of

$$\theta(\tau) = 0.9964e^{-0.63\tau} + 0.00353e^{-5.09\tau}. \tag{2.51}$$

2.2.4.3. Solving the Nonlinear Differential Equation by Numerical Method and Its Comparison with AGM

In regard to the initial conditions and the given constant values, the differential equation in the given time domain $t \in \{0, 1\}$ has been solved by numerical method, and the obtained results have been compared with AGM in the following table:

Table 2.4. Comparing the Result of AGM Solution with Numerical Method

τ	*0*	*0.2*	*0.4*	*0.6*	*0.8*	*1.0*
$\theta(\tau)$ *NUM*	1.0	0.87982	0.77485	0.68238	0.6007	0.5286
$\theta(\tau)$ AGM	1.0	0.87984	0.7750	0.68314	0.6011	0.5301

After that, the charts of achieved solutions by numerical method and AGM are graphically compared in the following form:

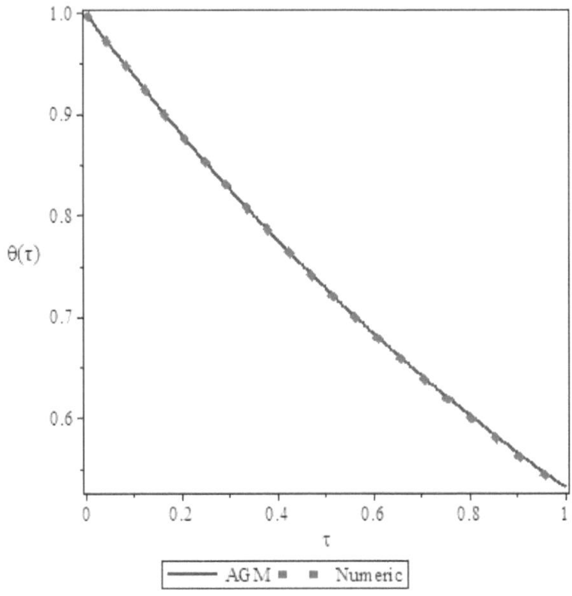

Fig. 2.4. Comparing the achieved solutions by AGM and numerical method

Example 2.2.5

The following equation is harmonic because of its trigonometric factor, $\cos(\omega t)$, which is mainly applied for harmonic vibrational systems. This kind of harmonic equation is basically used in electric circuits and electronics, control systems, and the systems that have vibrational responses. Also, the parameters used in this example show the ohmic resistance (R_0), the condenser capacity (C), the initial voltage (V_0), and the frequency of electrical circuit (ω_0).

$$f(t):\quad (R_0 + ki(t))\frac{di(t)}{dt} + \frac{i(t)}{c} = V_0\omega_0\cos(\omega_0 t) \tag{2.52}$$

And the related initial condition is expressed below:

$$i(0) = \mathrm{I}_0. \tag{2.53}$$

Consider the constant values of the given differential equation as

$$R_0 = 2 \quad , \quad c = 2 \quad , \quad k = \frac{1}{8} \quad , \quad \omega_0 = 1 \quad , \quad V_0 = 0.5 \quad , \quad I_0 = 0. \tag{2.54}$$

2.2.5.1. Solving the Differential Equation by AGM

In order to solve equation 2.52, we need an answer function with a trigonometric factor and a finite power series with the frequency (ω_0) in the following form:

$$i(t) = a e^{-bt} + d\cos(\omega_0 t + \varphi). \tag{2.55}$$

By applying the initial condition, the constant values a, b, d and φ can be computed.

2.2.5.2. Applying Initial Condition by AGM

The initial conditions are applied in the following two ways:

a. Applying the initial condition on the answer function is done as follows:

$$i = i(IC). \tag{2.56}$$

Therefore, the initial condition is applied on the answer function as follows:

$$i(0) = 0 \quad \rightarrow \quad a + d\cos\varphi = 0. \tag{2.57}$$

b. Applying the initial condition on the differential equation 2.52, shown by $f(t)$, is done after substituting the answer function into the differential equation 2.52 instead of the dependent parameter $i(t)$ as the following general form:

$$f(i(t)) \quad \rightarrow \quad f(i(IC)) = 0 \quad , \quad f'(i(IC)) = 0 \quad , \quad \ldots. \tag{2.58}$$

So according to the above equation, the initial condition is applied as follows:

$$f(i(0)): \quad \rightarrow \quad -(2+\frac{1}{8}a+\frac{1}{8}d\cos\varphi)(ab+d\sin\varphi)+\frac{1}{2}a+\frac{1}{2}d\cos\varphi = 0.5. \tag{2.59}$$

Applying the initial condition on the first derivative of the differential equation according to the right-hand side of equation 2.58 is done as

$$f'(i(0)):$$

$$\frac{1}{8}(ab+d\sin\varphi)^2+(2+\frac{1}{8}a+\frac{1}{8}d\cos\varphi)(ab^2-d\cos\varphi)--\frac{1}{2}(ab+d\sin\varphi)=0. \tag{2.60}$$

And for the second derivative, we will have

$$f''(i(0)):$$

$$(ab^2-d\cos\varphi)\{-\frac{1}{8}(ab+d\sin\varphi)-\frac{1}{4}(ab+d\sin\varphi)+\frac{1}{2}\}+(2+\frac{a}{8}+ \frac{d}{8}\cos\varphi)(-ab^3+d\sin\varphi)=0. \tag{2.61}$$

By solving the set of algebraic equations consisting of four equations with four unknowns, the constant coefficients of the answer function a, b, d and φ can be computed simply as

$$a = -0.061 \ , \ b = 0.297 \ , \ d = 0.2397 \ , \ \varphi = -1.3134 . \tag{2.62}$$

By substituting the resulting values in equation 2.55, the solution of the nonlinear differential equation 2.52 is achieved as follows:

$$i(t) = -0.061 e^{-0.297t} + 0.2397 \cos(t - 1.3134). \tag{2.63}$$

2.2.5.3. Numerical Solution of the Differential Equation and Comparing It with AGM

In regard to the given physical values from equation 2.54, the differential equation has numerically been solved in the certain domain $t \in \{0, 40\}$ and compared with AGM in the table below:

Table 2.5. The Resulted Comparison between AGM and Numerical Method

t	0.0	8	16	24	32	40
$i(t)$ NUM	0.0	0.21568	-0.12485	-0.188347	0.178611	0.1365212
$i(t)$ AGM	0.0	0.21485	-0.12571	-0.18712	0.17877	0.13510

The obtained solutions by AGM and numerical method are drawn as

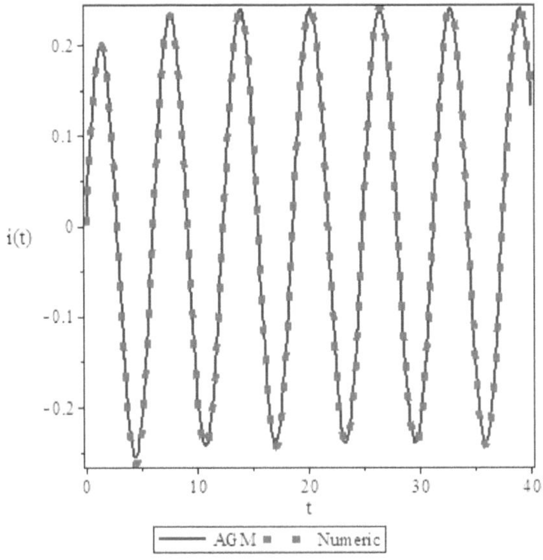

Fig. 2.5. Resulted charts from the achieved solutions by AGM and numerical method

It is important to mention that by choosing more sentences for equation 2.55, the precision of the solution can be significantly more.

Example 2.2.6

Solve the following nonlinear differential equation by AGM.

$$f(t): \quad (R_0 + kw)\frac{dw}{dt} + \frac{w}{c}\left(\frac{dw}{dt}\right) + w = \lambda\cos(\omega_0 t) \tag{2.64}$$

And the related initial condition is

$$w(0) = A. \tag{2.65}$$

Consider the physical values of the differential equation as follows:

$$R_0 = 10 \quad , \quad \omega_0 = 0.5 \quad , \quad A = 0.1 \quad , \quad c = 3 \quad , \quad k = \frac{2}{5} \quad , \quad \lambda = 0.05. \tag{2.66}$$

2.2.6.1. Solving the Differential Equation by AGM

This kind of equation is mostly used in electric circuits in the engineering fields such as electricity and control of circuits that are frequently variable. In order to solve the differential equation 2.64 by AGM, an answer function like the previous examples is needed by choosing a trigonometric and a power finite series as

$$w = ae^{-bt} + d\cos(\omega_0 t + \varphi). \tag{2.67}$$

2.2.6.2. Applying Initial Condition by AGM

As it is previously explained, initial condition is applied on equation 2.67 and the differential equation 2.64, shown by $f(\mathrm{t})$, as follows:

$$u(u(IC) \tag{2.68}$$

$$f(u(t)) \rightarrow f(u(IC)) = 0 \quad , \quad f'(u(IC)) = 0 \quad , \quad f''(u(IC)) = 0 \ , \ \ldots. \tag{2.69}$$

Therefore, applying the initial condition on the answer function is done with respect to equation 2.65 as

$$w(0) = 0.1 \rightarrow \qquad a + d\cos\varphi = 0.1. \tag{2.70}$$

And also on the differential equation 2.64 is done below.

$f(w(0))$:

$$(ab + 0.5d\sin\varphi)\{(10 + \frac{2}{5}a + \frac{2}{5}d\cos\varphi) + \frac{1}{3}(a + d\cos\varphi)\} - (\mathrm{a} + \mathrm{d}\cos\varphi) = -0.05 \tag{2.71}$$

Finally, applying initial condition on the higher derivatives of the differential equation according to equation 2.69 is done in the following form:

$$f'(w(t=0)) = 0 \tag{2.72}$$

$$f''(w(t=0)) = 0. \tag{2.73}$$

By solving the set of algebraic equations consisting of four equations with four unknowns from equation 2.70 to equation 2.73, the constant coefficients a, b, d and φ can easily be computed as follows:

$$a = 0.098 \quad , \quad b = 0.099 \quad , \quad d = 0.00973 \quad , \quad \varphi = -1.374. \tag{2.74}$$

By substituting the constant coefficients of equation 2.74 into the answer function, the solution of the nonlinear differential equation 2.64 is gained in the form of

$$w(t) = 0.98 e^{-0.099t} + 0.00973 \cos(0.5t - 1.374). \tag{2.75}$$

2.2.6.3. Solving the Differential Equation by Numerical Method and Comparing It with AGM

With regard to the physical values of equation 2.66, the differential equation has numerically been solved in $t \in \{0, 100\}$ and compared with AGM in the table below:

Table 2.6. Comparing the Achieved Solution of AGM with Numerical Method

t	0.0	20	40	60	80	100
$w(t)$ NUM	0.1	0.006523	0.01137	-0.00806	0.005915	-0.0006608
$w(t)$ AGM	0.1	0.00667	0.01134	-0.00888	0.00588	-0.0006647

The chart gained by numerical method and AGM are depicted as follows:

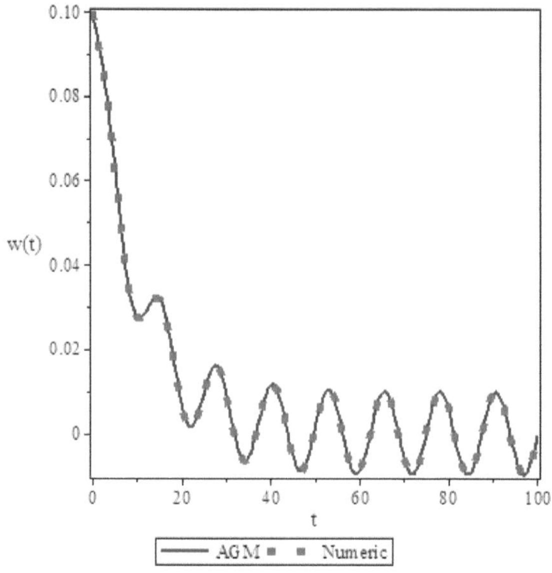

Fig. 2.6. Graphical comparison between AGM and numerical method

Example 2.2.7

Consider the following nonlinear differential equation and solve it.

$$f(t):\quad (5+u+u^2+u^3)\frac{du}{dt}+\frac{u}{1+u^2}=\cos(t) \tag{2.76}$$

And the related initial condition is presented as

$$u(0)=0. \tag{2.77}$$

2.2.7.1. Solving the Differential Equation by AGM

In order to solve the differential equation 2.76 by AGM, an answer function like the previous examples is needed as follows:

$$u=a e^{-bt}+d\cos(t+\varphi). \tag{2.78}$$

2.2.7.2. Applying Initial Condition by AGM

Initial condition is applied on the answer function and also on the differential equation after substituting the answer function, equation 2.78, into the differential equation 2.76 shown by $f(u(\mathrm{t}))$.

a. Applying the initial condition on the answer function is done as follows:

$$u(0)=0 \quad \rightarrow \quad a+d\cos\varphi=0. \tag{2.79}$$

b. Also applying the initial condition on the differential equation 2.76 is done, like the previous examples, as

$f(u(0))$:

$$\{[+(a+d\cos\varphi)]^3+(a+d\cos\varphi)^2+a+d\cos\varphi\}(-ab-d\varepsilon\varphi)+ \frac{a+d\cos\varphi}{1+(a+d\cos\varphi)^2}=0 \quad . \tag{2.80}$$

After that, for the first and second derivatives of the differential equation shown by $f'(\mathrm{u(t)})$ and $f''(\mathrm{u(t)})$, respectively, the initial condition is applied in the following forms:

$$f'(u(t=0))=0 \tag{2.81}$$

$$f''(u(t=0))=0\,. \tag{2.82}$$

By solving the set of algebraic equations consisting of four equations with four unknowns from equation 2.79 to 2.82, the constant coefficients a, b, d and φ can be gained as follows:

$$a=-0.0451 \quad , \quad b=0.2534 \quad , \quad d=0.194 \quad , \quad \varphi=-1.336. \tag{2.83}$$

By substituting the above constant values in the answer function, the solution of the nonlinear differential equation is obtained below:

$$u(t)=-0.0451e^{-0.2534t}+0.194\cos(t-1.336). \tag{2.84}$$

2.2.7.3. Solving the Differential Equation by Numerical Method and Comparing It with AGM

The nonlinear differential equation 2.76 can be solved numerically in $t \in \{0,\ 40\}$ and compared with AGM in the table below:

Table 2.7. The Compared Solutions of AGM and Numerical Method

t	0.0	8	16	24	32	40
$u(t)$ *NUM*	0.0	0.17511	-0.09233	-0.1594	0.13694	0.1192
$u(t)$ *AGM*	0.0	0.1740	-0.0982	-0.1517	0.1415	0.1111

The obtained chart by numerical method and AGM is depicted as follows:

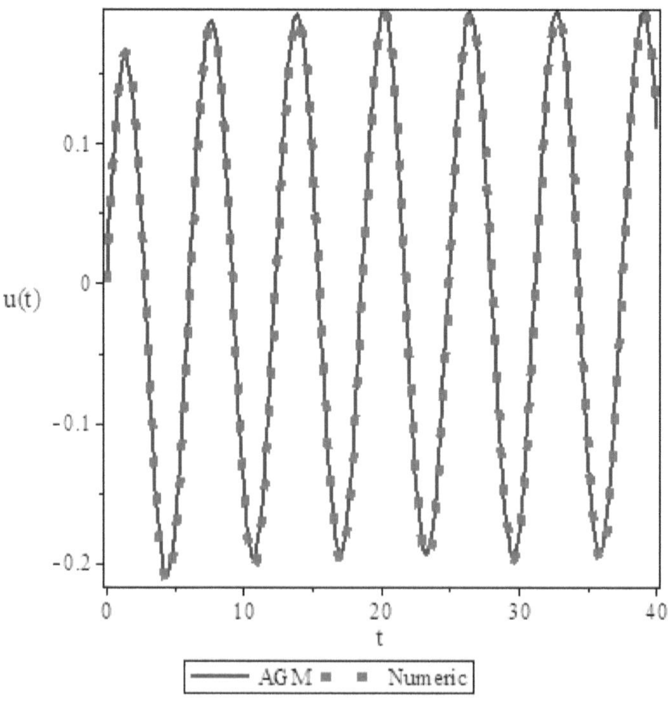

Fig. 2.7. The chart of obtained solutions by numerical method and AGM

In order to increase the precision of the achieved solution, more sentences from the answer function should be chosen, so the achieved answer by AGM is strongly tended toward the exact solution.

Example 2.2.8

Consider the following nonlinear differential equation and solve it by AGM.

$$\begin{cases} f(t): & \dfrac{du}{dt} = u(v-1) + u^2 v \\ g(t): & \dfrac{dv}{dt} = u - v^2 - u^2 - v \end{cases} \tag{2.85}$$

And the related initial conditions are defined as

$$u(0) = u_0 \ , \ \ v(0) = v_0. \tag{2.86}$$

Where, in the above equation, $u_0 = 0.15, v_0 = 0.1.$

2.2.8.1. Solving the Differential Equation by AGM

In this step, an answer function as an infinite power series is needed to solve the set of nonlinear differential equations 2.85 by AGM as follows:

$$u = \sum_{k=1}^{2} a_k e^{-b_k t} = a_1 e^{-b_1 t} + a_2 e^{-b_2 t}$$

$$v = \sum^{2} c_k e^{-d_k t} = c_1 e^{-d_1 t} + c_2 e^{-d_2 t}. \tag{2.87}$$

The constant coefficients a_1 , b_1 , a_2 , b_2 and also c_1 , d_1 , c_2 , d_2 can easily be gained by applying the initial conditions.

2.2.8.2. Applying Initial Conditions by AGM

Like the previous examples, initial conditions in AGM are applied in two ways.

a. Applying initial conditions on the given answer functions is done as follows:

$$u = u(IC) \quad , \quad v = v(IC). \tag{2.88}$$

IC is the abbreviation of applying initial equations. So the initial conditions are applied on the answer functions as

$$u(0) = 0.15 \quad \rightarrow \quad a_1 + d_1 = 0.15 \tag{2.89}$$

$$v(0) = 0.1 \quad \rightarrow \quad a_2 + d_2 = 0.1. \tag{2.90}$$

b. The initial conditions are applied on equation 2.85, shown by *f(t)* and *g(t)*, after substituting the answer functions instead of the dependent parameters u, v in the following general forms:

$$\begin{aligned} f(u,v):&\ \frac{d}{dt}[\sum_{k=1}^{2} a_k e^{-b_k t}] = [\sum_{k=1}^{2} a_k e^{-b_k t}]\{[\sum_{k=1}^{2} a_k e^{-d_k t}] - 1\} + [\sum_{k=1}^{2} a_k e^{-b_k t}]^2 [\sum_{k=1}^{2} c_k e^{-d_k t}] \\ g(u,v):&\ \frac{d}{dt}[\sum_{k=1}^{2} c_k e^{-d_k t}] = \sum_{k=1}^{2} a_k e^{-b_k t} - [\sum_{k=1}^{2} c_k e^{-d_k t}]^2 - [\sum_{k=1}^{2} a_k e^{-b_k t}]^2 - \sum_{k=1}^{2} c_k e^{-d_k t} \end{aligned} \tag{2.91}$$

Now, applying the initial conditions on equation 2.91 and also on their derivatives is done as follows:

$$f(u(0)\ ,\ v(0)): \rightarrow -a_1 b_1 - d_1 c_1 = (a_1 + d_1)[(a_2 + d_2 - 1) + (a_1 + d_1)(a_2 + d_2)] \tag{2.92}$$

$$g(u(0)\ ,\ v(0)): \rightarrow -a_2 b_2 - d_2 c_2 = (a_1 + d \ \rightarrow)[1 - (a_1 + d_1)] - (a_2 + d_2)[1 + (a_2 + d_2)]. \tag{2.93}$$

Then, the initial conditions are applied on the first derivative of the achieved equations in terms of the variable t as

$$f'[u(x=0) \quad , \quad v(x=0)] = 0 \tag{2.94}$$

$$g'[u(x=0) \quad , \quad v(x=0)]=0 . \tag{2.95}$$

Applying the initial conditions on the second derivatives is done as follows:

$$f''[u(x=0) \quad , \quad v(x=0)]=0 \tag{2.96}$$

$$g''[u(x=0) \quad , \quad v(x=0)]=0. \tag{2.97}$$

By solving the set of algebraic equations consisting of equation 2.89 to equation 2.97, except equation 2.91, the constant coefficients of the answer functions can be achieved as

$$\begin{aligned} &a_1=0.15 \quad , \ b_1=0.884 \quad , \ a_2=0.00000314 \ , \ b_2=-18.96 \\ &c_1=0.05+0.19i \ , \ d_1=-0.87-0.276i \ , \ c_2=0.05-0.189 \ , \ d_2=-0.871+0.276i. \end{aligned} \tag{2.98}$$

Where $\boldsymbol{i}$ is the imaginary part of the complex number. And also, by substituting the constant coefficients of equation 2.98 into the equation 2.87 and by using mathematical relations, the set of the nonlinear differential equations 2.85 can be solved as follows:

$$\begin{aligned} &u(t)=0.15e^{-0.284t}+0.0000314e^{-18.96t} \\ &v(t)=0.1e^{-0.87t}\cos(0.276t)+0.379e^{-0.87t}\sin(0.276t). \end{aligned} \tag{2.99}$$

2.2.8.3. A Comparison between the Obtained Answer by AGM and Numerical Solution

With regard to the amounts of boundary conditions in the time domain $t\in\{0, 10\}$, the set of nonlinear differential equations 2.85 has been solved by numerical method, and the obtained results have been compared with AGM in the following table:

Table 2.8. The Result of AGM and Numerical Solutions

t	0.0	2	4	6	8	10
u(t)	0.15	0.02418	0.003467	0.0004755	0.00006452	0.00000873
v(t)	0.1	0.05065	0.01304	0.002682	0.0004908	0.00008384

In order to increase the precision of answer functions by AGM, we can increase the number of the series sentences just by adding one sentence to each answer function; and as a result, the accuracy will be increased significantly.

The gained charts by AGM and numerical solution have been compared as follows:

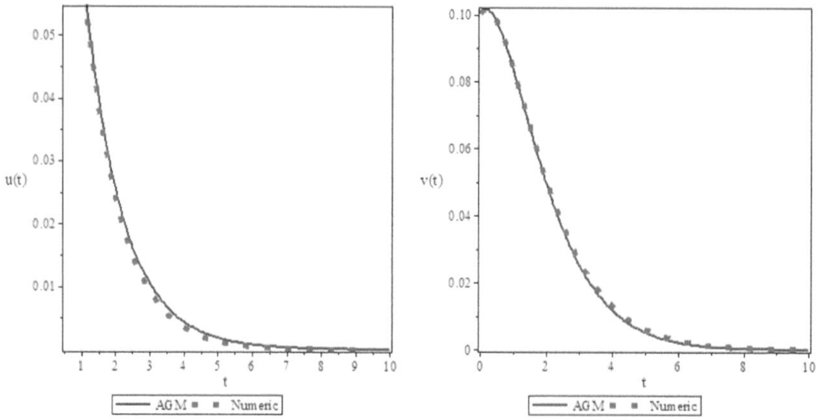

Fig. 2.8. A comparison between the obtained $u(t)$ by AGM and numerical method

Fig. 2.9. A comparison between the obtained $v(t)$ by AGM and numerical method

Example 2.2.9

A gas with Mach number (M) which is much smaller than 1, enters a tube with a friction coefficient of f in the adiabatic state as follows:

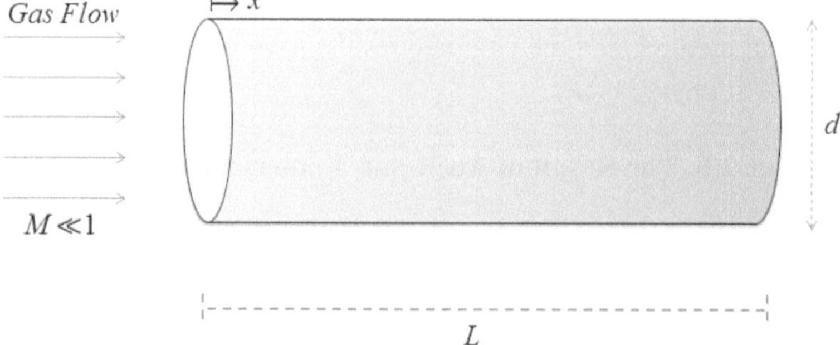

Fig. 2.10. The schematic diagram of the physical model

In the adiabatic state, the heat transferred to the circumstance is zero ($Q = 0$).

In the governing differential equation, the Mach number is a function of the tube length as follows:

$$g(x) : \frac{f}{d} = \frac{2}{M}\frac{dM}{dx}\left\{\frac{1-M^2}{(1-\frac{\gamma-1}{2}M^2)\gamma M^2}\right\} . \tag{2.100}$$

In the aforementioned equation, gamma is defined as $\gamma = \frac{C_p}{C_v}$, and the related initial condition for the above differential equation is expressed as

$$IC : M(0) = M_1. \tag{2.101}$$

2.2.9.1. Solving the Differential Equation by AGM

A power series is considered for the answer function of the presented problem as

$$M = \sum_{k=0}^{n} a_k e^{kb_k x} = a_0 + a_1 e^{b_1 x} + a_2 e^{2b_2 x} + \dots . \tag{2.102}$$

To reduce the size of operations, consider the first two series sentences from equation 2.102. It is citable that more sentences can be chosen in order to increase the precision of the solution.

$$M = a_0 + a_1 e^{b_1 x} \tag{2.103}$$

2.2.9.2. Applying the Initial Conditions

The initial condition is applied on the answer function and also on the differential equation 2.100 shown by g(M(x)) in the following forms.

a. Applying the initial condition on the answer function is done as follows:

$$M(0) = M_1 \quad \rightarrow \quad a_0 + a_1 = M_1. \tag{2.104}$$

b. Applying it on the main differential equation is done after substituting equation 2.103 into equation 2.100 and also on its derivative as follows:

$$g(M(x=0)) \ : \ \frac{f}{d} = \frac{2a_1b_1[1-(a_0+a_1)^2]}{\gamma(a_0+a_1)^3\{1+\frac{1}{2}(\gamma-1)(a_0+a_1)^2\}}. \tag{2.105}$$

Then, for the derivative of the differential equation shown by $g'(M(x=0))$, we will have

$$g'(M(x=0)) \ : \ \frac{2a_1b_1[1-(a_0+a_1)^2]}{\gamma(a_0+a_1)^3\{1+\frac{1}{2}(\gamma-1)(a_0+a_1)^2\}}\{1-\frac{3}{(a_0+a_1)}\} - \frac{2a_1^2b_1^2}{\gamma(a_0+a_1)^2\{1+\frac{1}{2}(\gamma-1)(a_0+a_1)^2\}}\{2+\frac{1-(\gamma-1)[1-(a_0+a_1)^2]}{1+\frac{1}{2}(\gamma-1)(a_0+a_1)^2}\} = 0. \tag{2.106}$$

By solving the set of algebraic equations consisting of three equations with three unknowns from equation 2.104 to equation 2.106, the constant coefficients of the answer function can be calculated in the following form.

To simplify, consider the following new variable:

$$\psi_1 = M_1^2(7-3M_1^2+3\gamma M_1^2-5\gamma)-6 \ , \quad \psi_2 = M_1^2(M_1^2-2)(\gamma-1)-2$$
$$\psi_3 = (M_1^2-1)(\gamma M_1^2 - M_1^2+2) \ , \quad \Delta = \frac{f\gamma M_1^2}{\alpha(1-M_1^2)^2} \ . \tag{2.107}$$

Therefore, the constant coefficients of the answer function can be obtained as

$$a_0 = \frac{2M_1\psi_2}{\psi_1} \ , \quad a_1 = \frac{M_1\psi_3}{\psi_1} \ , \quad b_1 = -\frac{1}{4}\Delta. \tag{2.108}$$

By substituting the values of equation 2.108 into equation 2.103, the answer of the nonlinear differential equation, which is the exact value of the Mach number in the tube length, can be achieved by AGM as follows:

$$M(x) = \frac{2M_1\psi_2}{\psi_1} + \frac{M_1\psi_3}{\psi_1} e^{-\frac{1}{4}\Delta\psi_1 x} . \tag{2.109}$$

By choosing the physical values for the system such as

$$d = 30(cm) \;\; ; \;\; \gamma = 1.4 \;\; , \;\; f = 0.05 \;\; , \;\; L = 100(\mathrm{m}) \;\; , \;\; \mathrm{M}_1 = 0.1 \tag{2.110}$$

the final solution of the presented problem is acquired in the following form:

$$\mathrm{M(x)} = 0.06693333866 + 0.03306666133\mathrm{e}^{0.0035709992755\mathrm{x}} . \tag{2.111}$$

2.2.9.3. Solving the Differential Equation by Numerical Method

With regard to the given physical values, the differential equation 2.100 has been solved numerically in the domain $x \in (0\ ,\ 100)$ in the following table:

Table 2.9. Various Values of Mach Number Have Been Obtained in Terms of the Tube Length by Numerical Method

$x(m)$	*0.00*	*20*	*40*	*60*	*80*	*100*
$M(x)$ NUM	0.10	0.10245	0.10509	0.10794	0.111054	0.11445

The charts resulting from AGM and numerical method have been drawn and compared below:

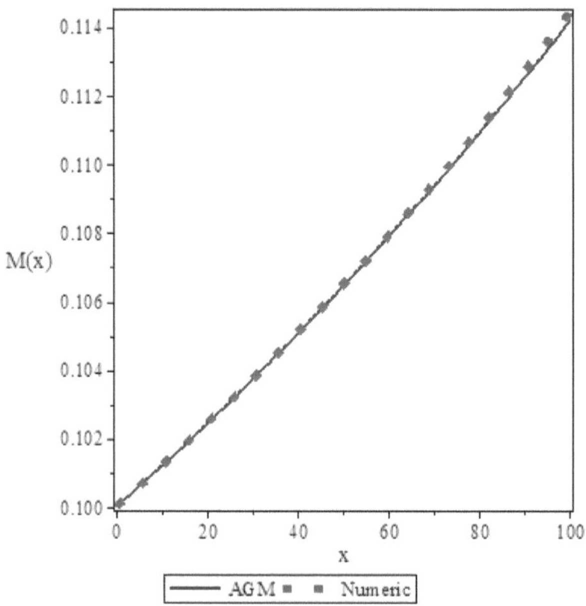

Fig. 2.11. A comparison between the acquired trend of Mach number in terms of tube length by AGM and numerical method

2.2.9.4. Computing the Gas Pressure and Temperature in the Length of the Tube

According to reference books [1-5], the gas pressure and temperature by having the Mach number in the tube length can be obtained as follows:

$$P = P_0 \exp[-\int_0^x \{[\frac{1+(\gamma-1)M^2}{M(1+\frac{\gamma-1}{2}M^2)}]\frac{dM}{dx}\}dx] \tag{2.112}$$

$$T = T_0 \exp\{-\int_0^x \frac{(\gamma-1)M\frac{dM}{dx}}{1+\frac{\gamma-1}{2}M^2}dx\}. \tag{2.113}$$

P_0 and T_0 are the entrance pressure and temperature to the tube length. By selecting the physical values of P_0=1 *atm* and T_0=303° k, the charts of pressure and temperature in the tube length due to equation 2.112 and equation 2.113 have been depicted below:

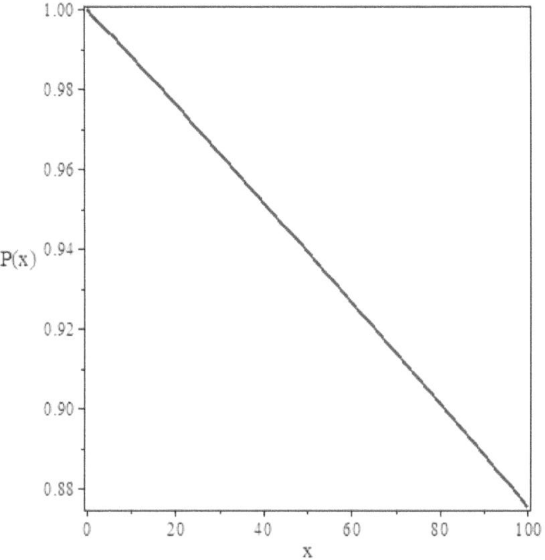

Fig. 2.12. Variation of fluid pressure in the length of the tube

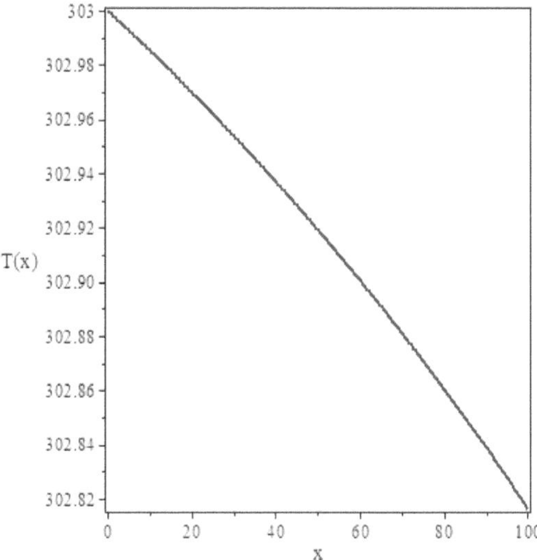

Fig. 2.13. Fluid temperature changes in the tube length

Example 2.2.10

Solve the nonlinear differential equation governing the gas flow in the tube length in general manner by AGM according to the following figure.

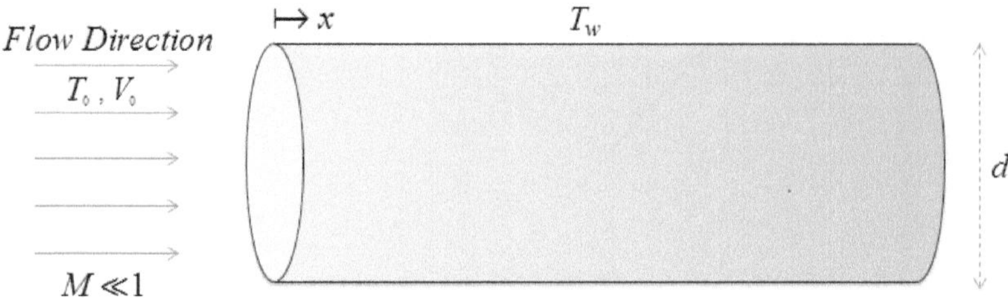

Fig. 2.14. The geometry of the introduced system

It is supposed that the tube has been exchanging heat, and the friction coefficient in the tube is considered as f . And also, the gas with the Mach number which is much smaller than 1, enters the tube. Moreover, T_w is the surface temperature of the tube. Like the previous example, P_0 and T_0 are the entrance pressure and temperature to the tube.

2.2.10.1. Solution Procedure

In general manner (without any presupposing), the nonlinear differential equation governing the system in the dynamic of gas books can be calculated as follows:

$$g(x): \frac{T_w - T_t}{T_t}\frac{(1+\gamma M^2)f}{4d} + \frac{1}{2}\gamma M^2 \frac{f}{d} = \{\frac{1-\gamma M^2}{M} + \frac{(\gamma-1)(1+\gamma M^2)\frac{M}{2}}{1+\frac{\gamma-1}{2}M^2}\}\frac{dM}{dx}. \tag{2.114}$$

Where T_t is the record temperature of the fluid entering the tube, which can be obtained as

$$T_t = \frac{v_0^2}{\gamma g R T_0 M^2}(1+\frac{\gamma-1}{2}M^2). \tag{2.115}$$

It is noteworthy that R is the general constant of gases.

By substituting the values of equation 2.115 into equation 2.114, the differential equation governing the system can be achieved. Afterward, the following initial condition is presented as

$$M(0) = M_1 = \frac{V_0}{\sqrt{\gamma g R(T_0 + 273)}}. \tag{2.116}$$

2.2.10.2. Solving Differential Equation by AGM

Choose an answer function as a power series with constant coefficients in the following form:

$$M = a_0 + e^{b_1 x} M = a_0 + e^{b_1 x}. \tag{2.117}$$

2.2.10.3. Applying the Initial Conditions

The initial condition is applied, like the previous example, on the answer function as follows:

$$M(0) = M_1 \quad \rightarrow \quad a + b = M_{1.} \tag{2.118}$$

And applying it on the differential equation is done as

$$g(M(x=0)):. \tag{2.119}$$

And also for the derivative of equation 2.114, we will have

$$g'(M(x=0)):. \tag{2.120}$$

By solving the set of algebraic equations consisting of three equations with three unknowns from equation 2.118 to equation 2.120, the constant coefficients of the answer function can easily be calculated.

To simplify, consider the following new variables:

$$
\begin{aligned}
\psi_1 &= 2T_w\gamma gRT_0M_1^2(3\gamma M_1^4-5\gamma M_1^2+M_1^2-3)+\gamma M_1^2V_0^2(3\gamma M_1^4-3M_1^4-\\
&\quad 5\gamma M_1^2+6M_1^2-3)+V_0^2(M_1^4-M_1^2+2)\\
\psi_2 &= (\gamma M_1^4-2\gamma M_1^2-1)(2T_w\gamma gRT_0+\gamma V_0^2-V_0^2)\\
\psi_3 &= (M_1^2-1)(V_0^2\gamma M_1^2+2T_w\gamma^2gRT_0M_1^4-\gamma V_0^2M_1^4+\gamma V_0^2M_1^2+2T_w\gamma gT_0M_1^2V_0^2M_1^2-2V_0^2)\\
\Delta &= \frac{f}{(1-M_1^2)^2dv_0^2} \quad .
\end{aligned}
\tag{2.121}
$$

The constant coefficients of the answer function can be calculated in the following form:

$$a_0=\frac{2M_1^3\psi_2}{\psi_1} \quad , \quad a_1=\frac{M_1\psi_3}{\psi_1} \quad , \quad b_1=-\frac{1}{8}\Delta. \tag{2.122}$$

Then, by substituting equation 2.122 into equation 2.117, the answer of nonlinear differential equation 2.114 can be obtained by AGM as follows:

$$M(x)=\frac{2M_1^3\psi_2}{\psi_1}+\frac{M_1\psi_3}{\psi_1}e^{-\frac{1}{8}\Delta\psi_1x}. \tag{2.123}$$

By considering the following physical values of the system as

$$
\begin{aligned}
&d=0.4 \quad , \quad \gamma=1.4 \quad , \quad f=0.00045 \quad , \quad L=50\\
&T_w=100 \quad , \quad R=8.3 \quad , \quad V_0=5 \quad , \quad T_0=25
\end{aligned}
\tag{2.124}
$$

the final solution of the problem is achieved below:

$$M=0.01920289138+0.008443587296\,e^{0.006818983152x}. \tag{2.125}$$

2.2.10.4. Comparing the Resulted Solution by Numerical Method and AGM

The differential equation 2.114 in the domain $x\in[0,50]$ has been solved numerically in the following table:

Table 2.10. Mach Number Obtained by Numerical Method

x	*0.0*	*10*	*20*	*30*	*40*	*50*
$M(x)$	M_1=0.02764	0.02824	0.028883	0.029573	0.03032	0.03113

The charts of the solutions by numerical method and AGM have been depicted as follows:

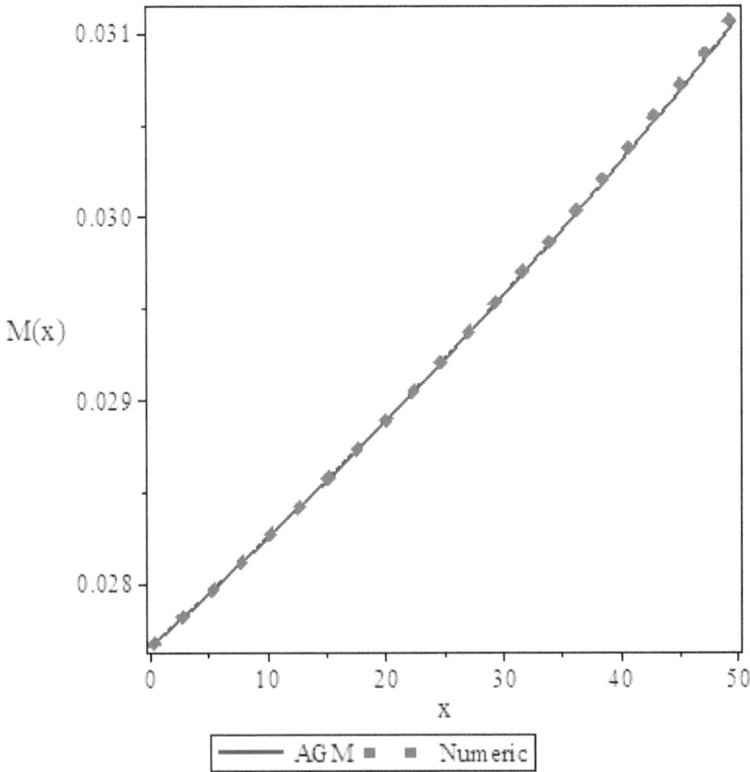

Fig. 2.15. Comparing the profiles of Mach number in the tube length acquired by AGM and numerical method

2.3. Problems

2.3.1. Suppose the following function in the domain $x \in \{0,\ 2\}$ and solve it by AGM.

$$f\ :\ \frac{du}{dx} = u^3 + u^2\sinh(u) = 0$$

And the related boundary condition is introduced as $u(0)\ =\ 0.1$.

Moreover, compare the obtained solution graphically by numerical method in the mentioned domain.

2.3.2. Consider the nonlinear differential equation as follows:

$$g\ :\ \frac{du}{dx} =\ \cosh(xu) - u + x + 1.$$

And the related boundary condition is expressed as $u(0)\ =\ 0$. Solve the mentioned differential equation in $x \in \{0,\ 1\}$.

2.3.3. Obtain the absolute values of the difference between AGM and numerical solution in the domain $t \in \{0, 40\}$ according to the introduced parameters and depict them.

$$\omega_0 = 1\ ,\ v_0\ =\ 0.2\ ,\ c\ =\ 5$$

$$f\ :\ \frac{dv}{dt} + \frac{v^5}{c} + v = v_0.\sin(\omega_0 t)$$

And the related boundary condition is $v(0) = 0.1$.

2.3.4. Solve the following differential equation by AGM in $t \in \{0,\ 20\}$ when $v(0)=0$, and also, the physical parameters of the system are $R_o = 5, \omega_0 =\ 1,\ v_0\ = 0.5\ ,\ \alpha = 0.2$. Then based on the given information, acquire the absolute values of $\frac{v_{AGM} - v_{NUM}}{v_{NUM}}$.

$$h : (v^2\alpha + R_0)(\frac{dv}{dt}) + \frac{v}{v^2+1} = v_0 \cos(\omega_0 t)$$

2.3.5. Solve the differential equation below by AGM when $\omega_0 = 1$, $v_0 = 0.2$, $n = 4$, $m = 3$.

$$f : (1 + \sqrt{1 + v^n + v^m})(\frac{dv}{dt}) + v = v_0 \sin(\omega_0 t)$$

And the boundary conditions are expressed as $v(0) = 0$. Finally, obtain the error function of the aforementioned problem in $x \in \{0, 20\}$ and define the maximum error of the obtained solution.

2.3.6. Consider the following first-order nonlinear differential equation in $t \in \{0, 20\}$.

$$g : (v^3 + v^2 + 1)(\frac{dv}{dt}) + \frac{v}{v^2+1} = v_0(\sin(\omega_0 t) + v^2 \cos(\omega_0 t))$$

And the boundary condition is expressed as $v(0) = 0$. Find a suitable answer function for the solution procedure.

HELP = Compare each of the solutions gained by different answer functions with numerical method.

2.4. References

1. Robert. D. Zucker, Oscar Biblarz, *Fundamentals of Gas Dynamics,* 2nd ed. (John Wiley and Sons, 2002).

2. Patrick. H. Oosthuizen, William E. Carscallen, *Introduction to Compressible Fluid Flow,* 2nd ed. (CRC Press, 2013).

3. S. C. Gupta, *Fluid Mechanics and Hydraulic Machines,*(Pearson Education India, 2006).

4. Theodore Gresh, *Compressor Performance: Aerodynamics for the User*, 2nd ed. (Newnes, 2001).

5. Michael. J. Moran et al., *Fundamentals of Engineering Thermodynamics,* 7th ed.(John Wiley and Sons, 2010).

Chapter 3

The Second- and Higher-Order Nonvibrational Differential Equations

3.1. Physical Examples

Example 3.1.1

Consider the following differential equation in $x \in \{1,\ 2\}$ and solve it by AGM.

$$f(x): xy''+xy'+y=0 \tag{3.1}$$

And the related boundary conditions are defined as

$$y(1)=0 \quad , \quad y(2)=3. \tag{3.2}$$

In this step, an answer function as a finite polynomial series is needed to solve differential equations by AGM as follows:

$$y(x)=\sum_{k=0}^{3} a_k x^k = a_0 + a_1 x + a_2 x^2 + a_3 x^3. \tag{3.3}$$

3.1.1.1. Applying Boundary Conditions by AGM

Boundary conditions in AGM are applied on the answer function as

$$y(1)=0 \quad \rightarrow \quad a_0 + a_1 + a_2 + a_3 = 0 \tag{3.4}$$

$$y(2)=3 \quad \rightarrow \quad a_0 + 2a_1 + 4a_2 + 8a_3 = 3. \tag{3.5}$$

Then, after substituting equation 3.3 into equation 3.1, the boundary conditions are applied on the obtained equation as follows:

$$f(y(1)): \quad \rightarrow \quad 5a_2 + 10a_3 + 2a_1 + a_0 = 0 \tag{3.6}$$

$$f(y(2)): \quad \rightarrow \qquad 16a_2 + 56a_3 + 4a_1 + a_0 = 0. \tag{3.7}$$

By solving the set of algebraic equations consisting of equation 3.4 to equation 3.7, the constant coefficients a_0 to a_3 are computed in the form of

$$a_0 = \frac{-90}{11} \quad , \quad a_1 = \frac{225}{22} \quad , \quad a_2 = \frac{-42}{11} \quad , \quad a_3 = \frac{9}{22}. \tag{3.8}$$

Afterward, by substituting the above constant coefficients into equation 3.3, the solution of the differential equation is obtained as

$$y(x) = \frac{-90}{11} + \frac{225}{22}x - \frac{42}{11}x^2 + \frac{9}{22}x^3. \tag{3.9}$$

3.1.1.2. A Comparison between the Obtained Answer by AGM and the Exact Solution

The differential equation 3.1 in the given domain is answered by exact solution as follows:

$$y(x) = \frac{3e^{2-x}[e^x + Ei(1,x)x - ex - Ei(1,-1)x]}{e^2 + 2Ei(1,-2) - 2\mathrm{e} - 2\,\mathrm{Ei}(1,-1)}. \tag{3.10}$$

Where Ei is defined as the gamma function (Γ) in the following form:

$$Ei(a,z) = z^{a-1}\Gamma(1-a,z). \tag{3.11}$$

Here, the obtained charts by AGM and exact solution are compared as

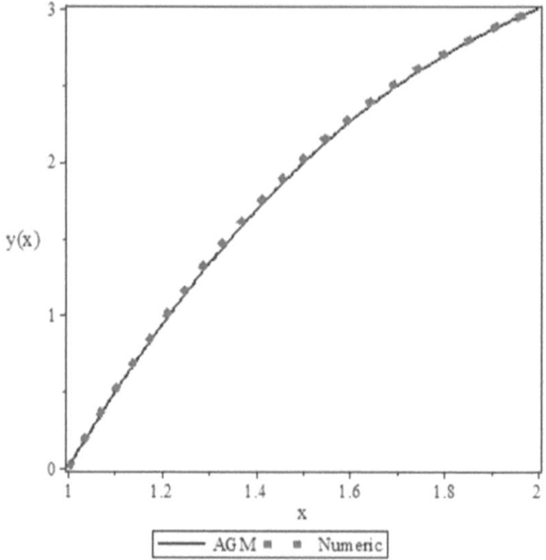

Fig. 3.1. The result of obtained solutions by AGM and exact solution

Example 3.1.2

Solve the nonlinear differential equation with the following initial conditions by AGM.

$$f(x): \quad u''' = 2e^{2x} + u'u^2 - u' + u + 1 \tag{3.12}$$

And the initial conditions of this equation are

$$u(0) = 0 \quad , \quad u'(0) = 0 \quad , \quad u''(0) = 1. \tag{3.13}$$

3.1.2.1. Choosing the Answer Function of the Equation by AGM

In order to solve the differential equation 3.12, an answer function is considered as a finite series in the form of

$$u(x) = \sum_{n=0}^{6} a_n x^n = a_0 + a_1 x + a_2 x^2 + a_3 x^3 + a_4 x^4 + a_5 x^5 + a_6 x^6. \tag{3.14}$$

The given answer function has the constant coefficients a_0 to a_6, which can easily be computed by applying the initial conditions from equations 3.13. It is notable that the more numbers of series sentences of equation 3.14, the more precise the answer, and the answer is tended to the exact solution.

3.1.2.2. Applying Initial or Boundary Conditions by AGM

In AGM, initial or boundary conditions are applied in the following two general ways.

a. Initial conditions are applied on the answer function as the following general equation:

$$u = u(IC, BC). \tag{3.15}$$

BC and IC are the abbreviations of boundary and initial conditions in differential equations, respectively. Then with respect to the given initial conditions, their application on the given answer function is shown as

$$u(0) = 0 \quad \rightarrow \quad a_0 = 0 \tag{3.16}$$

$$u'(0) = 0 \quad \rightarrow \quad a_1 = 0 \tag{3.17}$$

$$u''(0) = 1 \quad \rightarrow \quad 2a_2 = 1. \tag{3.18}$$

b. Applying boundary or initial conditions on the differential equation and their derivatives obtained by substituting the answer function, which is equation 3.14 into the differential equation 3.12, is done as the following general equation:

$$f(u(x)) \quad \rightarrow \quad f(u(IC\ ,\ BC)) = 0. \tag{3.19}$$

And for its derivative, we will have

$$f'(u(x)) \quad \rightarrow \quad f'(u(IC\ ,\ BC)) = 0. \tag{3.20}$$

The above equation is the first derivative of equation 3.19, and also, we follow the above procedure for higher orders.

Based on the above explanations, we have

$$f(u(0)): \quad \rightarrow \qquad 6a_3 = 3 + a_1 a_0^2 - a_1 - a_0 \tag{3.21}$$

$$f'(u(0)): \quad \rightarrow \qquad 24a_4 = 4 + 2a_2 a_0^2 + 2a_1^2 a_0 - 2a_2 + a_1 \tag{3.22}$$

$$f''(u(0)): \quad \rightarrow \qquad 120a_5 = 8 + 6a_3 a_0^2 + 12a_2 a_0 a_1 + 2a_1^3 - 6a_3 + 2a_2 \tag{3.23}$$

$$f'''(u(0)): \quad \rightarrow \qquad 720a_6 = 16 + 24a_4 a_0^2 + 48a_3 a_0 a_1 + 24a_2 a_1^2 + 24a_2^2 a_0 - 24a_4 + 6a_3 \tag{3.24}$$

Since the answer function of the considered differential equation consists of seven constant coefficients, a_0 to a_6, by solving the set of algebraic equations from equation 3.16 to equation 3.18 and from equation 3.21 to equation 3.24, the constant coefficients can be acquired as follows:

$$a_0 = 0 \quad , \quad a_1 = 0 \quad , \quad a_2 = \frac{1}{2} \quad , \quad a_3 = \frac{1}{2} \quad ,$$
$$a_4 = \frac{1}{8} \quad , \quad a_5 = \frac{1}{20} \quad , \quad a_6 = \frac{1}{45} \quad . \tag{3.25}$$

By substituting the obtained values from equation 3.25 into the answer function, we can obtain the solution of nonlinear differential equation as

$$u(x) = \frac{1}{2}x^2 + \frac{1}{2}x^3 + \frac{1}{8}x^4 + \frac{1}{20}x^5 + \frac{1}{45}x^6. \tag{3.26}$$

3.1.2.3. Numerical Solution of the Differential Equation and Comparing It with AGM

In comparison with AGM, the differential equation 3.12 has numerically been solved in the certain domain of $x \in \{0, 1\}$ as:

Table 3.1. The Achieved Results of AGM and Numerical Method

x	0	0.2	0.4	0.6	0.8	1.0
$u(x)$ NUM	0	0.024217	0.1158145	0.3093576	0.6515807	1.2109248
$u'(x)$	0	0.264445	0.6799758	1.293707	2.1837853	3.5007142
$u'(x)$ AGM	0	0.024217	0.115803	0.3091248	0.6494094	1.1972222

Comparing the obtained solutions by AGM and numerical method

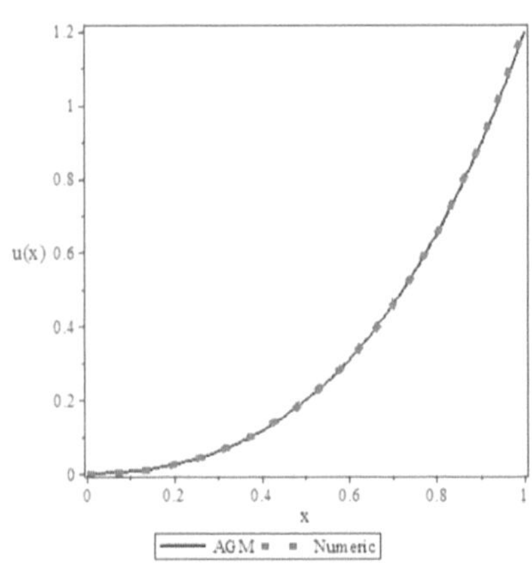

Fig. 3.2. Comparing the achieved solutions by AGM and numerical method

Fig. 3.3. Graphical results for the first derivative of the obtained solution by AGM and numerical method

Example 3.1.3

In general, all the physical problems with definite boundary conditions at the beginning and at the end will be led to the fourth-order differential equations, which are very applicable in engineering and basic sciences like the following example:

$$f(x):\ u^{IV}(x)+u^2(x)+u(x)u''(x)+u(x)=0 \tag{3.27}$$

And the related boundary conditions are

$$u(0)=0 \quad , \quad u(1)=0 \quad , \quad u'(0)=1 \quad , \quad u'(1)=0. \tag{3.28}$$

Like the previous examples for this equation, an answer function is considered as

$$u(x)=\sum_{k=0}^{7} a_k x^k = a_0 + a_1 x + a_2 x^2 + a_3 x^3 + a_4 x^4 + a_5 x^5 + a_6 x^6 + a_7 x^7. \tag{3.29}$$

By applying the boundary conditions, the constant coefficients a_0 to a_7 are easily computed by solving the set of algebraic equations.

3.1.3.1. Applying Boundary Conditions by AGM

As it has been said in the previous examples, boundary conditions in AGM are applied in two ways.

a. Applying boundary conditions on the answer function is done as follows:

$$u(x)=u(BC). \tag{3.30}$$

Where BC is the abbreviation of applying boundary conditions on the given answer function. So the boundary conditions are applied on equation3.29 as

$$u(0)=0 \quad \rightarrow \quad a_0=0 \tag{3.31}$$

$$u(1)=0 \quad \rightarrow \quad a_0+a_1+\ldots+a_7=0 \tag{3.32}$$

$$u'(0)=0 \quad \rightarrow \quad a_1=1 \tag{3.33}$$

$$u'(1)=0 \quad \rightarrow \quad a_1+2a_2+3a_3+4a_4+5a_5+6a_6+7a_7=0. \tag{3.34}$$

b. Boundary conditions are applied on equation 3.27 shown by $f(\mathrm{x})$ in the following form:

$$f(u(x)) \quad \rightarrow \quad f(u(\mathrm{B.C}) = 0 \quad , \quad f'(u(b.C)) = 0 \quad , \quad .... \tag{3.35}$$

The above procedure indicates that equation 3.29 should be substituted into equation 3.27 instead of the dependent variable u, and its derivatives should be calculated in relevance to the independent variable x. Finally, the boundary conditions are applied as

$$f(u(0)): \quad \rightarrow \quad 24a_4 + a_0^2 + 2a_0a_2 + a_0 = 0 \tag{3.36}$$

$$\begin{aligned} f(u(1)):& \\ & 25a_4 + 121a_5 + 361a_6 + 841a_7 + (a_0 + a_1 + ... + a_7)\{(a_0 + a_1 + ... + a_7) + (2a_2 + \\ & 6a_3 + 12a_4 + 20a_5 + 30a_6 + 42a_7)\} + a_0 + a_1 + a_2 + a_3 = 0 \end{aligned} \quad . \tag{3.37}$$

And like before, the boundary conditions are applied on the first-order derivative of the differential equation as

$$f'(u(0)): \quad \rightarrow \quad 120a_5 + 2a_0a_1 + 2a_1a_2 + 6a_0a_3 + a_1 = 0 \tag{3.38}$$

$$\begin{aligned} f'(u(1)):& \\ & 125a_5 + 726a_6 + 2527a_7 + (a_0 + a_1 + ... + a_7)\{2(a_1 + 2a_2 + 3a_3 + 4a_4 + 5a_5 + 6a_6 + \\ & 7a_7) + (6a_3 + 24a_4 + 60a_5 + 120a_6 + 210a_7)\} + (a_1 + 2a_2 + 3a_3 + 4a_4 + 5a_5 + 6a_6 + \\ & 7a_7)(2a_2 + 6a_3 + 12a_4 + 20a_5 + 30a_6 + 42a_7) + a_1 + 2a_2 + 3a_3 + 4a_4 = 0 \end{aligned} \quad . \tag{3.39}$$

By solving the set of algebraic equations consisting of eight equations with eight unknowns, the constant coefficients a_0 to a_7 can easily be computed as follows:

$$\begin{aligned} & a_0 = 0 \quad , \quad a_1 = 1 \quad , \quad a_2 = -1.98585 \quad , \quad a_3 = 0.974056 \\ & a_4 = 0.0247641 \quad , \quad a_5 = 0 \quad , \quad a_6 = -0.0165094 \quad , \quad a_7 = 0.0035377 \, . \end{aligned} \tag{3.40}$$

By substituting the constant coefficients from equation 3.40 into equation 3.29, the solution of the nonlinear differential equation 3.27 can be gained as

$$u(x) = x - 1.98585x^2 + 0.974956x^3 + 0.0247641x^4 - 0.0165094x^6 + 0.0035377x^7 . \tag{3.41}$$

3.1.3.2. Solving the Differential Equation by Numerical Method and Its Comparison with AGM

As regards the given boundary conditions, the nonlinear differential equation 3.27 in $x \in \{0, 1\}$ has numerically been solved in the table below and compared with AGM as follows:

Table 3.2. The Result of AGM and Numerical Solutions

x	*0*	*0.2*	*0.4*	*0.6*	*0.8*	*1*
$u(x)$ *NUM*	0.0	0.1280574	0.1440755	0.0960276	0.03199755	0.0
$u'(x)$	1.0	0.3203158	-0.1201289	-0.3202526	-0.2800366	0.0
$u''(x)$	-3.99403	-2.801661	-1.601871	-0.399349	0.80100163	1.99921
$u(x)$ AGM	0.0	0.1283654	0.1447955	0.096745	0.0323023	0.0

Comparison between the obtained charts by numerical method and AGM

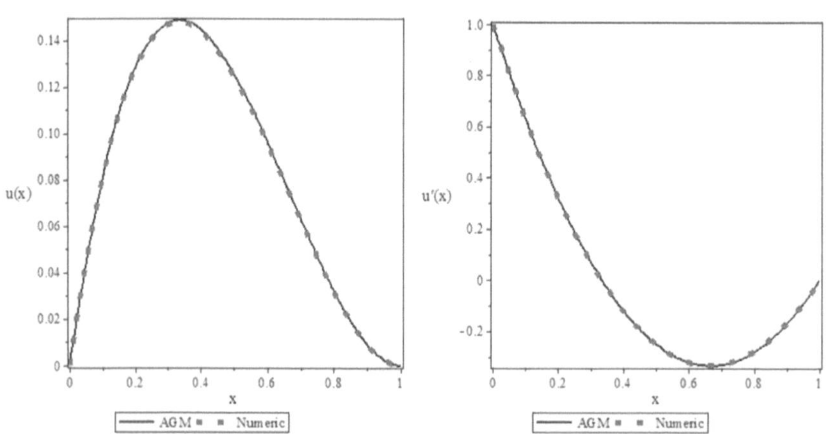

Fig. 3.4. The achieved results by AGM and numerical method for $u(x)$

Fig. 3.5. The chart of the first derivative for the obtained solutions by AGM and numerical method

With regard to the above charts, by choosing an answer function consisting of eight sentences, we can say that it has been a suitable approximation for the solution of the nonlinear differential equation.

These problems are very applicable in systems involved with heat and fluids, which have two boundary conditions at the beginning and two at the end, and are also applied for structural systems such as beams or columns.

Example 3.1.4

Consider the nonlinear differential equation in the cylindrical coordinates and solve it.

$$g(r)=\frac{1}{r}\frac{d}{dr}(r\mu(r)\frac{du(r)}{dr})=f(r) \tag{3.42}$$

Where $\mu(r)$ and $f(r)$ are defined as follows:

$$\mu(r)=1+\alpha u+\beta u^2 \quad ; \quad f(r)=-(u+u^2) \tag{3.43}$$

And the boundary conditions are

$$u(R_i)=u_i \quad , \quad u(R_0)=u_0. \tag{3.44}$$

And the constant coefficients in cylindrical coordinates are expressed as

$$\begin{array}{llll} R_i=1 & , \quad R_0=2 & , \quad u_i=1 & , \\ u_0=0 & , \quad \alpha=0.1 & , \quad \beta=0.1 & . \end{array} \tag{3.45}$$

3.1.4.1. Solution

In order to solve the differential equation by AGM, an answer function is needed by choosing a finite polynomial series with the following constant coefficients:

$$u(r)=\sum_{k=0}^{5}a_k r^k=a_0+a_1r+a_2r^2+a_3r^3+a_4r^4+a_5r^5. \tag{3.46}$$

The constant coefficients of the answer function, a_0 to a_5, can easily be computed by applying boundary conditions.

3.1.4.2. Applying Boundary Conditions by AGM

Like before, boundary conditions are applied in two ways:

a. Applying boundary conditions on the answer function is done as

$$u(r) = u(BC). \tag{3.47}$$

Where BC is the abbreviation of "boundary conditions"; its application on the presented answer function is done as

$$u(r=1) = 1 \tag{3.48}$$

$$u(r=2) = 0. \tag{3.49}$$

b. Applying boundary conditions on equation 3.42 shown by $g(r)$ and its derivatives is done as follows:

$$g(u(r)) \quad \rightarrow \quad g(u(BC)) = 0 \quad , \quad g'(u(BC)) = 0 \quad , \quad \ldots. \tag{3.50}$$

The above equation indicates that the answer function should be substituted in the main differential equation instead of the dependent variable U, and then its derivatives can be acquired. After that, the boundary conditions are applied on the obtained equations as

$$g(u(r)): \ \frac{1}{r}\frac{d}{dr}\{r\mu(r)\frac{d}{dr}[\sum_{k=0}^{5} a_k r^k]\} = f(r). \tag{3.51}$$

Now, the boundary conditions are applied on equation 3.51 as follows:

$$g(u(r=1)) = 0 \tag{3.52}$$

$$g(u(r=2)) = 0. \tag{3.53}$$

Applying boundary conditions on the derivative of equation 3.51 is done in the following form:

$$g'(u(r=1))=0 \tag{3.54}$$

$$g'(u(r=2))=0. \tag{3.55}$$

By solving a set of algebraic equations from equation 3.48 to equation 3.55, except equation 3.50 and equation 3.51, the constant coefficients a_0 to a_5 can simply be computed by the Maple software as follows:

$$a_0=0.56374 \quad , \quad a_1=2.49593 \quad , \quad a_2=-3.083 \quad ,$$
$$a_3=1.23136 \quad , \quad a_4=-0.22336 \quad , \quad a_5=0.015616 \quad . \tag{3.56}$$

By substituting the achieved values from equation 3.56 into equation 3.46, the solution of the nonlinear differential equation 3.42 is obtained as

$$u(r)=0.56347+2.49593r-3.083r^2+1.23136r^3-0.22336r^4+0.015616r^5. \tag{3.57}$$

3.1.4.3. Comparing the Obtained Solution of Differential Equation by AGM and Numerical Method

With regard to the given physical values and the boundary conditions, we can solve the differential equation 3.42 in $r \in \{1,\ 2\}$ numerically and compare it with AGM in the table below:

Table 3.3. Comparison between the Achieved Solution by AGM and Numerical Method

r	*1.0*	*1.2*	*1.4*	*1.6*	*1.8*	*2*
$u(r)$ *NUM*	1.0	0.82158	0.61844	0.406761	0.198064	0.0
$u'(r)$	-0.7969	-0.9695	-1.04848	-1.05877	-1.02189	-0.95496
u(r) AGM	1.0	0.82252	0.61981	0.40798	0.19873	0.0

The obtained charts by numerical method and AGM are depicted as follows:

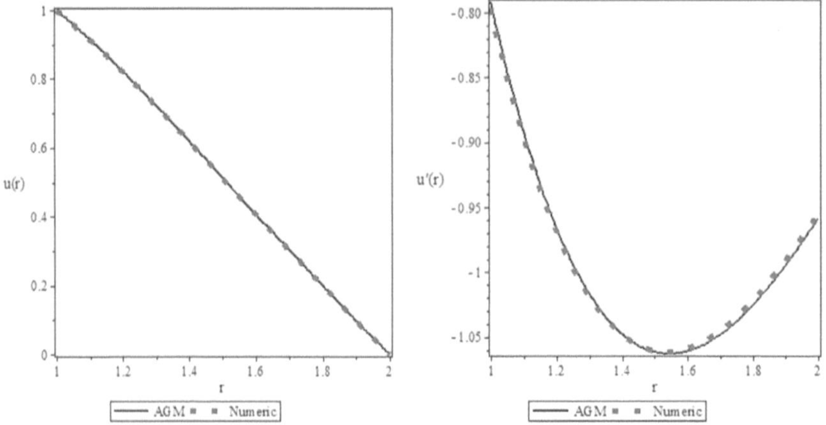

Fig. 3.6. The results of AGM and numerical solution for $u(\mathrm{r})$

Fig. 3.7. Comparing the first derivative of the obtained solution by AGM and numerical method

These types of nonlinear equations, like equation 3.42, are widely used in different fields of study such as strength of materials, elasticity, chemistry, and mechanics (heat transfer and fluid mechanics), especially equations that are defined in polar coordinates.

Example 3.1.5

It is notable that most approaches in engineering and basic sciences are investigated in determined domains, so the following example is solved and analyzed in a specified domain.

$$f(x): \ \frac{d^2u(x)}{dx^2}+x\frac{du(x)}{dx}+\alpha\, u(x)=0 \tag{3.58}$$

And the related boundary conditions are introduced as

$$u(1)=0 \qquad , \qquad u(2)=1. \tag{3.59}$$

α in the above equation is a constant and real parameter.

3.1.5.1. Solution

The nonlinear differential equation 3.58 is an erf(x), which can suitably be solved in the specified domain (boundary conditions) with negligible errors.

Based on the previous examples, in order to solve this problem, an answer function is selected as a finite polynomial series as follows:

$$u(x)=\sum_{k=0}^{5} a_k x^k = a_0 + a_1 x + a_2 x^2 + a_3 x^3 + a_4 x^4 + a_5 x^5. \tag{3.60}$$

It is citable that when the answer function is considered as the infinite series $u(x)=\sum_{k=0}^{\infty} a_k x^k$, the obtained answer in any domain of boundary conditions is similar to the exact solution.

3.1.5.2. Applying Boundary Conditions by AGM

Boundary conditions in AGM are applied in the following two manners:

a. Applying the boundary conditions on the answer function of equation 3.60 is done as follows:

$$u(1)=0 \quad \rightarrow \quad a_0 + a_1 + \ldots + a_5 = 0 \tag{3.61}$$

$$u(2)=1 \quad \rightarrow \quad a_0 + 2a_1 + 4a_2 + 8a_3 + 16a_4 + 32a_5 = 1. \tag{3.62}$$

b. Boundary conditions are applied on equation 3.58, shown as f(x), after substituting equation 3.60 into equation 3.58 instead of the dependent parameter u and on its derivatives as

$$f(u(x)) \quad \rightarrow \quad f(u(BC))=0 \quad , \quad f'(u(BC))=0 \quad , \quad \ldots. \tag{3.63}$$

Therefore due to equation 3.63, the boundary conditions are applied in the following form:

$f(u(1))$:

$$4a_2 + 9a_3 + 16a_4 + 25a_5 + a_1 + \alpha(a_0 + a_1 + ... + a_5) = 0 \quad (3.64)$$

$f(u(2))$:

$$10a_2 + 36a_3 + 112a_4 + 320a_5 + 2a_1 + \alpha(a_0 + 2a_2 + 4a_3 + 8a_4 + 16a_4 + 32a_5) = 0. \quad (3.65)$$

Then, boundary conditions are applied on the derivative of the achieved equation, which is the right-hand side of equation 3.63 as follows:

$f'(u(1))$:

$$15a_3 + 40a_4 + 85a_5 + a_1 + 4a_2 + \alpha(a_1 + 2a_2 + 3a_3 + 4a_4 + 5a_5) = 0 \quad (3.66)$$

$f'(u(2))$:

$$42a_3 + 176a_4 + 640a_5 + a_1 + 8a_2 + \alpha(a_1 + 4a_2 + 12a_3 + 32a_4 + 80a_5) = 0. \quad (3.67)$$

By solving the set of algebraic equations consisting of equation 3.61 to equation 3.67, with the exception of equation 3.63, the constant coefficients a_0 to a_5 can easily be computed as

$$a_0 = -2.305084746 \ , \ a_1 = 1.144067797 \ , \ a_2 = 3.186440678$$
$$a_3 = -2.754237288 \ , \ a_4 = 0.8135593220 \ , \ a_5 = -0.8474576271e\text{-}1. \quad (3.68)$$

By substituting the constant coefficients from equation 3.68 into equation 3.60, the solution of the nonlinear differential equation 3.58 can be obtained as follows:

$$u = -0.08474576271x^5 + 0.8135593220x^4 - 2.754237288x^3 + 3.186440678x^2 + 1.144067797x - 2.305084746 \quad . \quad (3.69)$$

3.1.5.3. A Comparison between the Obtained Answer by AGM and the Exact Solution

As it is explained previously, the exact solution of the nonlinear differential equation 3.58 for $\alpha = 1$ as erf(x) is presented as follows:

$$u(x) = \frac{e^{2-\frac{1}{2}x^2}}{erf(\frac{i}{\sqrt{2}}) - erf(i\sqrt{2})}[erf(\frac{i}{\sqrt{2}}) - erf(\frac{i}{\sqrt{2}}x)]. \tag{3.70}$$

Where the parameter i is the imaginational part of the complex number.

The obtained charts by AGM and exact solution have been compared as follows:

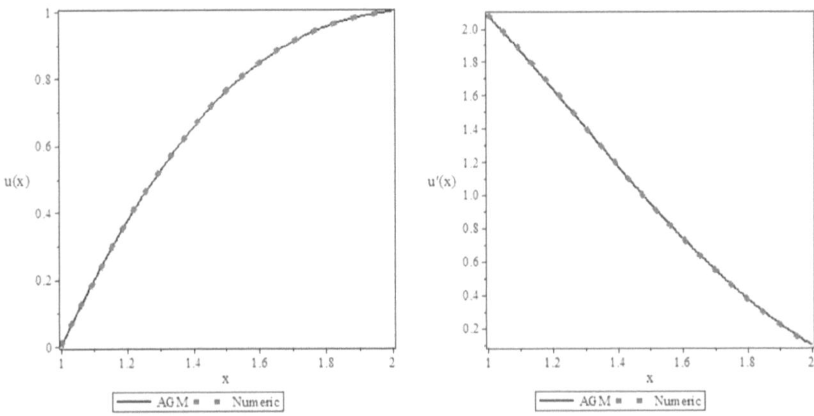

Fig. 3.8. The chart of obtained solutions by AGM and exact solution for $u(x)$

Fig. 3.9. The chart of obtained solutions by AGM and exact solution for $u'(x)$

It is notable that erf(x) is defined in all the mathematical books as

$$erf(x) = \frac{2}{\sqrt{\pi}}\int_0^x e^{-t^2}\,dt. \tag{3.71}$$

Differential equations, whose solutions are functions of erf(x), have abundant applications in certain domains of engineering and basic sciences that we can say the gained solutions by AGM can be an acceptable approximation for the existent exact solution.

Example 3.1.6

In the theory of big changing figures for computing critical load in columns, the differential equation governing the beam column in regard to its form as a fixed end and hinged end is introduced as

$$\frac{d^2\theta}{ds^2}+k^2\sin(\theta)=0. \tag{3.72}$$

And the boundary conditions are explained as

$$\theta(0)=\alpha \quad , \quad \theta(L)=0. \tag{3.73}$$

The parameter k in equation 3.72 is defined as follows:

$$k^2=\frac{p}{EI}. \tag{3.74}$$

In the above equations, the slope, along with the length of column, is θ, and the arch and the slope from the base are S , α , respectively.

The term $\sin(\theta)$ in equation 3.72 can be written by Taylor series expansion in the form of $\theta-\frac{\theta^3}{3!}$, and so we will have

$$f(s):\ \frac{d^2\theta}{ds^2}+k^2(\theta-\frac{\theta^2}{6})=0. \tag{3.75}$$

3.1.6.1. Solution

Like the previous examples, an answer function is needed as a finite series in the following form:

$$\theta=\sum_{k=0}^{5}a_k s^k=a_0+a_1s+a_2s^2+a_3s^3+a_4s^4+a_5s^5. \tag{3.76}$$

The constant coefficients a_0 to a_5 can easily be computed by applying the boundary conditions.

3.1.6.2. Applying Boundary Conditions by AGM

The boundary conditions are applied in two ways:

a. Applying the boundary conditions on the answer function shown by θ(BC)is done as

$$\theta(s=0)=\alpha \quad \rightarrow \quad a_0=\alpha . \tag{3.77}$$

Which means the initial slope at the hinged end is α and is computed in terms of radian, and then we have

$$\theta(s=L)=0 \quad \rightarrow \quad a_0+a_1L+a_2L^2+a_3L^3+a_4L^4+a_5L^5=0. \tag{3.78}$$

b. Applying the boundary conditions on the differential equation and on its derivatives, after substituting the answer function into differential equation 3.75, is done as the following general equation shown by $f(\theta(\mathrm{BC}))$:

$$f(\theta(s=0)): \quad \rightarrow \quad 2a_2+k^2(a_0-\frac{1}{6}a_0^3)=0 \tag{3.79}$$

$f(\theta(\mathrm{s}=L))$:

$$2a_2+6a_3L+12a_4L^2+20a_5L^3+k^2(a_0+a_1L+a_2L^2+a_3L^3+a_4L^4+ a_5L^5)\{1-\frac{1}{6}(a_0+a_1L+a_2L^2+a_3L^3+a_4L^4+a_5L^5)^2\}=0 \tag{3.80}$$

$$f'(\theta(s=0)): \quad \rightarrow \quad 6a_3+k^2(a_1-\frac{1}{2}a_0^2a_1)=0 \tag{3.81}$$

$f'(\theta(s=L))$:

$$6a_3+24a_4L+60a_5L^2+k^2(a_0+a_1L+2a_2L^2+3a_3L^3+4a_4L^4+5a_5L^5)\{1- \frac{1}{2}(a_0+a_1L+a_2L^2+a_3L^3+a_4L^4+a_5L^5)^2\}=0 \quad . \tag{3.82}$$

By solving the obtained algebraic equations composed of six equations with six unknowns from equation 3.77 to equation 3.82, the constant coefficients of the answer function can be obtained as follows:

To simplify, consider the following new variables:

$$\begin{aligned}
&\psi = k^2L^2\{2(72-k^2L^2)-\alpha^2(36-k^2L^2)\}-1440\\
&\Delta_1 = k^2L^2\{2(104-k^2L^2)-\alpha^2(28-k^2L^2)\}-480\\
&\Delta_2 = k^2L^2\{\alpha^2(106-11\alpha^2)-240\\
&\Delta_3 = 18k^2L^2(10\alpha^2-\alpha^4-8)-144\alpha^2\\
&\xi = k^4L^4(\alpha^4-8\alpha^2+12)
\end{aligned} \quad . \tag{3.83}$$

And finally, the constant coefficients are obtained as

$$a_0=\alpha \quad , \quad a_1=-\frac{3\alpha\Delta_1}{L\psi} \quad , \quad a_2=\frac{1}{12}k^2\alpha^3-\frac{1}{2}k^2\alpha \ ,$$
$$a_3=\frac{k^2\alpha\Delta_1(2-\alpha^2)}{4L\psi} \quad , \quad a_4=\frac{\alpha k^2(\Delta_2+\xi-240\alpha^2)}{4L^2\psi} \quad , \quad a_5=-\frac{\alpha k^2(\Delta_3+\xi)}{12L^3\psi} \quad . \tag{3.84}$$

After substituting the aforementioned constant coefficients into the answer function, the solution of nonlinear differential equation 3.75 is achieved by AGM in the form of

$$\theta(s)=\alpha-\frac{3\alpha\Delta_1}{L\psi}s+(\frac{1}{12}k^2\alpha^3-\frac{1}{2}k^2\alpha)s^2+\frac{k^2\alpha\Delta_1(2-\alpha^2)}{4L\psi}s^3+$$
$$6\frac{\alpha k^2(\Delta_2+\xi-240\alpha^2)}{4L^2\psi}s^4-\frac{\alpha k^2(\Delta_3+\xi)}{12L^3\psi}s^5 \quad . \tag{3.85}$$

By considering considering L=3, k=0.28, β=20 and α=(π/180) β, the final solution can be gained as

$$\theta(\mathrm{s})=-0.000006189369148s^5+0.00009915250428s^4+0.001080266225s^3-$$
$$0.01340550159s^2-0.08803695345s+0.3490658504 \quad . \tag{3.86}$$

3.1.6.3. Computing the Critical Load (P_{cr})

In regard to figure 3.10, at the hinged end $(s=0)$, we have $\theta=\alpha$. So $\frac{d\theta}{ds}=0$, which means the bending moment at the hinged end is equal to zero ($M=EI\frac{d\theta}{ds}=0$). Finally, we will have

$$\frac{d\theta(s)}{ds}=0 \xrightarrow{s=0} \Delta_1=0\rightarrow k^2L^2\{2(104-k^2L^2)-\alpha^2(28-k^2L^2)\}-480=0. \quad (3.87)$$

By solving equation 3.87, written in terms of k and substituted into equation 3.74, the critical load $P\rightarrow P_{cr}$ can be gained as follows:

$$p_{cr}=\frac{2EI}{(6-\alpha^2)L^2}(52-7\alpha^2-\sqrt{1984-608\alpha^2+49\alpha^4}). \quad (3.88)$$

It is notable that the angular α is in terms of radian, obtained by $\alpha=(\pi/180)\ \beta$, where the angular β is in terms of degree. As regards equation 3.88, the values of critical loads in terms of α have been obtained in the table below:

Table 3.4. Various Amounts of Critical Loads in Terms of α

α	0°	10°	20°	30°	40°	50°	60°
p_{cr}	$\frac{2.486EI}{L^2}$	$\frac{2.4968EI}{L^2}$	$\frac{2.53EI}{L^2}$	$\frac{2.5875EI}{L^2}$	$\frac{2.67228EI}{L^2}$	$\frac{2.7896EI}{L^2}$	$\frac{2.94766EI}{L^2}$

3.1.6.4. Numerical Solution (RKF45) and Comparing It with AGM

The solution procedure of nonlinear differential equation 3.75 by AGM is compared with the numerical method. In this particular case, we consider $\alpha=20°$, $k=0.28$ and $L=3m$, and as a result, the numerical solution in $s\in\{0,L\}$ can be presented as follows:

Table 3.5. Comparing the Achieved Solution by AGM and Numerical Method

$s(m)$	**0.0**	**0.6**	**1.2**	**1.8**	**2.4**	**3.0**
$\theta(s)$ $Rk45$	0.349066	0.2916887	0.2262166	0.1544312	0.07831652	0.00
$\theta(s)$ AGM	0.349065	0.2916634	0.2261744	0.15438895	0.07829187	0.00
$\theta'(s)$ $Rk45$	-0.087993	-0.102837	-0.1149072	-0.1238247	-0.12929805	-0.13114345
$\theta'(s)$ AGM	-0.088036	-0.102875	-0.1149222	-0.1238084	-0.1292603	-0.1311010
Error Percentage $\theta(s)\times 100$	0.00028	0.00867	0.0186	0.027	0.0314	-

The obtained charts by numerical method and AGM have been illustrated and compared below:

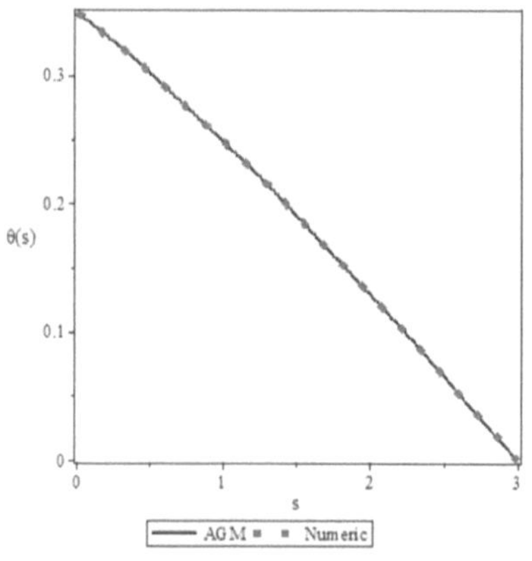

Fig. 3.10. The chart of obtained solutions by AGM and numerical method for θ(s)

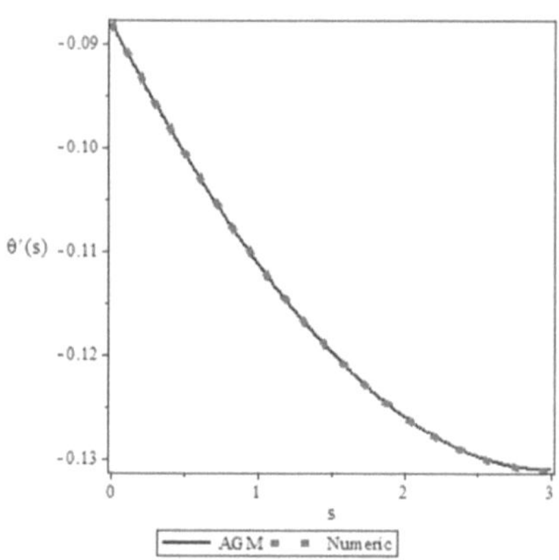

Fig. 3.11. The chart of obtained solutions by AGM and numerical method for θ'(s)

The difference of absolute value for $\theta(s)$ between solving by AGM and numerical method has been depicted as

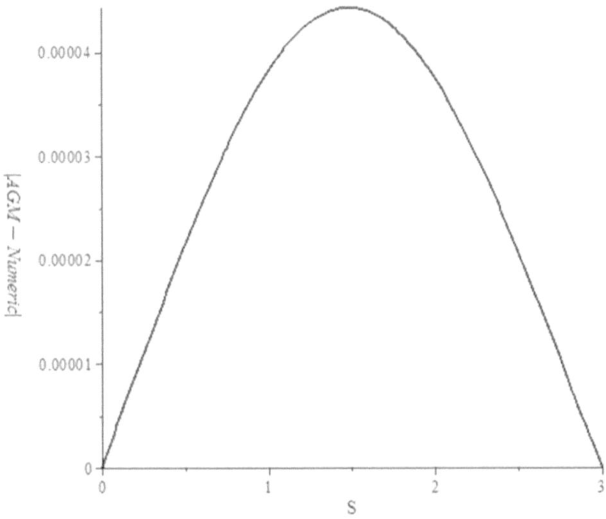

Fig. 3.12. Absolute difference for $\theta(s)$ between the gained values by AGM and numerical method

After being sure about the accuracy of the achieved solution by AGM, now we are able to depict the chart of $y(s)$ for $s = \frac{L}{2}$, which is equal to $\frac{E.Ie}{P_{cr}} \cdot \frac{d\theta(s)}{ds}$ for $\beta = 5^0$ as follows:

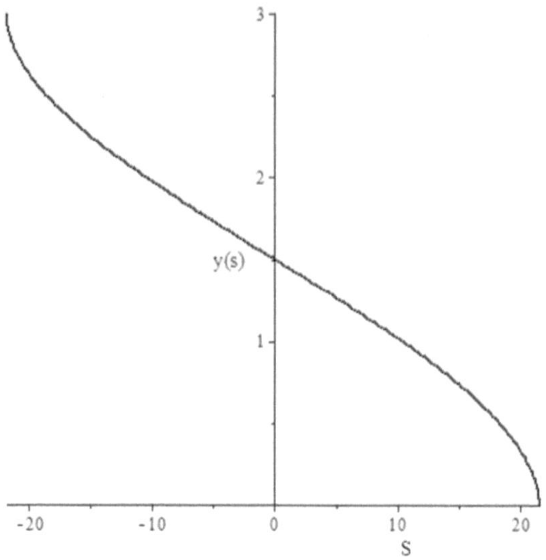

Fig. 3.13. The chart of $y(\text{s})$ for $\beta=5^0$ in $\text{s}=\frac{L}{2}$

Example 3.1.7

Solve the following parametric nonlinear differential equation:

$$f(x):\ \ u''-u'+e^{u^2}-u^2+\alpha Ln(1+\beta u)+\lambda\sin(u)=0. \tag{3.89}$$

And the related boundary conditions are defined as follows:

$$u(x)=0 \quad \text{at} \quad x=0 \qquad \text{and} \qquad u(x)=0 \quad \text{at} \quad x=1. \tag{3.90}$$

3.1.7.1. Solution

In order to analytically solve these kinds of equations by AGM, we do not need the Taylor expansion of ln , sin and e^{u^2}, etc., while this operation is necessary to solve differential equations by other methods. This issue can fairly be considered as one of the benefits of AGM. An answer function like the previous examples is required to solve the differential equation 3.89 as follows:

$$u(x)=\sum_{k=0}^{5}a_k x^k = a_0+a_1x+a_2x^2+a_3x^3+a_4x^4+a_5x^5 \tag{3.91}$$

Where the constant coefficients a_0 to a_5 can be computed by applying boundary conditions.

3.1.7.2. Applying Boundary Conditions by AGM

Boundary conditions are applied in two ways like the aforementioned examples as follows:

a. The boundary conditions are applied on the considered answer function, like the previous examples, as

$$u(x=0)=0 \quad \rightarrow \quad a_0=0 \tag{3.92}$$

$$u(x=1)=0 \quad \rightarrow \quad a_0+a_1+a_2+a_3+a_4+a_5=0. \tag{3.93}$$

b. Applying boundary conditions on the differential equation shown by $f(\mathrm{x})$is done by substituting equation 3.91 into equation 3.89 instead of the u as follows:

$$f(x):[\sum_{k=0}^{5}a_kx^k]''-[\sum_{k=0}^{5}a_kx^k]'+\exp([\sum_{k=0}^{5}a_kx^k])+\alpha Ln(1+\beta[\sum_{k=0}^{5}a_kx^k])+\lambda\sin([\sum_{k=0}^{5}a_kx^k])=0. \tag{3.94}$$

Now, the boundary conditions are applied on equation 3.94 and its derivative as

$$f(u(x=0)) \quad \rightarrow \quad 2a_2-a_1+e^{a_0^2}-a_0^2+\alpha Ln(1+\beta a_0)+\lambda\sin(a_0)=0. \tag{3.95}$$

And also

$$f(u(x=1))=0. \tag{3.96}$$

Applying boundary conditions on the first derivative of equation 3.94 is done in the following forms:

$$f'(u(x=0)) \quad \rightarrow \quad 6a_3 - 2a_2 + 2a_0a_1e^{a_0^2} - 2a_0a_1 + \frac{\alpha\beta a_1}{1+\beta a_0} + \lambda a_1 \cos(a_0) = 0. \tag{3.97}$$

And then

$$f'(u(x=1)) = 0. \tag{3.98}$$

By solving the set of algebraic equations from equation 3.92 to equation 3.98, except equation 3.94, we can compute the constant coefficients of the considered answer function.

To simplify, the following changing variables are considered as follows:

$$\Delta = \alpha^2\beta^2 + 2\lambda\alpha\beta \tag{3.99}$$

$$\psi = \Delta + \lambda^2 - 74\lambda + 721. \tag{3.100}$$

So the coefficients are obtained in terms of the given variables as

$$a_0 = 0 \quad , \quad a_1 = -\frac{7(\alpha\beta+\lambda-43)}{\psi} \quad , \quad a_2 = -\frac{\Delta+\lambda^2-67\lambda+420}{2\psi}$$
$$a_3 = \frac{\Delta+\lambda^2-39\lambda-70}{\psi} \quad , \quad a_4 = -\frac{\Delta+\lambda^2-\lambda+30}{2\psi} \quad , \quad a_5 = \frac{6(2\alpha\beta+2\lambda-1)}{\psi} \quad . \tag{3.101}$$

The solution of nonlinear differential equation 3.89 can be gained by substituting the constant coefficients of equation 3.101 into equation 3.91 as follows:

$$u(x) = -\frac{7(\alpha\beta+\lambda-43)}{\psi}x - \frac{\Delta+\lambda^2-67\lambda+420}{2\psi}x^2 + \frac{\Delta+\lambda^2-39\lambda-70}{\psi}x^3 - \frac{\Delta+\lambda^2-\lambda+30}{2\psi}x^4 + \frac{6(2\alpha\beta+2\lambda-1)}{\psi}x^5 \quad . \tag{3.102}$$

For $\alpha = 0.3$, $\beta=0.1$ and $\lambda=0.2$, the final solution of the presented problem is achieved as

$$u(x) = -0.004602057659x^5 - 0.02118004712x^4 - 0.1120929150x^3 - 0.2873749934x^2 + 0.4252500132x \quad . \quad (3.103)$$

In accordance with the above physical values and the domain $x \in \{0, 1\}$, the nonlinear differential equation has numerically been solved in the following table, and the obtained results have been compared with AGM as follows:

Table 3.6. Comparing the Obtained Results by AGM and Numerical Method

x	*0.0*	*0.2*	*0.4*	*0.6*	*0.8*	*1.0*
$u(x)$ *NUM*	0.0	0.072668	0.1164395	0.1244693	0.0887624	0.0
$u'(x)$	0.425463	0.296361	0.1356593	-0.0620096	-0.3028082	-0.5938345
$u(x)$ AGM	0.0	0.072623	0.116356	0.1243801	0.0887051	0.0

The obtained charts by numerical method and AGM are compared as

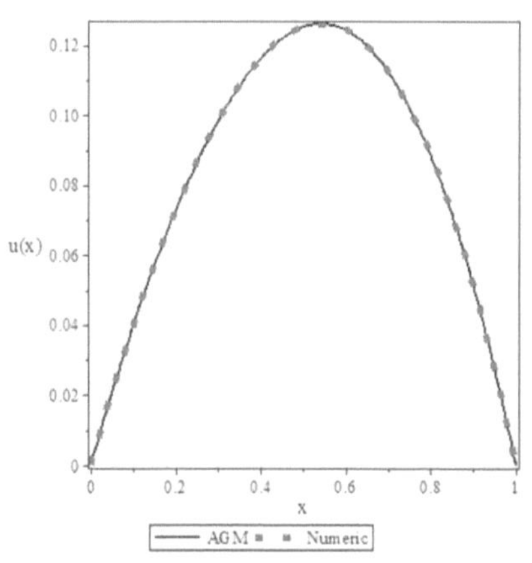

Fig. 3.14. The chart of obtained solution by AGM and numerical method for $u(x)$

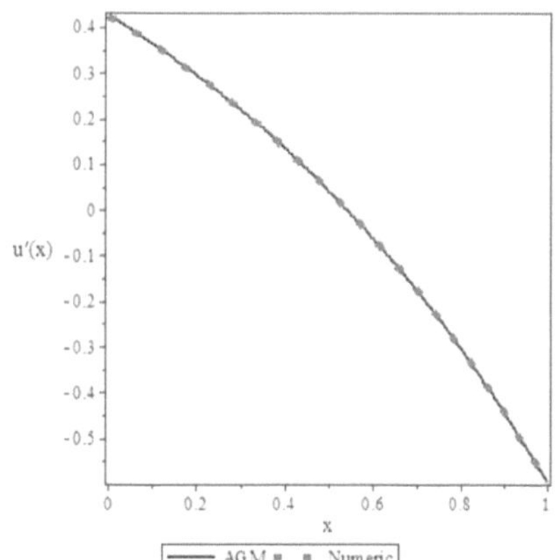

Fig. 3.15. The chart of obtained solution by AGM and numerical method for u'

According to the exponential and trigonometric factors (e and $\sin$) of the differential equation that makes the equation nonlinear, the differential equation can be solved by AGM without the Taylor expansion. Unlike the mentioned difficulty (e and $\sin$), the precision of the final solution is significantly acceptable, which can be observed in the above table.

Example 3.1.8

Consider the following set of coupled nonlinear differential equations and solve it.

$$\begin{cases} f(x): \quad u'''+u'w+u^2=0 \\ g(x): \quad v''+vu+v'v^2=0 \\ h(x): \quad w''+w^2+uw'+v=0 \end{cases} \tag{3.104}$$

And the related boundary conditions are expressed as

$$\begin{array}{llllll} u(x)=1, & u'(x)=0, & v(x)=1, & w(x)=0 & \text{at} & x=0 \\ u(x)=0, & v(x)=0, & w(x)=1 & & \text{at} & x=1. \end{array} \tag{3.105}$$

3.1.8.1. Solving the Set of Differential Equations by AGM

In order to solve the set of differential equations, consider the following answer functions as finite polynomial series in the forms of

$$\begin{aligned} u(x) &= \sum_{k=0}^{6} a_k x^k = a_0 + a_1 x + a_2 x^2 + a_3 x^3 + a_4 x^4 + a_5 x^5 + a_6 x^6 \\ v(x) &= \sum_{k=0}^{5} b_k x^k = b_0 + b_1 x + b_2 x^2 + b_3 x^3 + b_4 x^4 + b_5 x^5 \\ w(x) &= \sum_{k=0}^{5} c_k x^k = c_0 + c_1 x + c_2 x^2 + c_3 x^3 + c_4 x^4 + c_5 x^5 \end{aligned} \quad . \tag{3.106}$$

In the above equations, the constant coefficients a_0 to a_6, b_0 to b_5 and c_0 to c_5 are easily computed by applying boundary conditions.

3.1.8.2. Applying Boundary Conditions by AGM

Based on the given explanations in the previous examples, we will have the following:

a. Applying the boundary conditions on equation 3.106 is done as follows:

$$u = u(BC) \quad , \quad v = v(BC) \quad , \quad w = w(BC). \tag{3.107}$$

So the boundary conditions are applied with respect to equation 3.107 as follows:

$$u(0) = 1 \quad \rightarrow \quad a_0 = 1 \tag{3.108}$$

$$u'(0) = 0 \quad \rightarrow \quad a_1 = 0 \tag{3.109}$$

$$u(1) = 0 \quad \rightarrow \quad a_0 + a_1 + \ldots + a_6 = 0 \tag{3.110}$$

$$v(0) = 1 \quad \rightarrow \quad b_0 = 1 \tag{3.111}$$

$$v(1) = 0 \quad \rightarrow \quad b_0 + b_1 + \ldots + b_5 = 0 \tag{3.112}$$

$$w(0) = 0 \quad \rightarrow \quad c_0 = 0 \tag{3.113}$$

$$w(1) = 1 \quad \rightarrow \quad c_0 + c_1 + \ldots + c_5 = 1. \tag{3.114}$$

b. Boundary conditions are applied on equation 3.104, shown by $f(x)$, $g(x)$ and $h(x)$, and also on their derivatives as

$$\begin{aligned} &f(u(x) \;:\; f(u(BC)) = 0\,,\; f'(u(BC)) = 0, \ldots \\ &g(v(x)) \;:\; g(v(BC)) = 0\,,\; g'(v(BC)) = 0, \ldots \\ &h(w(x)) : h(w(BC)) = 0\,,\; h'(w(BC)) = 0, \ldots. \end{aligned} \tag{3.115}$$

Equation 3.115 means that the answer functions are substituted into the set of equation 3.104 instead of the dependent parameters u , v and w, and then the boundary conditions are applied on them as follows:

$$f(u(0)) \quad \rightarrow \quad 6a_3 + a_1c_0 + a_0^2 = 0. \tag{3.116}$$

And also in $x = 1$, we will have

$$f(u(x=1)) = 0 \tag{3.117}$$

$$g(v(0)) \quad \rightarrow \quad 2b_2 + b_0a_0 + b_1b_0^2 = 0. \tag{3.118}$$

Then in $x = 1$

$$g(v(x=1)) = 0 \tag{3.119}$$

$$h(w(0)) \quad \rightarrow \quad 2c_2 + a_0c_1 + b_0 = 0. \tag{3.120}$$

And finally in $x = 1$

$$h(w(x=1)) = 0. \tag{3.121}$$

Applying the boundary conditions on the derivatives of the set of differential equations is done in the following forms:

$$f'(u(0)) \quad \rightarrow \quad 24a_4 + 2a_2c_0 + a_1c_1 + 2a_0a_1 = 0. \tag{3.122}$$

And in $x = 1$

$$f'(u(x=1)) = 0 \tag{3.123}$$

$$g'(v(0)) \quad \rightarrow \quad 6b_3 + b_1a_0 + b_0a_1 + 2b_2b_0^2 + 2b_1^2b_0 = 0. \tag{3.124}$$

And then in $x = 1$

$$g'(v(x=1))=0 \tag{3.125}$$

$$h'(w(0)) \quad \rightarrow \quad 6c_3+2c_0c_1+a_1c_1+2a_0c_2+b_1=0. \tag{3.126}$$

After that, in $x=1$

$$h'(w(x=1))=0. \tag{3.127}$$

By solving the set of algebraic equations consisting of equation 3.108 to equation 3.114, and from equation 3.116 to equation 3.127, the constant coefficients of equation 3.106 can be obtained as

$$\begin{array}{lll}
a_0=1 \quad , & a_1=0 \quad , & a_2=-0.91466 \\
a_3=-0.1667 \quad , & a_4=0 \quad , & a_5=0.113416 \\
a_6=-0.0321 \quad , & b_0=1 \quad , & b_1=-0.94076 \\
b_2=-0.029618 \quad , & b_3=-0.12834 \quad , & b_4=0.143155 \\
b_5=-0.04427 \quad , & c_0=0 \quad , & c_1=2.053 \\
c_2=-1.5264 \quad , & c_3=0.6656 \quad , & c_4=-0.23756 \\
c_5=0.0755 & &
\end{array} \tag{3.128}$$

By substituting the achieved constant coefficients into equation 3.106, the solution of the set of coupled nonlinear differential equation is gained as follows:

$$\begin{aligned}
u(x)&=1-0.91466x^2-0.1667x^3+0.11341x^5-0.0321x^6 \\
v(x)&=1-0.94076x-0.029618x^2-0.12834x^3+0.143155x^4-0.04427x^5 \\
w(x)&=2.053x-1.5264x^2+0.6656x^3-0.23756x^4+0.0455x^5
\end{aligned} \tag{3.129}$$

3.1.8.3. Solving the Set of Coupled Nonlinear Differential Equations by Numerical Method and Its Comparison with AGM

In regard to the domain $x \in \{0,\ 1\}$, the set of nonlinear differential equations has been solved by numerical method, and the obtained results have been compared with AGM in the following table:

Table 3.7. Comparing the Obtained Results by AGM and Numerical Method

x	*0.0*	*0.2*	*0.4*	*0.6*	*0.8*	*1.0*
$u(x)$ *NUM*	1.0	0.96231	0.844691	0.643112	0.3589648	0.0
$u'(x)$	0.0	-0.38313	-0.796607	-1.218104	-1.616848	-1.961474
$u(x)$ *AGM*	1.0	0.96212	0.84402	0.642043	0.358035	0.0
$v(x)$ NUM	1.0	0.809518	0.6134765	0.411851	0.2065811	0.0
$v'(x)$	-0.94268	-0.96507	-0.995227	-1.01928	-1.03328	-1.033454
v(x) *AGM*	1.0	0.80985	0.613951	0.412254	0.206798	0.0
$w(x)$ *NUM*	0.0	0.358998	0.620853	0.805213	0.927488	1.0
$w'(x)$	2.0786	1.53344	1.1012366	0.755121	0.47788	0.255174
$w(x)$ *AGM*	0.0	0.35450	0.613913	0.79874	0.923790	1.0

After that, the charts of achieved solutions by numerical method and AGM are compared graphically in the following form:

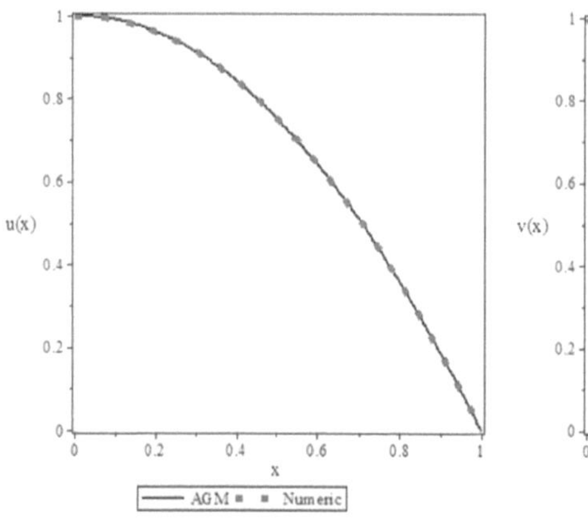

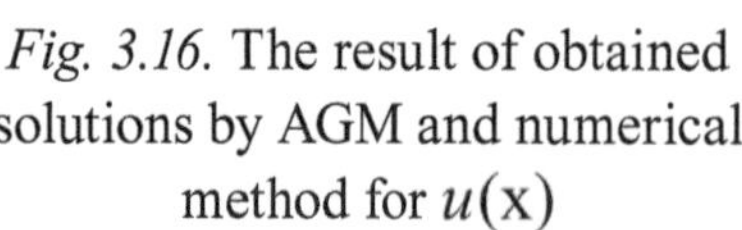

Fig. 3.16. The result of obtained solutions by AGM and numerical method for u(x)

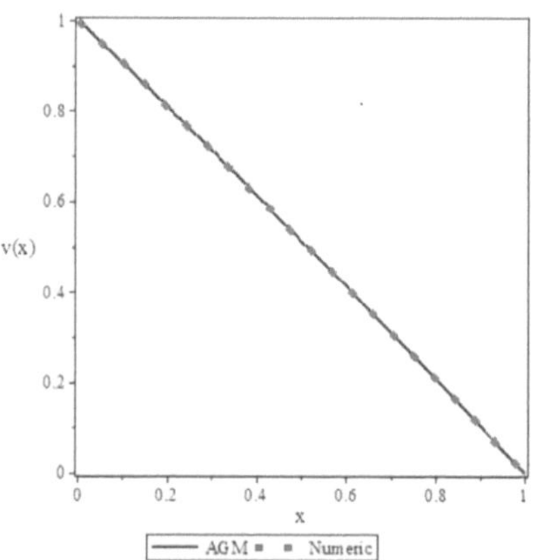

Fig. 3.17. The achieved chart by AGM and numerical method for $v(x)$

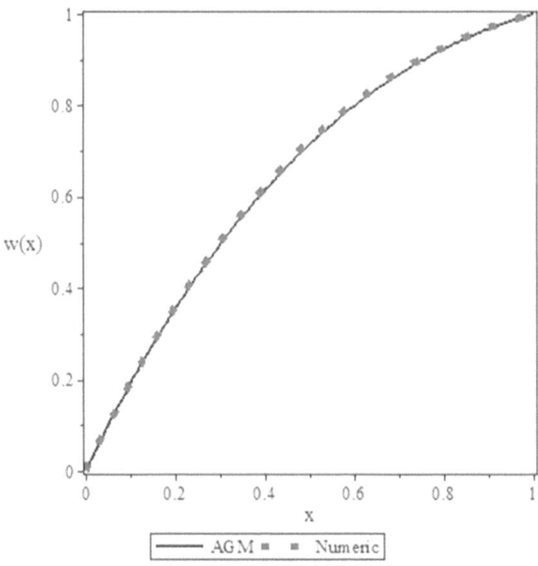

Fig. 3.18. The chart of obtained solutions by AGM and numerical method for $w(x)$

3.1.8.4. Investigating Computational Error

Generally, in order to understand what percentage of errors exists in the obtained solution of differential equations, the procedure below should be followed:

To start, the obtained solution of each set of differential equations should be substituted into the main set of differential equations $g(\eta)$, $h(\eta)$, and then it is necessary to depict the yielded equations in Cartesian coordinates. Afterward, the yielded errors of the given set of differential equations can be observed from the obtained charts.

To understand more, reading the following lines is recommended.

Consider a set of differential equations in the following form:

$$\begin{cases} g(\eta) = g(f(\eta)\ ,\ f'(\eta)\ , \dots) \\ h(\eta) = h(\theta(\eta)\ ,\ \theta'(\eta)\ , \dots). \end{cases} \tag{3.130}$$

And the answer of the aforementioned set of equations is assumed to be a function of η in the form of

$$f = h_1(\eta) \qquad , \qquad \theta = h_2(\eta). \tag{3.131}$$

Thus, by substituting equation 3.131 into equation 3.130, the computational error of the obtained solution by each analytic or semianalytic method can be achieved as follows:

$$\begin{aligned} g(\eta) &= g(f(h_1(\eta))\ ,\ f'(h_2(\eta))\ ,...) \\ h(\eta) &= h(\theta(h_1(\eta))\ ,\ \theta'(h_2(\eta))\ ,...). \end{aligned} \tag{3.132}$$

Then, the computational error of the presented problem can be depicted as

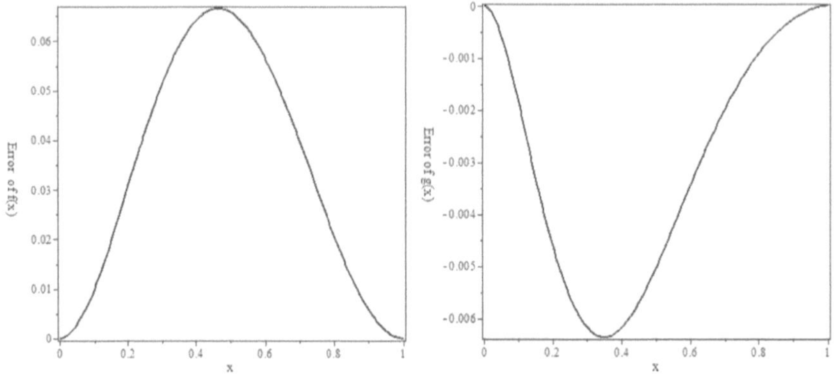

Fig. 3.19. The error of acquired solution for f(x)

Fig. 3.20. The error of acquired solution for $g(x)$

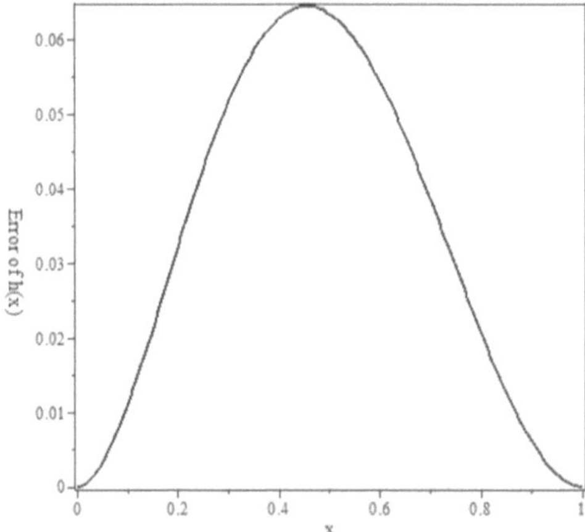

Fig. 3.21. The error of acquired solution for $h(x)$

Due to the computational error charts, the maximum error for differential equations *f(x), g(x)* and *h(x)* are 6.5%, 0.6%, and 6%, respectively, and the aforementioned errors can considerably be reduced by adding a sentence to the considered answer functions, but the action needs large mathematical operations.

Example 3.1.9

For the nonlinear differential equations, the number of their boundary conditions being more or less than the order of the differential equation can sometimes be seen in simulation of industrial processes. It is notable that in this situation, the boundaries of the system are available, but the number of the boundary conditions is not equal to the order of the differential equations governing the system. And so in order to solve this kind of problems, the balance and equality between boundary conditions and the order of the differential equation should be made. These two manners are comprehensively discussed below:

a. *The boundary conditions of the system are less than the order of the differential equation.* In AGM, differential equations with boundary conditions less than their order can easily be solved, because in the solution procedure, we come out of the atmosphere of differential equations and enter the algebraic equations. As a result, nonlinear differential equations can be answered by any shortage of the boundary conditions. Now, this is the question of how reliable and trusty the obtained answer of AGM is. Since the differential equation cannot numerically be answered with the shortage of the boundary conditions in relevance to the order of the differential equation, the numerical solution cannot be achieved. Then, the only way to check the accuracy of the obtained answer by AGM can be the chart of the error function that shows the precision of the achieved solution. The achieved accuracy can be investigated by AGM, because the process of the error chart means that if the gained solution by AGM or any other method is acceptable, then it should be satisfied in the differential equation.

Suppose that the differential equation has been defined as follows:

$$f(x): f\,(u^{n}, (u')^{m}, \ldots). \tag{3.133}$$

And consider the solution of the differential equation by AGM as

$$u = g(x). \tag{3.134}$$

Inasmuch as the achieved solution is the answer of the differential equation 3.133, the result should tend toward zero after substituting the achieved solution into the differential equation as follows:

$$f(u):\ f([g(x)]^{n}\ ,\ [g'(x)]^{m},...)\ \rightarrow\ f(u)\ \rightarrow 0. \tag{3.135}$$

By depicting the chart of $y = f(\mathrm{u})$ in Cartesian coordinates, the error percentage, which is the deflection of the curve from the horizontal axes, can be observed.

b. *When the boundary conditions of the system are more than the order of the differential equation.* When the number of boundary conditions is more than the order of the differential equation, the number of the boundary conditions and the order of the differential equation should be equaled by deriving from the differential equation. Therefore, in this manner, we can compare the solution of the problem by AGM with its numeral solution and make sure that we have acquired an acceptable and suitable answer in order to solve differential equations. Or like the previous technique, the precision of the achieved answer can be verified by the error chart.

Solve the following differential equation by AGM.

$$f(x):\ \frac{d^4u}{dx^4} + u\frac{d^3u}{dx^3} + \left(\frac{du}{dx}\right)^2 - u\left(\frac{du}{dx}\right) = 1 \tag{3.136}$$

And the related boundary conditions are expressed as

$$u(0) = 0 \quad , \quad u(1) = 1. \tag{3.137}$$

3.1.9.1. Solving the Differential Equation by AGM

As it is explained before, the order of the differential equation is four, and the number of the boundary conditions is two, which means two boundary conditions for solving the differential equation are not apparently enough, but it can be answered by AGM. For solution procedure, an answer function as a finite polynomial series is required in the form of

$$u=\sum_{k=0}^{5} a_k x^k = a_0 + a_1 x + a_2 x^2 + a_3 x^3 + a_4 x^4 + a_5 x^5. \quad (3.138)$$

3.1.9.2. Applying Boundary Conditions by AGM

Boundary conditions are applied in two ways that (first) is applied on the answer function and (second) on the differential equation and its derivatives shown by *f(x)*, after substituting equation 3.138 into the differential equation 3.136 shown by *f(u(x))*.

a. Applying boundary conditions on the answer function is done as follows:

$$u(0)=0 \quad \rightarrow \quad a_0 = 0 \quad (3.139)$$

$$u(1)=1 \quad \rightarrow \quad a_0 + a_1 + \ldots + a_5 = 1. \quad (3.140)$$

b. Applying boundary conditions on $f(u(x))$ is done as

$$f(u(0)): \quad \rightarrow \quad 24a_4 + 6a_0a_3 + a_1^2 - a_0a_1 = 1 \quad (3.141)$$

$f(u(1))$:

$$24a_6 + 120a_5 + (a_0 + a_1 + \ldots + a_5)\{(6a_3 + 24a_4 + 60a_5) - (a_1 + 2a_2 + 3a_3 + 4a_4 + 5a_5)\} + (a_1 + 2a_2 + 3a_3 + 4a_4 + 5a_5)^2 = 1 \quad . \quad (3.142)$$

And also applying boundary conditions on the first derivative of the differential equation $f'(u(x))$ is done in the form of

$$f'(u(0)): \rightarrow \qquad 120a_5 + 6a_1a_3 + 24a_0a_4 + 4a_1a_2 - a_1^2 - 2a_0a_2 = 0. \quad (3.143)$$

And also

$$f'(u(x=1)) = 0. \quad (3.144)$$

By solving the set of algebraic equations consisting of six equations with six unknowns from equation 3.139 to equation 3.144, the constant coefficients of the answer function can easily be computed as

$$a_0 = 0 \quad , \quad a_1 = 0.7606 \quad , \quad a_2 = 0.2304$$
$$a_3 = -0.00792 \quad , \quad a_4 = 0.1756 \quad , \quad a_5 = -0.00072. \quad (3.145)$$

By substituting the achieved constant values from equation 3.145 into equation 3.138, the solution of the nonlinear differential equation 3.136 is obtained as follows:

$$u(x) = 0.7606x + 0.2304x^2 - 0.00792x^3 + 0.01756x^4 - 0.00072x^5. \quad (3.146)$$

Then, the charts of the answer function and its derivative by AGM are graphically shown below:

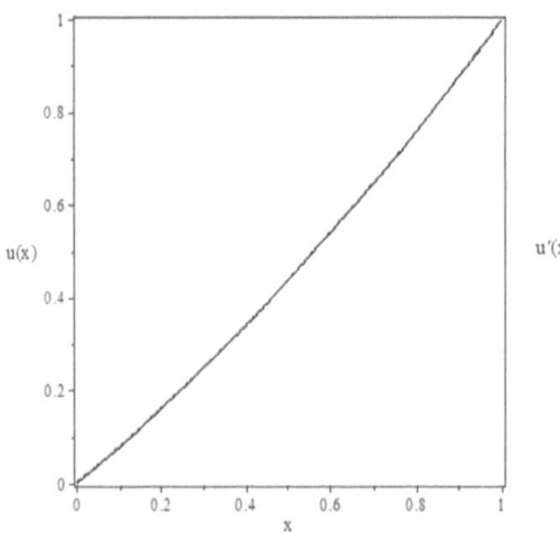

Fig. 3.22. The chart of obtained solution by AGM

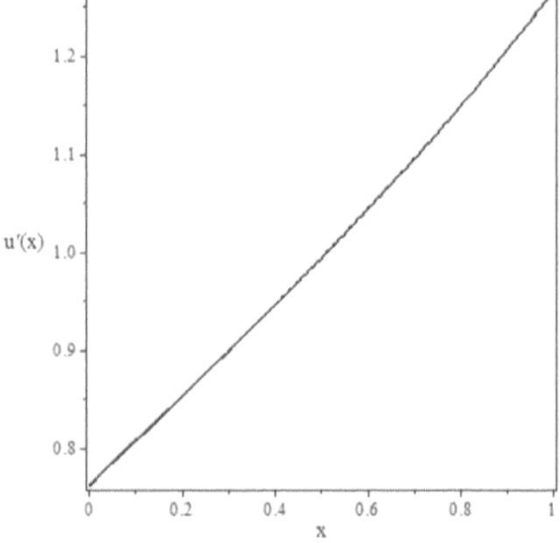

Fig. 3.23. The chart of the first derivative for the achieved solution (u')

It is important to be sure whether the answer function is the real answer of the nonlinear differential equation or not. And if so, with what percent of the error has been computed? Substituting the answer function 3.146, considered as the answer function into the differential equation instead of the dependent parameter (u), is the best way to verify that the final solution is the answer of equation 3.136, and then its result can be depicted as a function in the Cartesian coordinates. Thus, by substituting the answer function into the differential equation, we will have

$$y = f(u(x)) = 2.6\times10^{-6}x^{9} + 1.27\times10^{-4}x^{8} + 0.0061x^{6} - 0.02x^{5} - 0.012x^{4} + 0.067x^{3} - 0.04x^{2} + 6\times10^{-10}x. \quad (3.147)$$

Indeed, if the answer function was the exact answer of the differential equation, the result would be tended toward zero. To check that the differential equation has been solved with what percent of error, equation 3.147 should be depicted in the Cartesian coordinates as

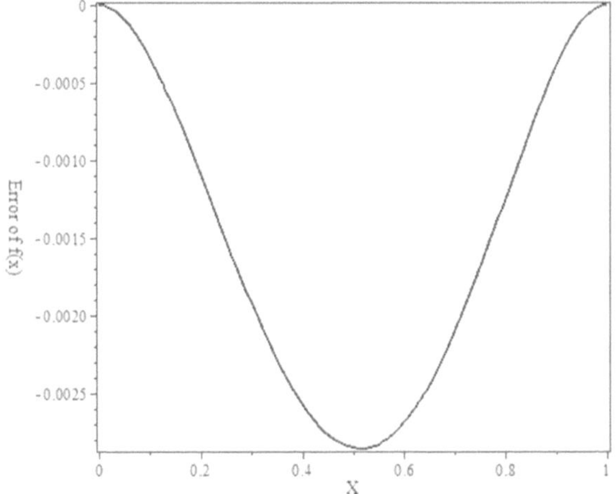

Fig. 3.24. Graphical results of the existing errors in the final solution by AGM

In regard to the error chart, the maximum error in solving the differential equation by AGM is approximately 0.25%, which is a negligible error in engineering; and for removing this error, we can add the number of sentences in the answer function of equation 3.138 and tend the answer of the differential equation toward the exact solution.

Example 3.1.10

Solve the following differential equation by AGM.

$$f(x):\ \frac{d^2u}{dx^2}+\frac{du}{dx}+u^2+u^4=0 \tag{3.148}$$

And the related boundary conditions are as follows:

$$u(0)=0 \quad , \quad u(1)=1 \quad , \quad u'(1)=0. \tag{3.149}$$

3.1.10.1. Solving the Differential Equation by AGM

Inasmuch as the number of boundary conditions is one more than the order of the presented problem, equation 3.148 should be derived so that the number of boundary conditions and the order of the yielded differential equation will be equaled. Thus, we will have

$$g(x):\ \frac{d^3u}{dx^3}+\frac{d^2u}{dx^2}+2u\frac{du}{dx}+4u^3\frac{du}{dx}=0. \tag{3.150}$$

Now the order of the differential equation 3.150 and the number of the boundary conditions have been equaled, and then the differential equation can be solved. To solve the nonlinear differential equation, an answer function is needed by choosing a finite series of polynomials in the following form:

$$u=\sum_{k=0}^{6}a_k x^k=a_0+a_1x+a_2x^2+a_3x^3+a_4x^4+a_5x^5+a_6x^6. \tag{3.151}$$

The constant coefficients a_1 to a_6 can easily be computed by applying boundary conditions.

3.1.10.2. Applying Boundary Conditions by AGM

As it is previously explained, boundary conditions are applied in two ways: first, on equation 3.151, and then on the differential equation 3.150 in the form of g(u).

a. Applying boundary conditions on equation 3.151 is done as

$$u(0)=0 \quad \rightarrow \quad a_0=0 \tag{3.152}$$

$$u(1)=1 \quad \rightarrow \quad a_0+a_1+a_2+...+a_6=1 \tag{3.153}$$

$$u'(1)=0 \quad \rightarrow \quad a_1+2a_2+3a_3+4a_4+5a_5+6a_6=0. \tag{3.154}$$

b. Applying the boundary conditions on the compound function of the differential equation shown by $g(u(BC))$ is done as follows:

$$g(u(0))'' \quad \rightarrow \quad 6a_3+2a_2+2a_0a_1+4a_0^3a_1=0. \tag{3.155}$$

And also in $x=1$, we will have

$g(u(1))$:

$$12a_3+36a_4+80a_5+2a_2+2(a_0+a_1+...+a_6)(a_1+2a_2+3a_3+4a_4+5a_5+6a_6)\{1+2(a_0+a_1+...+a_6)^2\}=0 \tag{3.156}$$

Applying boundary conditions on the first derivative of the differential equation shown by $g'(u(BC))$ is presented as

$$g'(u(0)): \quad \rightarrow \quad 24a_4+6a_3+2a_1^2+4a_0a_2+12a_0^2a_1^2+8a_0^3a_2=0. \tag{3.157}$$

And also in $x=1$, we will have

$$g'(u(x=1))=0. \tag{3.158}$$

By solving the set of algebraic equations consisting of seven equations with seven unknowns from equation 3.152 to equation 3.158, the constant coefficients a_1 to a_6 can easily be gained as follows:

$$\begin{aligned} &a_0 = 0 \quad , \quad a_1 = 2.015 \quad , \quad a_2 = -1.127 \\ &a_3 = 0.376 \quad , \quad a_4 = -0.4323 \quad , \quad a_5 = 0.1718 \\ &a_6 = -0.0029 \quad . \end{aligned} \tag{3.159}$$

By substituting the constant values of equation 3.159 into the answer function, the solution of the nonlinear differential equation 3.148 is acquired by AGM as follows:

$$u(x) = 2.015x - 1.127x^2 + 0.376x^3 - 0.4323x^4 + 0.1718x^5 - 0.0029x^6. \tag{3.160}$$

3.1.10.3. Solving the Differential Equation by Numerical Method and Comparing It with AGM

With regard to the given boundary conditions in $x \in \{0, 1\}$, the differential equation has been solved numerically and compared with AGM in the following table:

Table 3.8. Comparison between the Acquired Solution by AGM and Numerical Method

x	*0.0*	*0.2*	*0.4*	*0.6*	*0.8*	*1.0*
$u(x)$ *NUM*	0.0	0.3459	0.6236	0.8302	0.9583	1.0
$u'(x)$	1.9094	1.5567	1.2184	0.8450	0.42343	0.0
$u(x)$ AGM	0.0	0.351	0.637	0.8410	0.9601	0.0

The obtained charts by numerical method and AGM are drawn as follows:

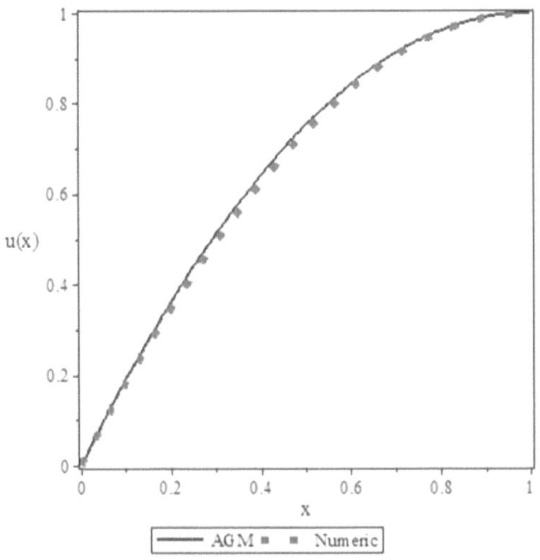

Fig. 3.25. Comparing the acquired solution by AGM and numerical method

Example 3.1.11

The equation governing the fluid system is presented in the following form:

$$g(y):\frac{d^4 f(y)}{dy^4}+\alpha(y\frac{d^3 f(y)}{dy^3}+3\frac{d^2 f(y)}{dy^2})+\mathrm{Re}\,(f(y)\frac{d^3 f(y)}{dy^3}-\frac{df(y)}{dy}\frac{d^2 f(y)}{dy^2})=0. \qquad (3.161)$$

And the related boundary conditions are introduced as

$$\begin{aligned} y=0 \quad &\rightarrow \quad f''=0\ ,\ f=0 \\ y=1 \quad &\rightarrow \quad f'=0\ ,\ f=1\,. \end{aligned} \qquad (3.162)$$

3.1.11.1. Solving the Differential Equation by AGM

Choose an answer function for the given problem in the following form:

$$f=\sum_{k=0}^{5} a_k y^k = a_0 + a_1 y + a_2 y^2 + \ldots + a_5 y^5. \qquad (3.163)$$

3.1.11.2. Applying the Boundary Conditions

a. The boundary conditions are applied on the answer function as follows:

$$f(0)=0 \quad \rightarrow \quad a \; =0 \tag{3.164}$$

$$f''(0)=0 \quad \rightarrow \quad a_2=0 \tag{3.165}$$

$$f(1)=1 \quad \rightarrow \quad a_0+a_1+...+a_5=1$$ 3.166)

$$f'(1)=0 \quad \rightarrow \quad a_1+2a_2+3a_3+4a_4+5a_5=0. \tag{3.167}$$

b. Applying the boundary conditions on the differential equation shown by $g\,(f\,(\mathrm{BC}))$ is done as follows:

$$g(f(0)):24a_4+6\alpha a_2+R_e(6a_0a_3-2a_1a_2)=0 \tag{3.168}$$

$$\begin{aligned} g(f(1)):\ & 24a_4+120a_5+\alpha(24a_3+60a_3+60a_4+120a_5+6a_2)+\mathrm{Re}\ \{(a_0+...+a_5)(6a_3+ \\ & 24a_4+60a_5)-(a_1+2a_2+3a_3+4a_4+5a_5)(2a_2+6a_3+12a_4+20a_5)\}=0 \quad . \end{aligned} \tag{3.169}$$

By solving the set of algebraic differential equations consisting of six equations with six unknowns from equation 3.164 to equation 3.169, the constant coefficients of equation 3.163 can be achieved as

$$\psi=5+3\alpha+2\,\mathrm{Re}$$

$$a_0=0 \quad , \quad a_1=\frac{5}{8\psi}(12+8\alpha+5\,\mathrm{Re}) \quad , \quad a_2=0$$

$$a_3=\frac{-5}{4\psi}(2+2\alpha+\mathrm{Re}) \quad , \quad a_4=0 \quad , \quad a_5=\frac{1}{8\psi}(4\alpha+\mathrm{Re}). \tag{3.170}$$

By substituting the values of equation 3.170 into the answer function, the solution of the problem can be acquired as follows:

$$f(y)=\frac{5}{8\psi}(12+8\alpha+5\,\mathrm{Re})y-\frac{5}{4\psi}(2+2\alpha+\mathrm{Re})\,y^3+\frac{1}{8\psi}(4\alpha+\mathrm{Re})y^5. \tag{3.171}$$

3.1.11.3. Solving the Differential Equation by Numerical Method and Comparing It with AGM

By choosing the physical values, we will have

$$\alpha = 0.1 \qquad , \qquad \text{Re} = 50. \tag{3.172}$$

The differential equation in the domain $y \in (0\ ,\ 1)$ has been solved in the following table:

Table 3.9. Comparing the Obtained Results by AGM and RKF45

y	*0.0*	*0.2*	*0.4*	*0.6*	*0.8*	*1.0*
$f(y)$	0.0	0.307544	0.58579	0.807523	0.950524	1.00
$f'(y)$	1.56238	1.4848	1.27102	0.927605	0.48993	0.0
$f''(y)$	-0.7365	-0.73651	-1.42332	-1.98408	-2.35727	-2.50168

In this part, the gained results by AGM and Numerical method have been depicted graphically in the following figures:

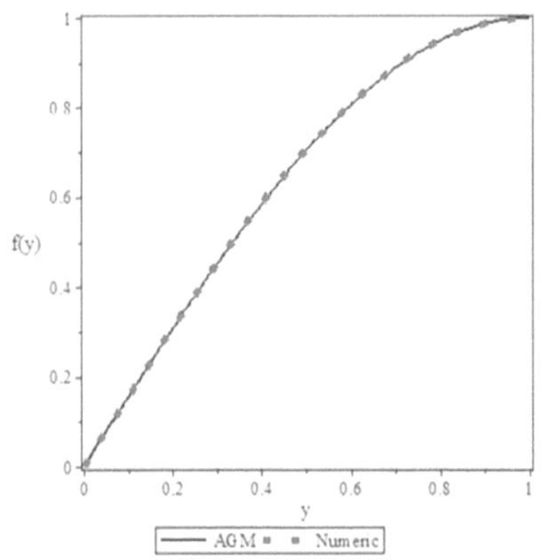

Fig. 3.26. Comparing the obtained charts by AGM and numerical method for $f(y)$

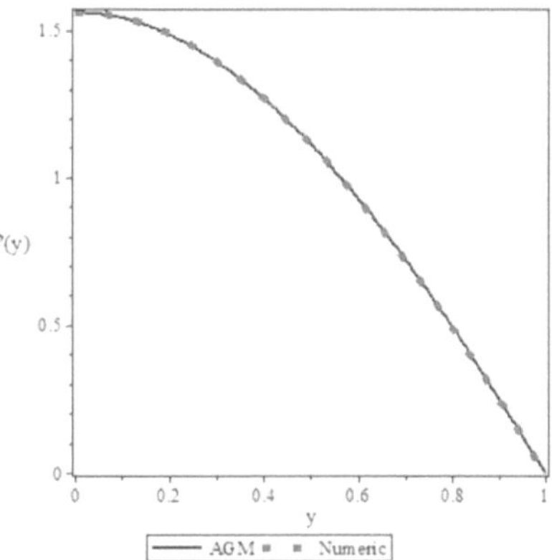

Fig. 3.27. Comparing the obtained charts by AGM and numerical method for $f'(y)$

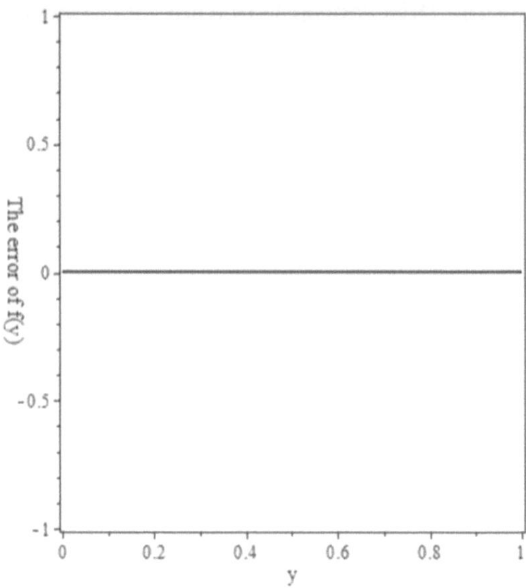

Fig. 3.28. The error of acquired solution by AGM

Example 3.1.12

Consider a convective fin according to the following figure:

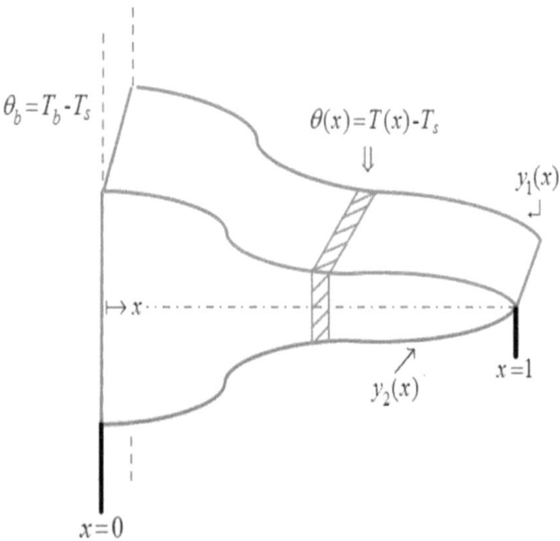

Fig. 3.29. The geometry of the presented problem

The nonlinear differential equation governing the heat transfer of the above fin is expressed as follows:

$$f(x):\ \frac{d}{dx}\{k(x)[y_1(x)-y_2(x)]\frac{d\theta(x)}{dx}\}=h(x)\theta(x)\{(1+y_1'^2(x))^{\frac{1}{2}}+(1+y_2'^2(x))^{\frac{1}{2}}\}. \quad (3.173)$$

Where y_1(x) and y_2(x) are the upper and lower curves of the fin, respectively, and T_s is the ambient temperature, and T_b is the base temperature of the fin.

By supposing the following boundary conditions

$$\theta(0)=1 \quad , \quad \theta(1)=0. \quad (3.174)$$

And also by considering that conductivity (k) and the convective heat transfer (h) are functions of temperature θ(x), we will have

$$k(x)=k_0e^{-0.1\theta(x)} \quad , \quad h(x)=h_0(1+0.02\theta(x)). \quad (3.175)$$

It is notable that the Taylor series expansion of $k(x)$ is written as

$$\mathrm{k(x)}=\mathrm{k}_0(1-0.1\theta(\mathrm{x})+...). \quad (3.176)$$

And the upper and lower fin curves are considered as follows:

$$y_1(x)=-x+e^{-0.05x} \quad (3.177)$$

$$y_2(x)=-0.2-x+e^{0.06x}. \quad (3.178)$$

After that, consider the following physical values of the system in (SI) as

$$\mathrm{k}_0=3.05 \quad , \quad h_0=1.2. \quad (3.179)$$

3.1.12.1. Solution

In order to solve the differential equation 3.173 by AGM, an answer function like the previous examples is needed by choosing a series of finite polynomials below:

$$\theta(x)=\sum_{k=0}^{5}a_k x^k=a_0+a_1x+a_2x^2+a_3x^3+a_4x^4+a_5x^5. \tag{3.180}$$

Where the constant coefficients a_0 to a_5 can be achieved by applying boundary conditions.

3.1.12.2. Applying Boundary Conditions by AGM

As it is previously explained, boundary conditions are applied in two approaches.

a. Boundary conditions are applied on equation 3.180 as follows:

$$\theta(0)=1 \quad \rightarrow \quad a_0=1 \tag{3.181}$$

$$\theta(1)=0 \quad \rightarrow \quad a_0+a_1+a_2+a_3+a_4+a_5=0. \tag{3.182}$$

b. Applying the boundary conditions on equation 3.173 is done as follows:

$$f(\theta(\mathrm{x}=0)):\ (1-0.1a_0)(a_2-0.3355a_1)-0.061a_1^2=3.48a_0(1+0.02a_0). \tag{3.183}$$

And also in $\mathrm{x}=1$, we will have

$$f(\theta(\mathrm{x}=1)):. \tag{3.184}$$

And then for $f'(\theta(\mathrm{BC}))$, we can have

$$f'(\theta(\mathrm{x}=0)):$$

$$(1-0.1a_0)(3.66a_3-1.342a_2-0.003355a_1)-0.366a_1a_2+0.0671a_1^2= \\ 0.0969a_1a_0+(1+0.02a_0)(3.48a_1-0.004345a_0) \quad . \tag{3.185}$$

As before, the derivative of the main differential equation in $x=1$ is obtained as follows:

$$f'(\theta(\mathrm{x}=1)):. \tag{3.186}$$

By solving the set of algebraic equations including six equations with six unknowns, the constant coefficients a_0 to a_5 can easily be computed in the following form:

$$a_0 = 1 \quad , \quad a_1 = -2.39986 \quad , \quad a_2 = 2.80633$$
$$a_3 = -2.40671 \quad , \quad a_4 = 1.46815 \quad , \quad a_5 = -0.46790. \tag{3.187}$$

By substituting the constant coefficients of equation 3.187 into the equation 3.180, the answer of the nonlinear differential equation 3.173 is acquired as follows:

$$\theta(x) = -0.46790x^5 + 1.46815x^4 - 2.40671x^3 + 2.80633x^2 - 2.39986x + 1. \tag{3.188}$$

3.1.12.3. Solving the Differential Equation by Numerical Method and Comparing It with AGM

With regard to the given physical values of equation 3.179 in $x \in \{0, 1\}$, we solve the differential equation numerically and compare it with AGM in the table below:

Table 3.10. Comparing the Obtained Temperature Profile of the Introduced Fin by AGM and Numerical Method

θ	*0.0*	*0.2*	*0.4*	*0.6*	*0.8*	*1.0*
$u(\theta)$ *NUM*	1.0	0.61210	0.36510	0.20171	0.09138	0.00
$u'(\theta)$	-2.7135	-1.5309	-0.98630	-0.65476	-0.65476	-0.48175
$u(\theta)$ AGM	1.0	0.61184	0.36418	0.20157	0.09035	0.000

The obtained charts by numerical method and AGM are drawn as follows:

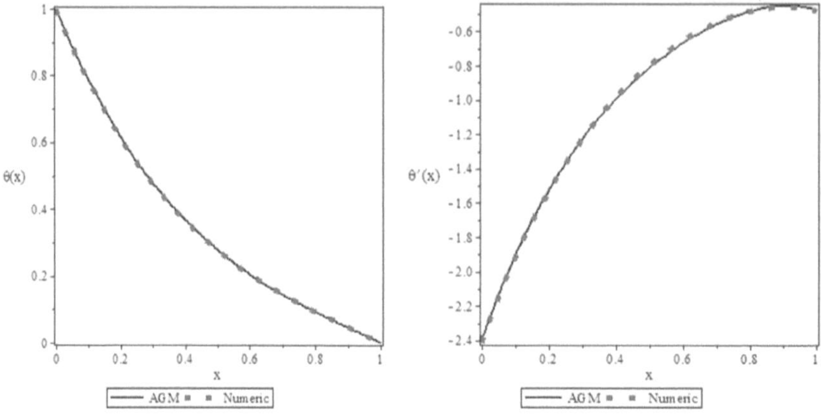

Fig. 3.30. Comparing the acquired solution by AGM with numerical method

Fig. 3.31. Comparing the first derivative of the acquired solution by AGM with numerical method

Example 3.1.13

The equations governing the nanofluid system are presented in the following form:

$$\begin{cases} F(\eta): & f''''-S(\eta f'''+3f''-2f.f''')-M^2 f''=0 \\ G(\eta): & \theta''+\Pr.S(2f\theta'-\eta\theta'+\Pr.N_b\theta'\phi'+\Pr N_t\theta'^2=0 \\ H(\eta): & \phi''+L_eS(2f-\eta)\phi'+\dfrac{N_t}{N_b}\theta''=0 \end{cases} \quad . \tag{3.189}$$

And the related boundary conditions are expressed as

$$\eta=0 \quad \rightarrow \quad f(\eta)=A \quad , \quad f'(\eta)=0 \quad , \quad \theta(\eta)=\phi(\eta)=1$$

$$\eta=1 \quad \rightarrow \quad f(\eta)=\frac{1}{2} \quad , \quad f'(\eta)=0 \quad , \quad \theta(\eta)=\phi(\eta)=0. \tag{3.190}$$

3.1.13.1. Solution

By choosing the physical values of the system, we will have

$$A = 0.2 \quad , \quad M = S = \Pr = L_e = N_b = 1 \quad , \quad N_t = 0.1. \tag{3.191}$$

Choose answer functions for the set of differential equations 3.189 in the following form:

$$f = \sum_{k=0}^{5} a_k \eta^k = a_0 + a_1\eta + a_2\eta^2 + \ldots + a_5\eta^5$$

$$\theta = \sum_{k=0}^{3} b_k \eta^k = b_0 + b_1\eta + b_2\eta^2 + \ldots + b_3\eta^3$$

$$\phi = \sum_{k=0}^{3} c_k \eta^k = c_0 + c_1\eta + c_2\eta^2 + \ldots + c_3\eta^3 . \tag{3.192}$$

3.1.13.2. Applying the Boundary Conditions

a. The boundary conditions are applied on the introduced answer functions as follows:

$$f(0) = 0.2 \quad \rightarrow \quad a_0 = 0.2 \tag{3.193}$$

$$f'(0) = 0 \quad \rightarrow \quad a_1 = 0 \tag{3.194}$$

$$f(1) = \frac{1}{2} \quad \rightarrow \quad a_0 + a_1 + \ldots + a_5 = \frac{1}{2} \tag{3.195}$$

$$f'(1) = 0 \quad \rightarrow \quad a_1 + 2a_2 + 3a_3 + 4a_4 + 5a_5 = 0 \tag{3.196}$$

$$\theta(0) = 1 \quad \rightarrow \quad b_0 = 1 \tag{3.197}$$

$$\theta(1) = 0 \quad \rightarrow \quad b_0 + b_1 + b_2 + b_3 = 0 \tag{3.198}$$

$$\phi(0)=1 \quad \rightarrow \quad c_0=1 \tag{3.199}$$

$$\phi(1)=0 \quad \rightarrow \quad c_0+c_1+c_3=0 \tag{3.200}$$

b. The boundary conditions are applied on the set of differential equations as follows:

$$F(f\ (\eta=0)): \quad 24a_4-8a_2+12a_0a_3=0 \tag{3.201}$$

$$F(f\ (\eta=1)): \quad -48a_4-20a_5-30a_3-8a_2+2(a_0+a_1+...+a_5)(6_3+24a_4+60a_5)=0 \tag{3.202}$$

$$G(\theta(\eta=0)): \quad 2b_2+2a_0b_1+b_1c_1+0.1b_1^2=0 \tag{3.203}$$

$$G(\theta(\eta=1)): \quad 2b_2+6b_3+(b_1+2b_2+3b_3)\{2(a_0+a_1+...+a_5)-1+(c_1+2c_2+3c_3) \\ +0.1(b_1+2b_2+3b_3)\}=0 \tag{3.204}$$

$$H(\phi(\eta=0)): \quad 2c_2+2a_0c_1+0.2b_2=0 \tag{3.205}$$

$$H(\phi(\eta=1)): \quad 2c_2+6c_3+[2(a_0+a_1+...+a_5)-1](c_1+2c_2+3c_3)+0.2b_2+0.6b_3=0 \tag{3.206}$$

By solving the set of algebraic differential equations consisting of fourteen equations with fourteen unknowns from equation 3.193 to equation 3.206, the constant coefficients of the answer functions can be obtained as follows:

$$\begin{array}{llll} a_0=0.2 & , a_1=0 & , a_2=1.0553 & , a_3=-1.062 \\ a_4=0.457 & , a_5=-0.15133 & , b_0=1 & , b_1=-0.5609 \\ b_2=-0.24136 & , b_3=-0.19772 & , c_0=1 & , c_1=-1.20451 \\ c_2=0.26504 & , c_3=-0.06053 & & \end{array} \tag{3.207}$$

By substituting the above values into the considered answer functions, the solution of the differential equation can be acquired as

$$\begin{aligned} f\ (\eta) &= 0.2+1.0553\eta^2-1.062\eta^3+0.4579\eta^4-0.15133\eta^5 \\ \theta(\eta) &= 1-0.5609\eta-0.24136\eta^2-0.19772\eta^3 \\ \phi(\eta) &= 1-1.20451\eta+0.26504\eta^2-0.06053\eta^3 \end{aligned} \tag{3.208}$$

3.1.13.3. Solving the Differential Equation by Numerical Method and Comparing Them with AGM

The set of differential equations in the domain $\eta \in (0\ ,\ 1)$ has been solved numerically in the following table:

Table 3.11. Comparing between the Velocity and Temperature Profiles and Volume Fraction of the Nanofluid Acquired by AGM with Numerical Solution

η	*0.0*	*0.2*	*0.4*	*0.6*	*0.8*	*1.00*
$f(\eta)$	0.2	0.23284	0.30785	0.395253	0.468564	0.50
$f'(\eta)$ *NUM*	0.0	0.29700	0.42866	0.423668	0.286667	0.00
$f'(\eta)$ *AGM*	0.0	0.2955	0.4281	0.4245	0.2878	-6×10^{-10}
$\theta(\eta)$ *NUM*	1.0	0.86435	0.70228	0.508518	0.277348	0.00
$\theta(\eta)$ *AGM*	1	0.8592	0.6930	0.4985	0.2709	0
$\phi(\eta)$ *NUM*	1.0	0.774823	0.56420	0.365653	0.178126	0.00
$\phi(\eta)$ *AGM*	1	0.7771	0.5679	0.3692	0.1802	2×10^{-10}

Finally, the resulting charts by AGM and the numerical method have been compared in the following form:

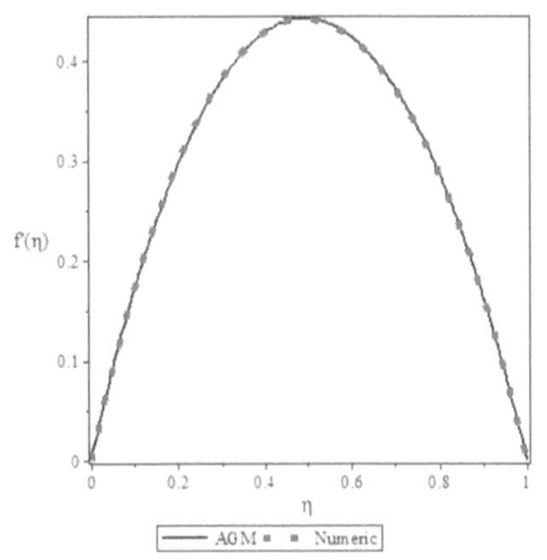

Fig. 3.32. Comparing the acquired velocity profile by AGM and numerical method

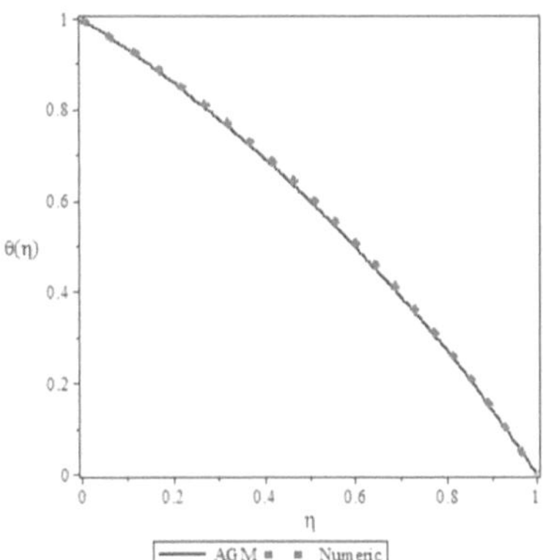

Fig. 3.33. Comparing the obtained temperature profile by AGM and numerical method

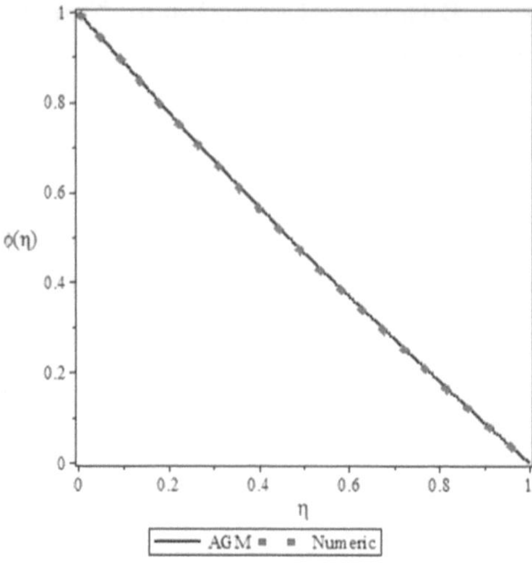

Fig. 3.34. A comparison between the volume fraction of nanofluid gained by AGM and RKF45

3.2. Problems

3.2.1. Consider the Bessel function with the order n in $x \in \{1, 2\}$ and solve it by AGM.

$$f(x):\ \ x^2u''+xu'+(u^2-n^2)=0$$

And the related boundary conditions are introduced as $u(1)=1$, $u(2)=0$. Moreover, the exact solution of the problem in the form of Bessel function with the order n is presented as follows:

$$u(x)=\frac{Y_n(2).J_n(x)-J_n(2).Y_n(x)}{Y_n(2).J_n(1)-Y_n(1).J_n(2)}.$$

3.2.2. Consider the parametric nonlinear differential equation as follows:

$$f(x):\ \ \frac{d^2y}{dx^2}\alpha y^n(\frac{dy}{dx})+\beta y^m+\lambda.y^p=\eta,\ \ y(0)=1\ ,\ y(1)=0\,.$$

The physical parameters α , β , λ , η , n , m and p are constant and are dependent upon the physical entity of the system. They appear with the differential equations governing the system as the effective parameters. The following parameters are defined in the following forms:

$$\alpha=0.7\quad,\quad\beta=0.5\quad,\quad\lambda=0.4\quad,\quad\eta=0.4$$
$$n=3\quad,\quad m=2.2\quad,\quad p=4\quad.$$

3.2.3. Suppose the following third-order nonlinear differential equation and solve it.

$$f(\eta):\ \ \frac{d^3w(\eta)}{d\eta^3}+\eta w(\eta)\frac{d^2w(\eta)}{d\eta^2}-w^2(\eta)\frac{dw(\eta)}{d\eta}+w(\eta)=\sin(\eta)$$

And the related boundary conditions are as $w(0)=0$, $w'(0)=1$, $w''(1)=2$

3.2.4. Consider the following differential equation in $x\in\{1,\ 2\}$ and solve it by AGM.

$$f(x):\quad \frac{d^2u}{dx^2}+x\frac{du}{dx}+x^2u=0$$

And the boundary conditions are introduced as $u(1)=1$, $u(2)=0$. Also, the exact solution of the mentioned problem in $x\in\{1,\ 2\}$ is given as follows:

$$u(x)=x(-1.4+1.42i)\exp[(-0.25-0.433i)x^2]\{(0.189-0.261i)kummerm(0.75-0.144i,1.5,0.866\,ix^2)+$$
$$(0.1193-0.7334\,i)\,kummermU(0.75-0.1443i,1.5,0.866ix^2)$$

.

The parameter i in the above equation is defined as the imaginational part of complex number. Moreover, in mathematics, the Kummer function is computed as $kummer\,M(mu,nu,z)$, and $kummer\,U(mu,nu,z)$ is achieved by solving the following differential equation:

$$zy''+(v-z)y'-\mu y=0.$$

3.2.5. Some of the differential equations in AGM, like the presented example here, do not need to convert their trigonometric, logarithmic, and exponential terms by Taylor series expansion. Solve the differential equation below by AGM.

$$f(x):\quad \frac{d^2u}{dx^2}+u^2-\frac{du}{dx}+u^3.e^{\frac{du}{dx}}+\ln(1+u^2)=0$$

,

and the boundary conditions are expressed as

$$u(0)=1,\ u(1)=0.$$

3.2.6. Solve the following set of coupled nonlinear differential equations.

$$\begin{cases} f(x): & u''-uv'+u^2u'=0 \\ g(x): & v''+uu'-v'v^2=0 \end{cases}$$

And the boundary conditions are expressed as:

$u(0)=1\ ,\ \ u(1)=v(1)=0$.

3.2.7. Consider the following set of parametric nonlinear differential equations:

$$\begin{cases} f(x): & \dfrac{d^2\theta}{\partial y^2}+\varepsilon\mu(\dfrac{du}{dy})^2=0 \\ g(x): & \dfrac{d}{dy}(\mu\dfrac{du}{dy})=0 \end{cases}$$

Where the parameter μ is a function of θ, which is equal to $\mu=1-\alpha\theta$. Afterward, α and ε are constant physical parameters. Then, the boundary conditions of the set of the equations are expressed as

$$\begin{cases} y=0 & \rightarrow & u=0 & , & \theta=0 \\ y=1 & \rightarrow & u=1 & , & \theta=0. \end{cases}$$

3.2.8. Solve the following nonlinear differential equation by AGM.

$$\mathrm{f}\,(\eta): \quad \frac{d^2\phi(\eta)}{d\eta^2}+\phi(\eta)\frac{d\phi(\eta)}{d\eta}+\phi^2(\eta)=0$$

And the related boundary conditions are expressed as $\varphi(0)=0$, $\varphi(1)=1$, $\varphi'(1)=1$.

Chapter 4

Nonlinear Vibrational Differential Equations

4.1. Applied examples

Example 4.1.1

Solve the following nonlinear differential equation by AGM.

$$\frac{d^2x}{dt^2}+\frac{1}{2}(\frac{dx}{dt})+x+\frac{1}{10}[x^3+(\frac{dx}{dt})^3]=0 \tag{4.1}$$

And the initial conditions are expressed as follows:

$$x(0)=0 \quad , \quad x'(0)=I_0. \tag{4.2}$$

The above nonlinear differential equation is governed on an electric circuit with the initial flow (I_0) in the systems of electric circuits in the field of power engineering.

4.1.1.1. Solving the Differential Equation by AGM

In order to solve the nonlinear differential equation, an answer function is needed as follows:

$$x=e^{-bt}\{a\cos(\omega t+\varphi)\}. \tag{4.3}$$

The constant coefficients $(a\ ,\ b\ ,\ \omega$ and $\phi\ )$ can be computed by applying the initial conditions.

4.1.1.2. Applying the Initial Conditions by AGM

The initial conditions can be applied on the answer function shown by x(IC)and on the differential equation 4.1 shown by $f(\mathrm{x}(\mathrm{I}C))$ and on its derivatives in two ways.

a. Applying the initial conditions on the answer function is done as follows:

$$x(IC): \quad \begin{cases} a\cos\varphi = 0 \\ -ab\cos\varphi - a\omega\sin\varphi = I_0. \end{cases} \tag{4.4}$$

b. The initial conditions are applied on the differential equation $f(\mathrm{x}(\mathrm{I}C))$ as

$f(x(t=1))$:

$$a(b^2-\omega^2-\frac{1}{2}b+a+\frac{1}{10}a^2\cos^2\varphi)\cos\varphi+\frac{1}{2}a\omega(4b-1)\sin\varphi-\frac{1}{10}a^3(b\cos\varphi+\omega\sin\varphi)^3=0 \tag{4.5}$$

And also applying the initial conditions on the derivative of the differential equation $f'(\mathrm{x}(\mathrm{I}C))$ is done in the following form:

$$f'(x(t=0))=0. \tag{4.6}$$

By solving the set of algebraic equations consisting of four equations with four unknowns from equation 4.4 to equation 4.6, the constant coefficients of the answer function can be gained very easily.

To simplify, the following new variable is considered as

$$\psi=\frac{1}{20}\sqrt{375-I_0^2(50+I_0^2)}. \tag{4.7}$$

Therefore, the constant coefficients of the answer function can be achieved as follows:

$$a=20I_0\sqrt{\psi} \quad , \quad b=(\frac{1}{4}+\frac{1}{20}I_0^2) \quad , \quad \varphi=-1.57 \quad , \quad \omega=\frac{1}{20}\sqrt{\psi}. \tag{4.8}$$

And after substituting the aforementioned constant values into the answer function, the solution of the nonlinear differential equation 4.1 is obtained by AGM as

$$x(t) = 20I_0\sqrt{\psi}e^{-(\frac{1}{4}+\frac{1}{20}I_0^2)t}\cos(\frac{1}{20}\sqrt{\psi} - 1.57). \tag{4.9}$$

4.1.1.3. Numerical Solution of the Nonlinear Differential Equation and Comparing It by AGM

Differential equation 4.1 by choosing the value (I_0=0.1) has numerically been solved in the time domain $t \in \{0, 20\}$ in the table below and compared by AGM.

Table 4.1. Comparing the Achieved Solutions by AGM and Numerical Method

x	*0.0*	*4*	*8*	*12*	*16*	*20*
x(t) Num	0.0	-0.02531	0.013871	-0.004169	0.0004101	0.0003321
x'(t)	0.1	-0.02099	-0.002023	0.003949	-0.001888	0.0004997
x(t) AGM	0.0	-0.02527	0.013841	-0.004175	0.0004214	0.0003319

The obtained charts have been depicted and compared as follows:

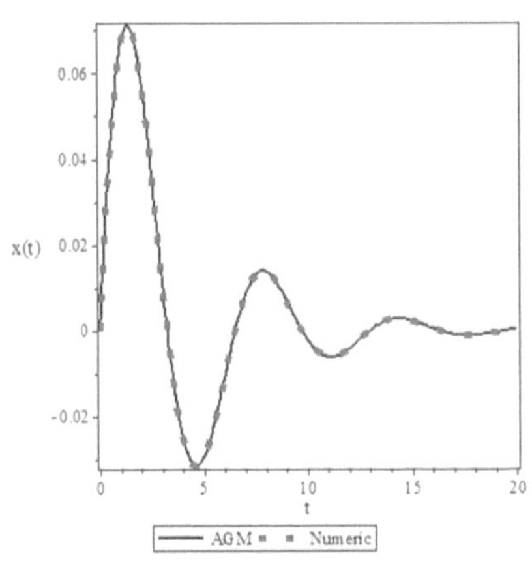

Fig. 4.1. The chart of obtained solution by AGM and numerical method for x(t)

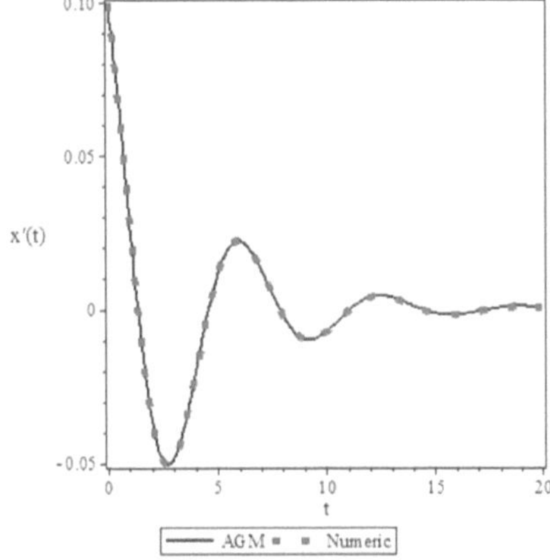

Fig. 4.2. The chart of obtained solution by AGM and numerical method for $x'(t)$

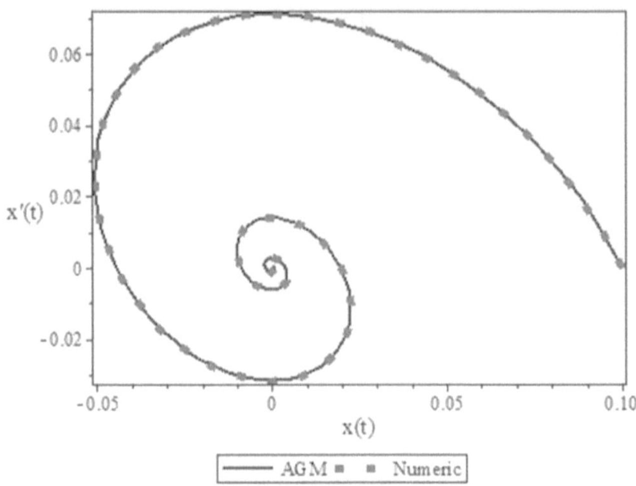

Fig. 4.3. The resulted phase plans by AGM and numerical method

Example 4.1.2

Solve the following nonlinear vibrational differential equation by AGM.

$$f(t):\ \ddot{\theta}(t)+\alpha(\dot{\theta}(\mathrm{t}))^{3}.\sin(\theta(t))+\beta.\theta(\mathrm{t}).\cos(\theta(t))+\eta.\theta(t)=0 \quad (4.10)$$

And the related initial equations are expressed as

$$\theta(0)=\mathrm{A} \quad , \quad \dot{\theta}(0)=0\cdot \quad (4.11)$$

The constant parameters α , β and η are constant values of the vibrational system.

4.1.2.1. Solving the Differential Equation by AGM

The differential equation 4.10 is exactly nonlinear because it has the trigonometric factors $\sin\theta$, $\cos\theta$, and also $(\dot{\theta})^{3}$. In order to solve the aforementioned differential equation, the Taylor expansion of the trigonometric factors is not needed. This can increase the precision of the obtained answer and is also for solving the differential equation by AGM. An answer function is needed as a finite series. Since there are no external forces in the vibrational system (without $\sin(\omega t)$), the best choice for the answer function is defined as

$$\theta = e^{-at}\{b\cos(\omega t+\varphi)\}. \quad (4.12)$$

The constant coefficients a and b, the angular frequency of vibration ω, and the initial phase ϕ are easily computed by applying the initial conditions. Inasmuch as the power of $(\dot{\theta})$ in the differential equation is 3, the vibrational system is non-damping, and then in the process of solving the problem by AGM, it is expected that the constant value (a) becomes zero $(a=0)$.

4.1.2.2. Applying Initial Conditions by AGM

Initial conditions are applied in two ways as follows:

a. Initial conditions are applied on the answer function in the following general form:

$$\theta(t) = \theta(IC) \quad (4.13)$$

Where (IC) is the abbreviation of initial conditions. Therefore, we have

$$\theta(0) = \mathrm{A} \quad \rightarrow \quad b\cos\varphi = A \quad (4.14)$$

$$\dot{\theta}(0) = 0 \quad \rightarrow \quad b(a\cos\varphi + \omega\varepsilon\varphi) = 0. \quad (4.15)$$

b. Applying initial conditions on the differential equation 4.10 and its derivatives is done after substituting the answer function instead of the dependent parameter θ into the differential equation shown by $f(\theta(\mathrm{t}))$ as the following general equation:

$$f(\theta(t)) \rightarrow f(\theta(IC) = 0 \quad , \quad f'(\theta IC)) = 0 \quad , \quad \ldots \quad (4.16)$$

The above equations mean that the answer function is substituted into the differential equation 4.10, and afterward, the initial conditions are applied on it as follows:

$f(\theta(0))$:

$$b(a^2-\omega^2+\eta)\cos\varphi+2ab\omega\sin\varphi-b\alpha(a\cos\varphi+\omega\sin\varphi)^3\sin\{b\cos\varphi\}+ \beta b\cos\varphi\cos\{b\cos\varphi\}=0 \quad . \tag{4.17}$$

Then for the derivative of differential equation according to equation 4.16, we will have

$f'(\theta(0))$:

$$-ab(a^2-3\omega^2+\eta)\cos\varphi-b\omega(3a^2+\eta-\omega^2)\sin\varphi-\beta b(a\cos\varphi+ \omega\sin\varphi)\cos\{b\cos\varphi\}+(a\cos\varphi+\omega\sin\varphi)^2\{3\alpha(a^2b\cos\varphi+2ab\sin\varphi- b\omega^2\cos\varphi)+\alpha ab\cos\varphi+b\omega\sin\varphi)^2\cos\{b\cos\varphi\}+\beta b(a\cos\varphi+ \omega\sin\varphi)\cos\varphi\sin\{b\cos\varphi\}=0 \quad . \tag{4.18}$$

By solving the set of algebraic equations consisting of four equations with four unknowns from equation 4.14 to equation 4.18, with the exception of equation 4.16, we can compute the constant coefficients of the considered answer function as

$$a=0 \quad , \quad b=A \quad , \quad \varphi=0. \tag{4.19}$$

And also, the angular frequency of vibration is obtained as follows:

$$\omega=\sqrt{\beta\cos(A)+\eta}\,. \tag{4.20}$$

By substituting the obtained constant coefficients into equation 4.12, the solution of nonlinear differential equation 4.10 can be gained by AGM in the form of harmonic and non-damping as

$$\theta(t)=A\cos\{\sqrt{\beta\cos(A)+\eta}\}. \tag{4.21}$$

Based on the given physical values in the foregoing part of this solution, we can show that

$$\theta(t)=\ 0.2\cos(0.8910779838t). \tag{4.22}$$

4.1.2.3. Solving the Differential Equation by Numerical Method and Comparing It with AGM

For solving the differential equation 4.10, it is necessary to determine the physical parameters of the differential equation. So the following physical values are considered for the vibrational system as follows:

$$\alpha = 0.2 \quad , \quad \beta = 0.3 \quad , \quad \eta = 0.5 \quad , \quad A = 0.2. \qquad (4.23)$$

In accordance with the above physical values, the nonlinear differential equation in the domain $t \in \{0, 30\}$ has numerically been solved by RKF45, and the obtained results and its comparison with AGM can be observed in the following table:

Table 4.2. Comparing the Obtained Solutions by AGM and Numerical Method

t	0.0	6	12	18	24	30
$\theta(t)$ NUM	0.2	0.11836	-0.0597	-0.1897	-0.16472	-0.00568
$\theta(t)$ AGM	0.2	0.1185	-0.0598	-0.18901	-0.1646	-0.00570

The charts achieved by numerical method and AGM are compared as

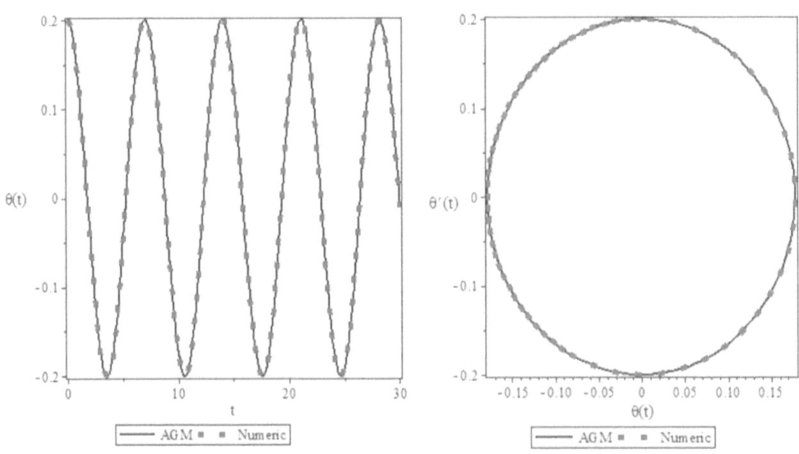

Fig. 4.4. The chart of obtained solution by AGM and numerical method for $\theta(t)$

Fig. 4.5. The chart of resulted phase plans by AGM and numerical method

It is necessary to mention that the chart of phase plan is the ratio of the answer function $\theta(t)$ to the derivative of the answer function $\theta'(t)$ in the coordinate system. According to the above charts, it is concluded that the achieved answer function is a suitable estimation for solving the nonlinear differential equation 4.10 and for increasing the precision in the solution of differential equation. The series sentences of the considered answer function can be increased, and so mathematical operations will be increased. Therefore, the solution is tended toward the exact answer. As it is explained before, the set of vibrational equations is applied in the earthquake simulation in civil engineering and engineering of solids and similar fields.

Example 4.1.3

Solve the following set of nonlinear differential equation by AGM.

$$\begin{cases} f(t) = \dfrac{d^2u}{dt^2} + \alpha u + \beta v^2 = 0 \\ g(t) = \dfrac{d^2v}{dt^2} + \lambda v + \mu u^3 = 0 \end{cases} \tag{4.24}$$

And the initial conditions are expressed as follows:

$$u(0) = u_0 \quad , \quad v(0) = v_0 \quad , \quad u'(0) = v'(0) = 0. \tag{4.25}$$

And α , β , λ and μ are the constant coefficients in the differential equation.

4.1.3.1. Solving the Differential Equation by AGM

Like the previous examples, in order to solve the set of nonlinear differential equation 4.24, two answer functions are needed as follows:

$$u = e^{-at}\{b\cos(\omega_1 t + \varphi_1)\} \quad ; \quad v = e^{-ct}\{d\cos(\omega_2 t + \varphi_2)\} \tag{4.26}$$

The constant coefficients a , b , c , d and also ω_1 , ω_2 , φ_1 and φ_2 can be computed by applying the initial conditions.

4.1.3.2. Applying the Initial Conditions by AGM

For the answer functions shown by $u(IC)$, $v(IC)$ and on the set of differential equation 4.24 shown by $f(u(IC)$, $g(v(IC))$ and then on its derivatives, we will have the following:

a. Applying the initial conditions on the answer functions is done as follows:

$$u(IC) = u_0 \quad \rightarrow \quad b\cos\varphi_1 = u_0 \tag{4.27}$$

$$u'(IC) = 0 \quad \rightarrow \quad b(b\cos\varphi_1 + \omega_1 \sin\varphi_1) = 0 \tag{4.28}$$

$$v(IC) = v_0 \quad \rightarrow \quad d\cos\varphi_2 = v_0 \tag{4.29}$$

$$v'(IC) = 0 \quad \rightarrow \quad d(c\cos\varphi_2 + \omega_2 \sin\varphi_2) = 0 \tag{4.30}$$

b. Applying the initial conditions on the set of differential equations $f(u(IC)$, $g(v(IC))$ is done in the following form:

$f(u(t=0)):$

$$b(a^2 - \omega_1^2 + \alpha)\cos\varphi_1 + 2ab\omega_1 \sin\varphi_1 + \beta d^2 \cos^2\varphi_2 = 0 \tag{4.31}$$

$g(v(t=0)):$

$$d(c^2 - \omega_2^2 + \lambda)\cos\varphi_2 + 2cd\omega_2 \sin\varphi_2 + \mu b^2 \cos^3\varphi_1 = 0 \tag{4.32}$$

And also applying the initial conditions on the derivative of the differential equations shown by $f'(u(t=0)$ and $g'(v(t=0))$ is done as follows:

$f'(u(t=0)):$

$$ab(3\omega_1^2 - a^2 - \alpha)\cos\varphi_1 - b\omega_1(3a^2 + \alpha - \omega_1^2)\sin\varphi_1 - 2\beta d^2(c\cos\varphi_2 + \omega_2 \sin\varphi_2)\cos\varphi_2 = 0 \tag{4.33}$$

$g'(v(t=0)):$

$$cd(3\omega_2^2 - c^2 - \lambda)\cos\varphi_2 - d\omega_2(3c^2 + \lambda - \omega_2^2)\sin\varphi_2 - 3\mu b^3(a\cos\varphi_1 + \omega_1 \sin\varphi_1)\cos^2\varphi_1 = 0 \tag{4.34}$$

By solving the set of algebraic equations consisting of eight equations with eight unknowns from equation 4.27 to equation 4.34, the constant coefficients of the answer functions can be gained as

$$a=0 \quad , \quad b=u_0 \quad , \quad c=0 \quad , \quad d=v_0 \quad , \quad \varphi_1=0 \quad , \quad \varphi_2=0$$
$$\omega_1=\sqrt{\frac{\alpha u_0+\beta v_0^2}{u_0}} \quad , \quad \omega_2=\sqrt{\frac{\lambda v_0+\mu u_0^3}{v_0}} \quad . \qquad (4.35)$$

By substituting the aforementioned constant values into the answer functions, the solution of the set of nonlinear differential equations 4.24 is gained by AGM as follows:

$$u(t)=u_0\cos\{\sqrt{\frac{\alpha u_0+\beta v_0^2}{u_0}}t\} \quad , \quad v(t)=v_0\cos\{\sqrt{\frac{\lambda v_0+\mu u_0^3}{v_0}}t\}. \qquad (4.36)$$

4.1.3.3. Numerical Solution of the Set of Differential Equations and Comparing It with AGM

In order to solve the given set of differential equations, it is required to define the following constant values as

$$u_0=0 \quad , \quad v_0=0.05 \quad , \quad \alpha=0.4$$
$$\beta=0.2 \quad , \quad \mu=0.2 \quad , \quad \lambda=0.2. \qquad (4.37)$$

Therefore, with respect to the above physical values, the set of equations has numerically been solved in the time domain $t\in\{0,\ 40\}$ in the table below:

Table 4.3. The Result of Obtained Solution by Runge-Kutta Method

t	0.0	8	16	24	32	40
$u(t)$	0.06	0.0202	-0.04683	-0.05296	0.009635	0.058344
$u'(t)$	0.0	0.03525	0.02374	-0.01943	-0.037133	-0.005726
$v(t)$	0.05	-0.04521	0.03204	-0.01301	-0.008612	0.028671
$v'(t)$	0.0	0.00947	-0.01706	0.02153	-0.02203	0.0182921

The obtained results have been depicted and compared graphically as follows:

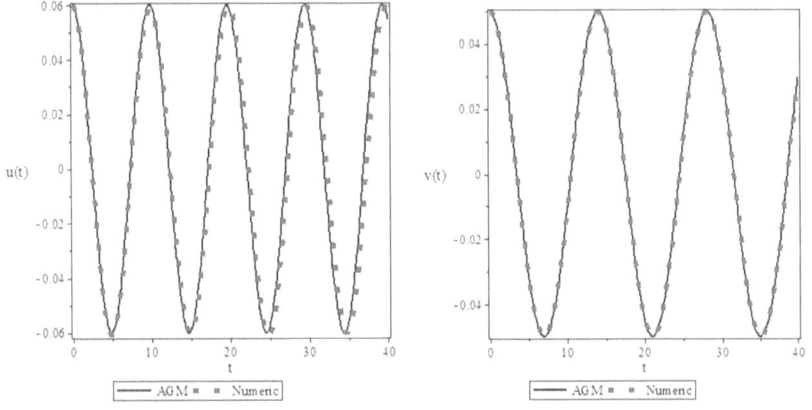

Fig. 4.6. The result of achieved solutions by AGM and numerical method for $u(t)$

Fig. 4.7. The chart of resulted solutions by AGM and numerical method for $v(t)$

And for their derivatives, we will have the following:

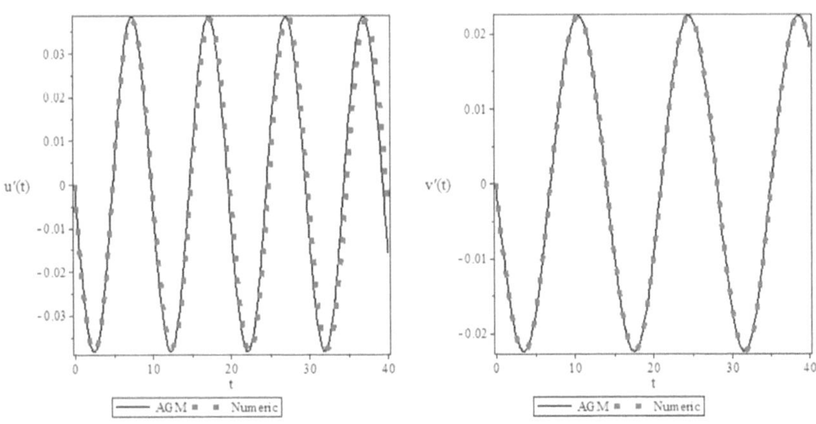

Fig. 4.8. The chart of obtained $u'(t)$ by AGM and numerical method

Fig. 4.9. The chart of resulted $v'(t)$ by AGM and numerical method

Finally, for their phase plans, we can show the following comparisons:

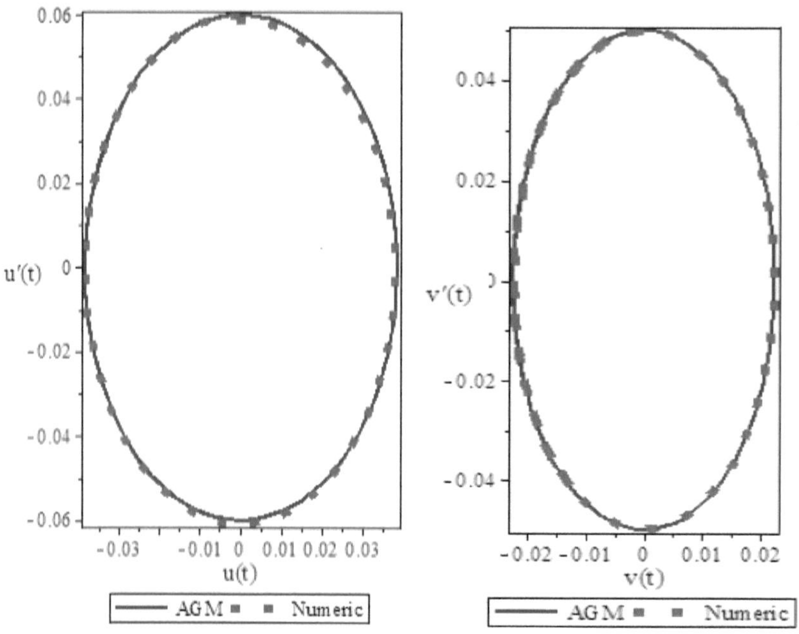

Fig. 4.10. The achieved phase plans by AGM and numerical method for u(t)

Fig. 4.11. The gained phase plans by AGM and numerical method for v(t)

It is citable that in AGM, there is no need to use the Taylor series expansion for trigonometric, exponential, and logarithmic functions—while in other methods for solving nonlinear differential equations, the Taylor expansion should be considered as a vital factor for solution procedure.

Example 4.1.4

Solve the following nonlinear differential equation (Duffing's equation) parametrically.

$$f(t):\ \frac{d^2u}{dt^2}+\beta^2(u+\mu^2u^3)=p\sin(\Omega t) \tag{4.38}$$

And the related initial conditions are expressed as

$$u(0) = A \quad , \quad u'(0) = 0. \tag{4.39}$$

The parameters β , μ , p , Ω and A are constant values for the vibrational system, which includes external forces $p\sin(\Omega t)$ with the frequency (Ω).

$$A = 0.15 \quad , \quad \Omega = 2 \quad , \quad \beta = 0.3 \quad , \quad \mu = 0.2 \quad , \quad p = 0.2 \tag{4.40}$$

4.1.4.1. Solving Differential Equations by AGM

An infinite power series is needed in order to solve equation 4.38, and a sentence from this function such as the following form as the answer function has been chosen to reduce the size of computational operations:

$$u = e^{-at}\{b\cos(\omega t + \varphi)\} + d\sin(\Omega t + \phi). \tag{4.41}$$

The constant coefficients (a , b , φ , d and ϕ) can easily be computed by applying the initial conditions.

4.1.4.2. Applying the Initial Conditions by AGM

The initial conditions can be applied like the previous examples in two ways: on the answer function of differential equation shown by $\mathrm{u}(\mathrm{I}C)$ and then on the differential equation 4.38 and its derivatives shown by $f(\mathrm{u}(\mathrm{I}C))$.

a. Applying the initial conditions on the answer function is done as follows:

$$u(IC): \quad \mathrm{b}\cos\varphi + \mathrm{d}\sin\varphi = 0.15 \tag{4.42}$$

$$u'(IC): \quad -b(a\cos\varphi + \omega\sin\varphi) + 2d\cos\phi = 0. \tag{4.43}$$

b. The initial conditions are applied on $f(\mathrm{u}(\mathrm{I}C))$ as

$$f(u(t=0)) = 0$$

$$b(a^2 - \omega^2 + 0.09)\cos\varphi + 2ab\omega\sin\varphi - 3.91d\sin\phi + 0.0036(b\cos\varphi + d\sin\phi)^3 = 0 \quad . \tag{4.44}$$

And also on the first derivative of the differential equation $f'(u(IC))$, we will have

$$f'(u(t=0))=0$$
$$ab(-a^2+3\omega^2-0.09)cos\varphi+b(-3a^2\omega+\omega^3-0.09\omega)\sin\varphi-$$
$$7.82d\cos\phi+0.0108(b\cos\varphi+d\sin\phi)^2(2d\cos\phi-ab\cos\varphi-b\omega\sin\varphi)=0.4. \quad (4.45)$$

And the initial conditions are applied on the second and third derivatives shown by $f''(u(IC))$ and $f'''(u(IC))$, respectively, like the aforementioned process, as follows:

$$f''(u(t=0))=0 \quad (4.46)$$

$$f'''(u(t=0))=0. \quad (4.47)$$

By solving the set of algebraic equations consisting of six equations with six unknowns from equation 4.42 to equation 4.47, the constant coefficients of the answer function can be acquired in the following form:

$$a=-0.0001 \quad , \quad b=0.372 \quad , \quad d=0.0511 \quad ,$$
$$\phi=3.141 \quad , \quad \varphi=-1.156 \quad , \quad \omega=0.3 \quad . \quad (4.48)$$

And by substituting the above constants into equation 4.41, the solution of nonlinear differential equation 4.38 is obtained by AGM as follows:

$$u(t)=0.372e^{0.0001t}\cos(0.372t-1.156)+0.0511\sin(2t+3.141). \quad (4.49)$$

4.1.4.3. Solving the Nonlinear Differential Equation by AGM and Numerical Method (RN45) and Comparing Them Together

The presented problem, with respect to the given physical values in the time domain $t\in\{0,\ 40\}$, has been solved by RKF45, and then the two methods have been compared with each other in the table below:

Table 4.4. The Values Resulted by Numerical Method and AGM

t	*0.0*	*8*	*16*	*24*	*32*	*40*
u(*t*) *Num*	0.15	0.13282	-0.3528	0.40221	-0.26011	-0.000198
u(*t*) *AGM*	0.0	0.13324	-0.3538	0.40248	-0.25837	-0.000225

The resulting charts by AGM and numerical method are drawn below:

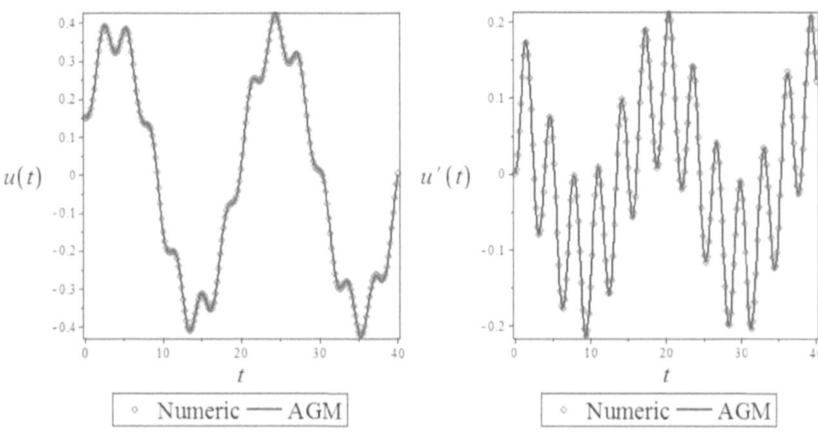

Fig. 4.12. The chart of achieved u(t) by AGM and numerical method

Fig. 4.13. Comparing the first derivative of the obtained solutions by AGM and numerical method

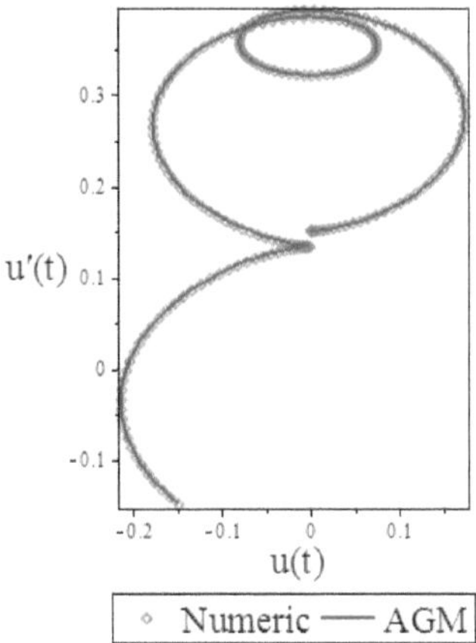

Fig. 4.14. The resulted phase plan by AGM and numerical method

In order to increase the precision and sensitivity of the answer, the number of the series sentences of equation 4.41 can be enhanced that we, here, could solve the nonlinear differential equation with high precision by choosing two sentences from the series of answer function.

Example 4.1.5

Solve the following nonlinear vibrational differential equation by AGM.

$$f(t):\ \frac{d^2x(t)}{dt^2}+\alpha\{x^n(t)+x(t)\}+\beta(\frac{dx(t)}{dt})=0 \tag{4.50}$$

And the initial conditions are expressed as

$$x(0)=A \quad , \quad x'(0)=0, \tag{4.51}$$

where α , β and n are the constant parameters of the vibrational system.

4.1.5.1. Solving the Differential Equation by AGM

Consider an answer function to solve the nonlinear vibrational differential equation 4.50 like the previous example.

$$x(t)=e^{-at}\{b\cos(\omega t+\varphi)\} \tag{4.52}$$

The constant coefficients a and b, the angular frequency of vibration ω , and the initial phase of damping vibration φ are easily computed by applying the initial conditions.

4.1.5.2. Applying Initial Conditions by AGM

As previously mentioned, two ways are used to apply initial conditions by AGM—that is, first, on the answer function $x(IC)$ and then on the differential equation $f(x(IC))$ as follows:

a. The initial conditions are applied on the answer function as

$$x(0) = A \quad \rightarrow \quad b\cos\varphi = A \tag{4.53}$$

$$x'(0) = 0 \quad \rightarrow \quad b(a\cos\varphi + \omega\sin\varphi) = 0. \tag{4.54}$$

b. Applying the initial conditions on the differential equation and its derivatives shown by $f(x(IC))$ and $f'(x(IC))$ is done as follows:

$$f(x(0)): \quad b(a^2 - \omega^2 - \beta a + \beta)\cos\varphi + b\omega(2a - \beta)\sin\varphi + \alpha b\cos^n\varphi = 0 \tag{4.55}$$

$f'(x(0))$:

$$b(3a\omega^2 + \beta a^2 - a^3 - \alpha a - \beta\omega^2)\cos\varphi - b\omega(3a^2 - \omega^2 - 2a\beta + \alpha)\sin\varphi -$$
$$\frac{\alpha nb}{\cos\varphi}(a\cos\varphi + \omega\sin\varphi)\cos^n\varphi) = 0 \tag{4.56}$$

By solving the set of algebraic equations consisting of four equations with four unknowns from equation 4.53 to equation 4.56, the constant coefficients of the considered answer function can be computed very easily.

To simplify, the following new variables are considered as

$$\psi = 4\alpha(A^n + A). \tag{4.57}$$

And the constant coefficients of the answer function can be obtained as follows:

$$a = \frac{1}{2}\beta \quad , \quad b = 2A\sqrt{\frac{\psi}{\psi - \beta^2 A}} \quad , \quad \varphi = -tg^{-1}(\beta\sqrt{\frac{A}{\psi}}). \tag{4.58}$$

And also, the damping angular frequency is gained below:

$$\omega = \sqrt{-\frac{A\beta^2 - 4A\alpha - 4e^{ln(A)n}\alpha}{4A}}. \tag{4.59}$$

After substituting the above constant coefficients into the answer function and considering $\alpha = 0.2$, $\beta = 0.15$, $n = 3$, $A = 0.2$, the solution of the differential equation 4.50 can be obtained by AGM as follows:

$$x = 0.2027604434e^{-0.075000000000t}\cos(0.4498610897t - 0.1651987502). \quad (4.60)$$

And then, for the frequency of the vibrational system, we will have

$$\omega = 0.4498610897. \quad (4.61)$$

4.1.5.3. Solving the Nonlinear Differential Equation by Numerical Method and Comparing It with AGM

By choosing the physical values of the vibrational system for the given differential equation based on the last part, the differential equation 4.50 in the domain $t \in \{0, 60\}$ has numerically been solved by RKF45 in the following table:

Table 4.5. The Results of Achieved Solution by Numerical Method

t	0.0	12	24	36	48	60
$x(t)$	0.2	0.03634	-0.017094	-0.013693	-0.003264	0.000813
$x'(t)$	0.0	0.03017	0.01409	0.001312	-0.001747	-0.000993

The charts gained by numerical method and AGM are compared as

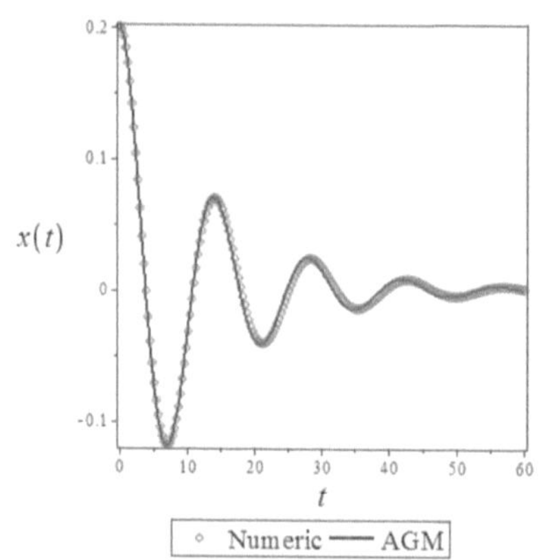

Fig. 4.15. The chart of achieved x(t) by AGM and numerical method

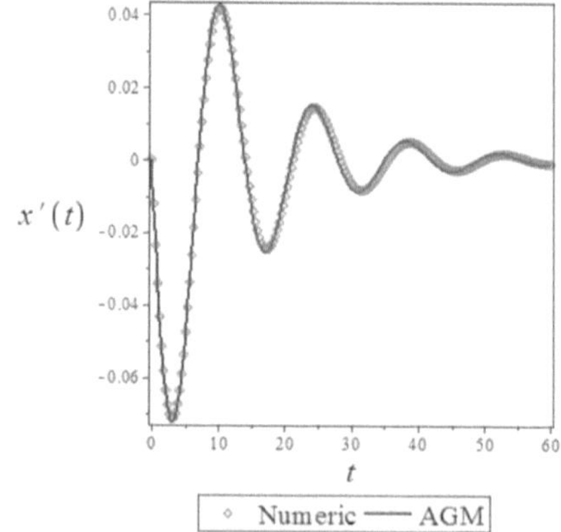

Fig. 4.16. Comparing the first derivative of the gained solutions by AGM and numerical method

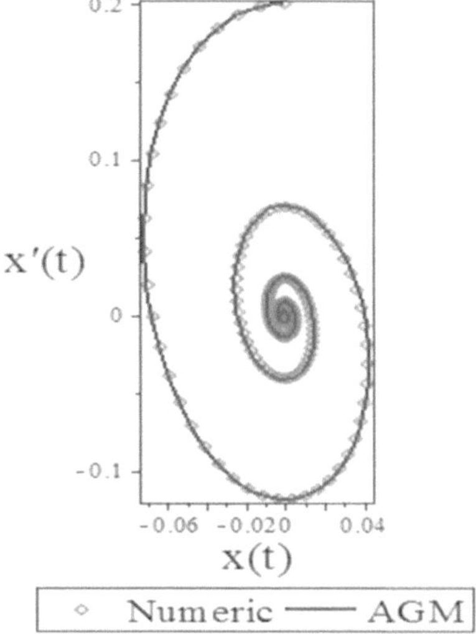

Fig. 4.17. The resulted phase plans by AGM and numerical method

With regard to the above obtained charts, solving the nonlinear differential equation with AGM by choosing the answer function like equation 4.52 can be considered as a suitable procedure. And for reaching the high-sensitivity answers or the exact solution, we can add a number of sentences to the answer function.

4.1.5.4 Investigation of Other Vibrational Parameters

As it is expressed in the first chapter, the equations of the locus of the maximum amplitude points for the displacement $x_m(t)$, velocity $\dot{x}_m(t)$, and acceleration $\ddot{x}_m(t)$ can be expressed for the vibration of the system in the following forms:

$$x_m(t) = be^{-at} \quad , \quad \dot{x}_m(t) = \text{b}\sqrt{a^2 + \omega^2 e^{-at}} \quad , \quad \ddot{x}_m(t) = \text{b}(a^2 + \omega^2)e^{-at}. \quad (4.62)$$

Since the values a , b and ω in the equation 4.58 and equation 4.59 have been computed, the locus charts of maximum displacement, velocity, and acceleration in terms of the amplitude of vibration with the related functions are depicted graphically as follows:

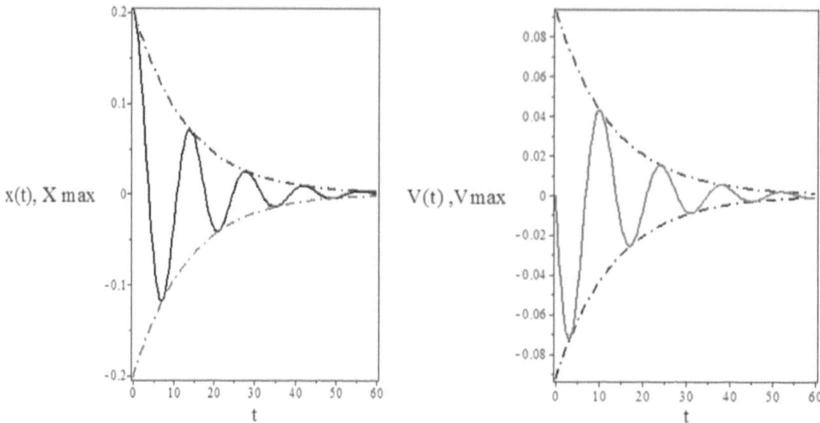

Fig. 4.18. The locus of maximum vibrational displacement points by AGM

Fig. 4.19. The locus of maximum vibrational velocity points by AGM

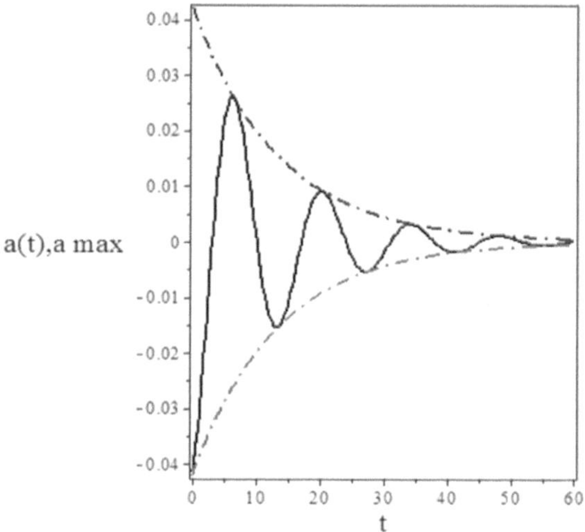

Fig. 4.20. The locus of maximum vibrational acceleration points by AGM

And also, the maximum points of displacement, velocity, and acceleration can be obtained from the following equations:

$$t_m(t) = \frac{k\pi}{\omega} \quad \rightarrow \quad x_k = be^{-a(\frac{k\pi}{\omega})} \quad , \quad \ddot{x}_k = \mathrm{b}(a^2+\omega^2)e^{-a(\frac{k\pi}{\omega})}$$

$$t_m(t) = \frac{(2k-1)^k}{2\omega} \quad \rightarrow \quad \dot{x}_k = b\sqrt{a^2+\omega^2}\, e^{-a(\frac{2k-1}{2\omega})} \quad . \tag{4.63}$$

4.1.5.5. Investigating the Charts of Damping Ratio (ξ), the Angular Frequency (ω), the Logarithmic Reduction Ratio (δ), and the Instantaneous Amplitude of Damping Vibration in Terms of the Initial Amplitude of Vibration (A)

On the basis of the first chapter, the damping ratio (ξ) and logarithmic reduction ratio (δ) have been defined as follows:

$$\xi = \frac{a}{\sqrt{a^2 + 4\omega^2}} \quad , \quad \delta = \frac{2\pi\xi}{\sqrt{1-\xi^2}}. \tag{4.64}$$

And also, the values of damping vibrational amplitude (b) and the damping angular frequency (ω) have been obtained due to the given physical values such as $\alpha = 0.2$, $\beta = 0.15$ and $n = 3$. Then, the following charts can be investigated in terms of the initial amplitude of the vibration $x(0) = \mathrm{A}$ in the charts below:

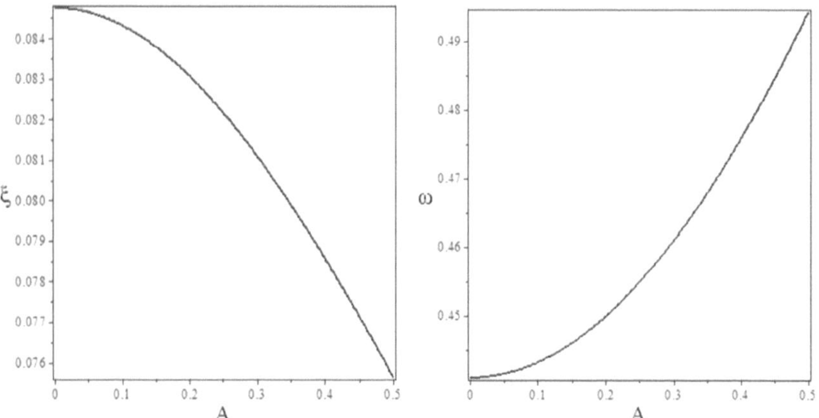

Fig. 4.21. Variation of damping ratio in terms of amplitude of vibration by AGM

Fig. 4.22. The result of changing angular frequency with amplitude of vibration

Then we will have:

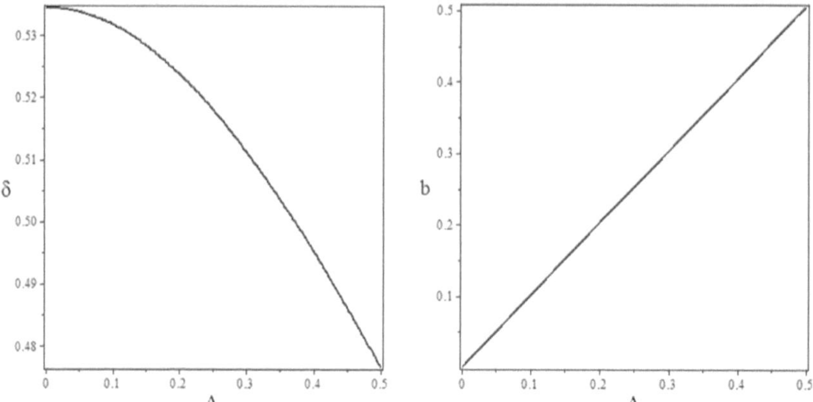

Fig. 4.23. Variation of logarithmic reduction ratio with amplitude of vibration

Fig. 4.24. Variation of damping vibrational amplitude in terms of amplitude of vibration

The above charts can be utilized for analyzing real vibrational systems applied in designing mechanisms consisting of vibrational components.

Example 4.1.6

Solve the following nonlinear vibrational differential equation by AGM.

$$f(t): \quad \frac{d^2u}{dt^2} + c\frac{du}{dt} + ku^2 = F_0 \sin(\Omega t) \tag{4.65}$$

And the related initial conditions are expressed as

$$u(0) = A \quad , \quad u'(0) = 0. \tag{4.66}$$

Consider the constant values of the differential equation as follows:

$$A = 0.15 \quad , \quad c = 1 \quad , \quad k = 0.01 \quad , \quad F_0 = 0.2 \quad , \quad \Omega = 2. \tag{4.67}$$

It is notable that the aforementioned differential equation can be applied for the vibration of systems, along with damping factor (C) and a spring with elasticity factor (K) with the external force $f(\mathrm{t}) = F.\sin(\Omega t)$, which are used in the earthquake, civil and mechanical engineering, etc.

4.1.6.1. Solving the Differential Equation by AGM

A sentence from an answer function with infinite power series is needed in order to solve the nonlinear differential equation 4.65 as

$$u = e^{-at}\{b\cos(\omega t+\varphi)\}+d\sin(\Omega t+\phi). \tag{4.68}$$

The constant values a, b, d, φ and ϕ can easily be computed, and the parameter (Ω) is the angular frequency of the external force exerting on the vibrational system.

4.1.6.2. Applying the Initial Conditions by AGM

Like the previous examples, applying the initial conditions is done on the answer function shown by $\mathrm{u}(IC)$ and on the differential equation shown by $f(\mathrm{u}(IC))$ as follows:

a. The initial conditions are applied on the answer function in the following form:

$$u(IC): \begin{cases} u(0) = A & \rightarrow \quad b\cos\varphi + d\sin\varphi = A \\ u'(0) = 0 & \rightarrow \quad b(a\cos\varphi+\omega\sin\varphi)-d\cos\phi = 0. \end{cases} \tag{4.69}$$

b. The initial conditions are applied on the differential equation as

$$f(u(t=0)) = 0 :$$

$$b(9a^2-\omega^2-ca)\cos\varphi + b\omega(2a-c)\sin\varphi -$$
$$d\Omega(\Omega\sin\phi - c\cos\phi) + k(b\cos\varphi + d\sin\phi)^2 = 0. \tag{4.70}$$

And also on the derivatives of the differential equation, we have

$$f'(u(IC): \rightarrow \qquad f'(u(t=0)) = 0 \tag{4.71}$$

$$f''(u(IC)):\rightarrow \qquad f''(u(t=0))=0 \tag{4.72}$$

$$f'''(u(IC)):\rightarrow \qquad f'''(u(t=0))=0. \tag{4.73}$$

By solving a set of algebraic equations consisting of six equations with six unknowns from equation 4.69 to equation 4.73 and by helping Maple software with respect to the physical values of the system in equation 4.67, the constant coefficients of the answer function can be computed as follows:

$$a=0.5 \quad , \quad b=-0.284i \quad , \quad d=0.0447$$
$$\phi=-tg^{-1}(0.447) \quad , \quad \varphi=-tg^{-1}(1.1647) \quad , \quad \omega=0.497i. \tag{4.74}$$

In the above equation (i) is the imaginary part of the complex number. After substituting the aforementioned values into equation 4.68, the solution of the nonlinear differential equation 4.65 can be obtained by utilizing mathematical simplification as

$$u(t)=0.2847e^{-0.5t}\sinh(0.497t+0.56)+0.044\sin(2t-2.6776)+(5.84)(10^{(-11)}I)e^{-0.4995t}+ \cosh(0.4974t+0.56635) \tag{4.75}$$

4.1.6.3. Solving the Nonlinear Differential Equation Numerically and Comparing It with AGM

Due to the given physical values, the differential equation 4.65 has numerically been solved in the domain $t\in\{0,\ 40\}$ in the table below:

Table 4.6. Comparing the Obtained Solution by AGM with Numerical Method

t	8	16	24	32	40
$u(t)$ NUM	0.27691	0.20379	0.28100	0.18995	0.26980
$u'(t)$ NUM	0.06454	-0.04528	0.01988	0.004947	-0.031524
$u(t)$ AGM	0.2773	0.20375	0.28203	0.18982	0.27255

In accordance with the above table, the charts of solving numerically and AGM have been depicted in the following form:

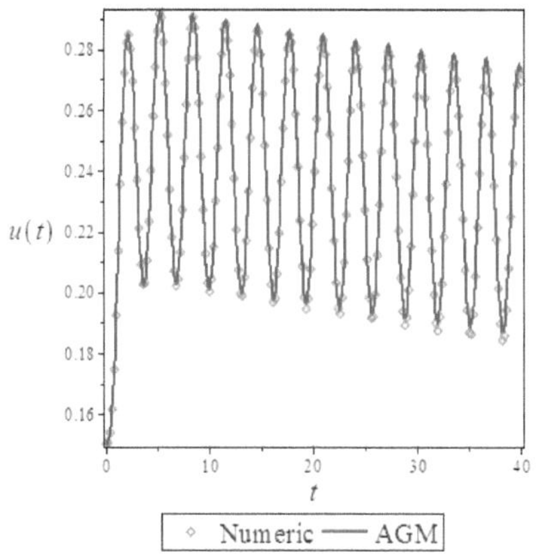

Fig. 4.25. Comparing the achieved solutions by AGM with numerical method

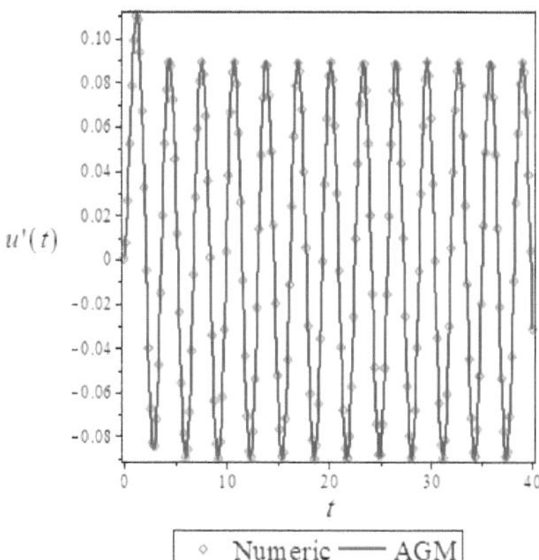

Fig. 4.26. A comparison between the first derivative of the obtained solutions by AGM and RKF45

And then for the phase plan, we will have

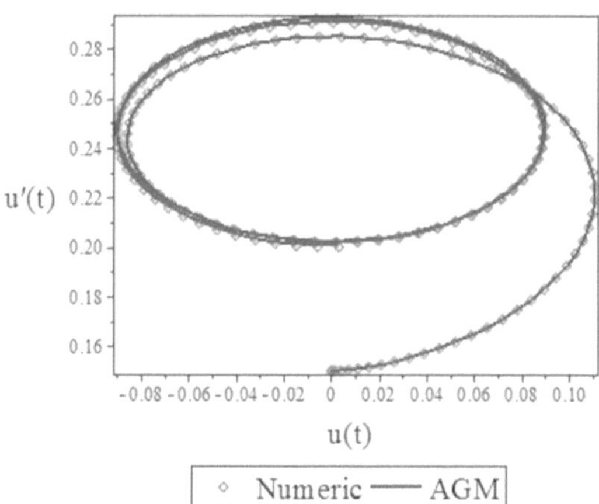

Fig. 4.27. The resulted phase plans by AGM and numerical method

Due to the aforementioned charts and solving by AGM and numerical method, choosing the answer function of equation 4.68 is a suitable approximation for solving the differential equation, and in order to answer the differential equation with high sensitivity and precision, we can increase the number of the series sentences.

Based on the explanations of example 4.2.5, the variations of vibrational parameters in terms of amplitude of vibration have been shown graphically as

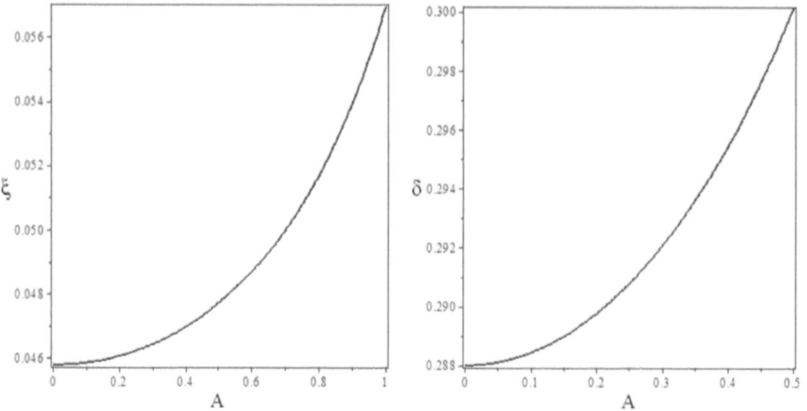

Fig. 4.28. The result of damping ratio variation in terms of amplitude of vibration

Fig. 4.29. Variation of logarithmic reduction ratio by vibrational amplitude

Moreover, the trend of both angular frequency and damping vibrational amplitude in terms of amplitude of vibration (A) has been obtained in the following forms:

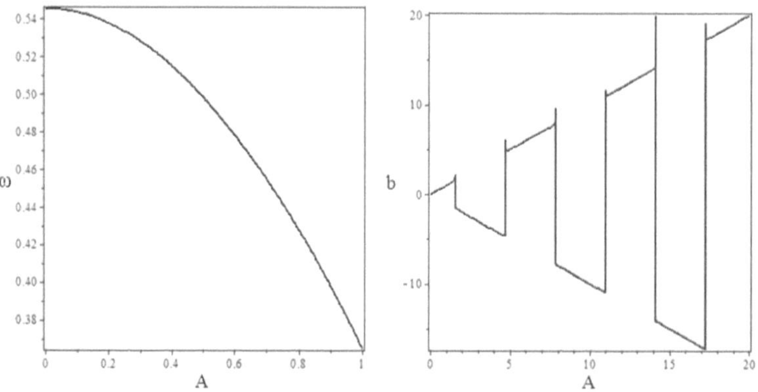

Fig. 4.30. Variation of angular frequency in terms of amplitude of vibration (A)

Fig. 4.31. The result of changing damping vibrational amplitude in terms of A

Example 4.1.7

Solve the following nonlinear differential equation by AGM.

$$f(t):\ (1+\mu u^{m})\frac{d^{2}u}{dt^{2}}+\alpha u^{n}\frac{du}{dt}+\gamma\frac{du}{dt}+\lambda\sin(u)=0 \tag{4.76}$$

And the initial conditions are expressed as follows:

$$u(0)=A \qquad , \qquad u'(0)=0. \tag{4.77}$$

4.1.7.1. Solution

In order to solve differential equation 4.76 by AGM, we do not need the Taylor expansion of $\sin(u)=u-\frac{u^{3}}{3}+\ldots$; thus, with respect to the previous examples, an answer function is considered to be shown by trigonometric and exponential series as follows:

$$u=e^{-at}\{b\cos(\omega t+\varphi)\}. \tag{4.78}$$

The constant values a , b , φ and ω are computed by applying the initial conditions.

4.1.7.2. Applying Initial Conditions by AGM

As it has been explained previously, the initial conditions can be applied in two ways: on the answer function of differential equation shown by $u(IC)$ and then on the differential equation and its derivatives as a general equation shown by $f(u(IC))$.

a. Applying the initial conditions on the answer function is done as

$$u(0)=A \quad\rightarrow\quad b\cos\varphi=A \tag{4.79}$$

$$u'(0) = 0 \quad \rightarrow \quad a\cos\varphi + \omega\sin\varphi = 0. \tag{4.80}$$

b. The initial conditions are applied on the differential equation and its derivatives after substituting equation 4.78 into differential equation 4.76 below:

$f(u(t=0))$:

$$(1 + mu(b\cos(\varphi))^m)(a^2 b\cos(\varphi) + 2ab\sin(\varphi)\omega - b\cos(\varphi)\omega^2)\alpha(b\cos(\varphi))^n(-ab\cos(\varphi) - b\sin(\varphi)\omega) + \gamma(-a\,b\cos(\varphi) - b\sin(\varphi)\,\omega) + \lambda\sin(b\cos(\varphi)) = 0 \tag{4.81}$$

$f'(u(t=0))$:

$$b[(a^2-\omega^2)\cos\varphi + 2a\omega\sin\varphi](\gamma + \alpha\, b^n\cos^n\varphi) - (a\cos\varphi + \omega\sin\varphi)[\lambda b\cos\{b\cos\varphi\} - n\alpha b^{n-1}(a\cos\varphi + \omega\sin\varphi)\cos^{n-1}\varphi] - \mu m b^{m+1}(a\cos\varphi + \omega\sin\varphi)[(a^2-\omega^2)\cos\varphi + 2a\omega\sin\varphi]\cos^{m-1}\varphi + b(1+\mu b^m\cos^m\varphi)(3a\omega^2\cos\varphi + \omega^3\sin\varphi - 3a^2\omega\sin\varphi - a^3\cos\varphi) = 0 \tag{4.82}$$

By solving the set of algebraic equations consisting of four equations with four unknowns from equation 4.79 to equation 4.82, the constant coefficients of the answer function can be achieved very conveniently.

For simplicity, consider the following new variable as

$$\psi_1 = 4\lambda(1+\mu A^m)\sin A$$
$$\varphi_2 = A(\gamma + \alpha A^n)^2 \quad . \tag{4.83}$$

Therefore, the constant coefficients of the answer function can be obtained as follows:

$$a = \frac{4\sqrt{A\psi_2}\sin A}{\psi_1}, \qquad b = A\sqrt{\frac{\psi_1}{\psi_1 - \psi_2}}$$
$$\omega = \frac{2\lambda}{\psi_1}\sqrt{A(\psi_1-\psi_2)}\sin A, \qquad \varphi = tg^{-1}\left(\frac{\sqrt{\psi_1-\psi_2}}{\psi_1}\right). \tag{4.84}$$

By substituting the above constant values into equation 4.78, the solution of nonlinear differential equation 4.76 is gained by AGM as

$$u(t) = A\sqrt{\frac{\psi_1}{\psi_1 - \psi_2}} e^{-\frac{4\sqrt{A\psi_2}\sin A}{\psi_1}t} \cos\{\frac{2\lambda}{\psi_1}\sqrt{A(\psi_1 - \psi_2)}\sin At + tg^{-1}(\frac{\sqrt{\psi_1 - \psi_2}}{\psi_1})\}. \quad (4.85)$$

By choosing physical values for the vibrational system such as

$$m = 3 \quad , \quad n = 4 \quad , \quad \mu = 0.2 \quad , \quad \alpha = 1.7 \quad , \quad \lambda = 0.3 \quad , \quad \gamma = 0.2 \quad , \quad A = 0.1, \quad (4.86)$$

we will have

$$\omega = -0.5379845352 \quad (4.87)$$

$$u = 0.1017150877e^{-0.10006498701t}\cos(0.5379845352t - 0.1838982552). \quad (4.88)$$

4.1.7.3. Numerical Solution of the Differential Equations and Comparing It with AGM

On the basis of the given physical values, differential equation 4.76 in the time domain $t \in \{0, 40\}$ has numerically been solved in the following table:

Table 4.7. Comparing the Achieved Solutions by AGM and Numerical Method

t	0.0	8	16	24	32	40
$u(t)$ NUM	0.1	-0.02541	-0.01119	0.00908	-0.00095	-0.00149
$u'(t)$ AGM	0.0	0.02298	-0.00814	-0.00176	0.00226	-0.00044

The charts obtained by numerical method and AGM have been depicted and compared as follows:

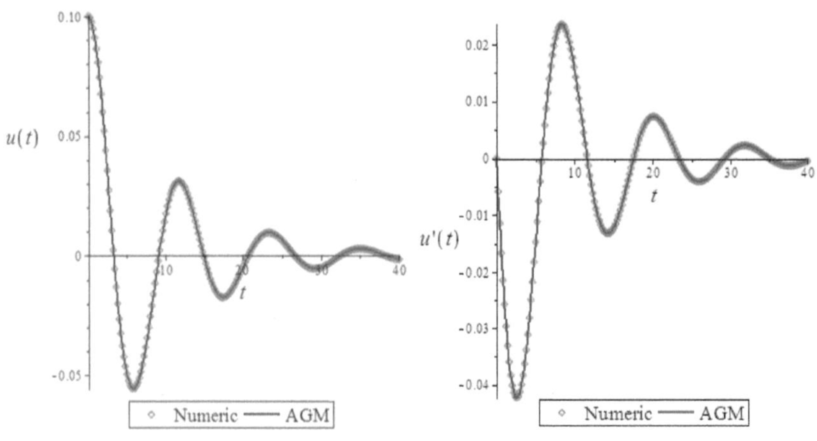

Fig. 4.32. The result of achieved solutions by AGM and numerical method

Fig. 4.33. The chart of first derivative of the gained solutions by AGM and numerical method

Then, the obtained phase plans have been compared graphically as

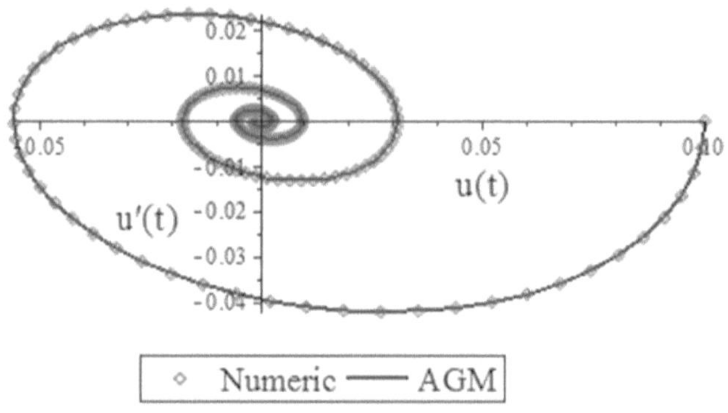

Fig. 4.34. A comparison between the acquired phase plans by AGM and RKF45

As regards table 4.7, the difference between gained solutions by AGM and numerical method can be presented as $\left| u_{num} - u_{AGM} \right|$ in terms of time below:

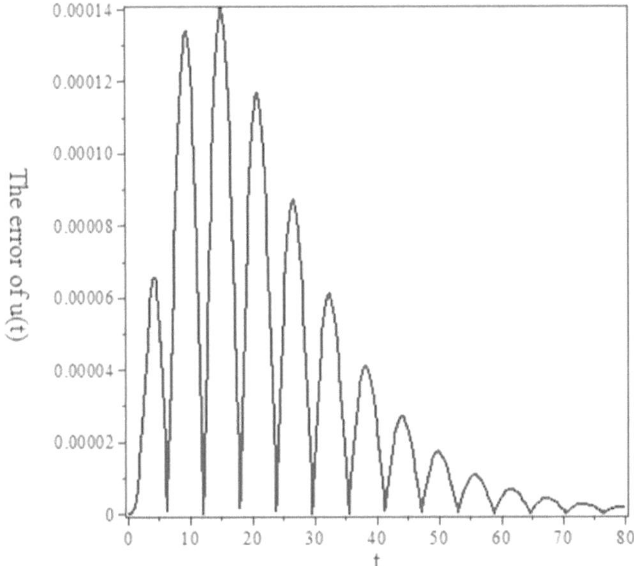

Fig. 4.35. The error chart of obtained solution by AGM in the given domain

Due to the aforementioned charts, it has been observed that the difference between obtained solutions by AGM and numerical method is very little. And also, we can conclude that the difference is tended toward zero for bigger times, or on the other hand, the system is converged in infinitive times.

Eventually, for the locus of maximum vibrational displacement, velocity, and acceleration points, we will have the following:

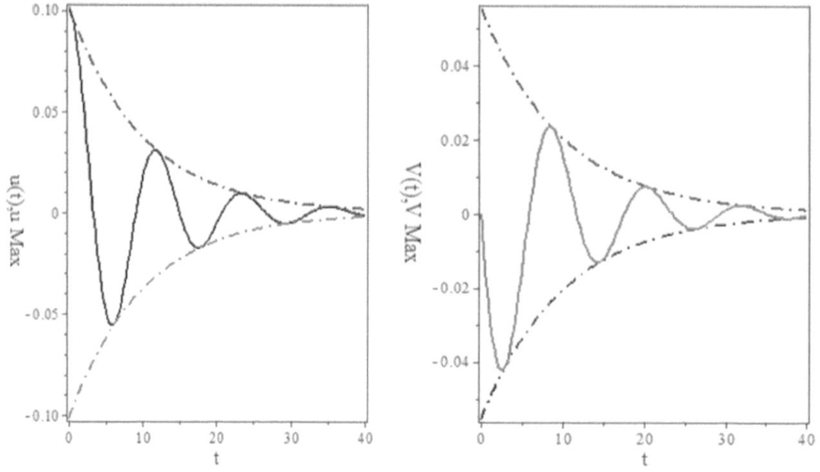

Fig. 4.36. The locus of maximum vibrational displacement points by AGM

Fig. 4.37. The locus of maximum vibrational velocity points by AGM

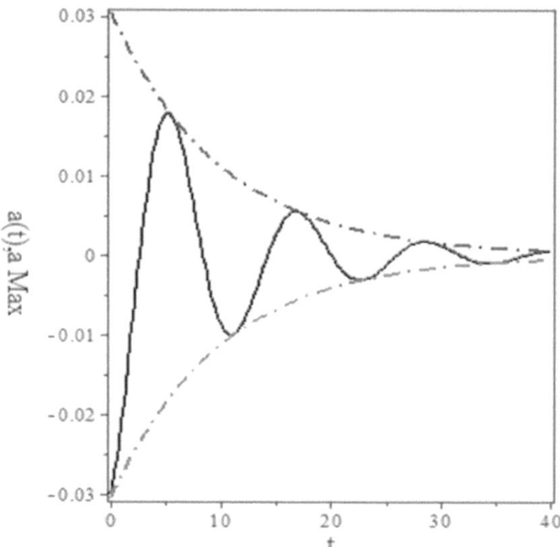

Fig. 4.38. The achieved locus of maximum vibrational acceleration points by AGM

Example 4.1.8

According to the following figure, a cable with negligible weight and a concentrated mass m in its center has been connected between two rigid materials with the initial tension T_0.

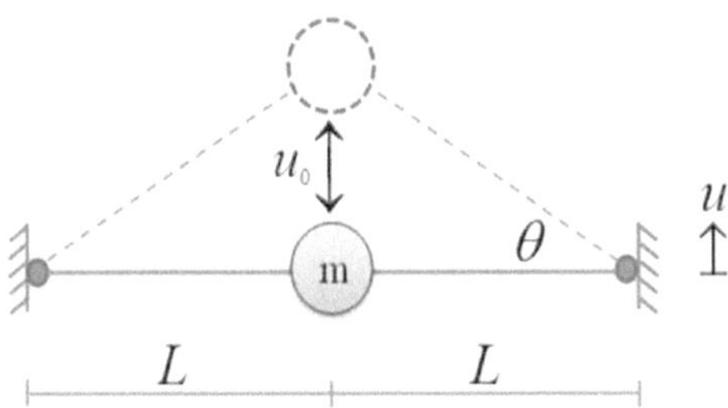

Fig. 4.39. Physical properties of the introduced system

Obtain the vibrational equation governing the system and solve it by AGM when the concentrated mass is deviated as u_0.

4.1.8.1. Simulation Procedure

The displacement of the cable can be achieved in the following form:

$$\delta = \sqrt{L^2 + u^2} - L. \tag{4.89}$$

Based on the Hooke's law $\delta = \varepsilon E$, we will have

$$\varepsilon = \frac{\delta}{L} \quad ; \quad \delta = \frac{T}{A} \quad \rightarrow \quad T = \frac{EA}{L}\delta. \tag{4.90}$$

Due to the balance of forces exerting on the concentrated mass m shown in the following figure, we can indicate that

$$\sum F = m\ddot{u} \quad \rightarrow \quad m\ddot{u} + 2TSin\theta = 0. \tag{4.91}$$

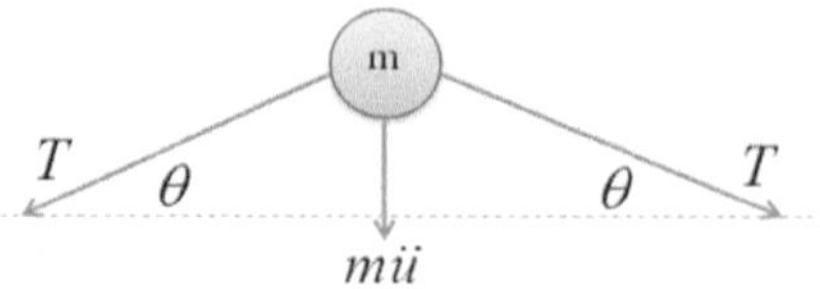

Fig. 4.40. The flowchart of forces exerting on the concentrated mass

The trigonometric relation between the angular θ, instantaneous displacement u, and the cable length L can be acquired as follows:

$$\sin\theta = \frac{u}{\sqrt{L^2 + u^2}}. \tag{4.92}$$

By substituting equation 4.89 into equation 4.90, and again after substituting equation 4.92 and equation 4.90 into the equilibrium equation 4.91, we will have

$$f(t):\quad m\ddot{u} + 2\{T_0 + \frac{EA}{L}[\sqrt{L^2 + u^2} - L]\}\frac{u}{\sqrt{L^2 + u^2}} = 0. \tag{4.93}$$

It is important to mention that differential equation 4.93 represents the nonlinear behavior of the mass m in the above system. Moreover, equation 4.93 can show the linear behavior of the system for small angles, whose linear form can be shown as follows:

$$u << L \qquad \rightarrow \qquad m\ddot{u} + 2T_0\frac{u}{L} = 0. \tag{4.94}$$

And the angular frequency of equation 4.94 can be obtained as follows:

$$\omega = \sqrt{\frac{2T_0}{mL}}. \tag{4.95}$$

It is citable that the answer of nonlinear differential equation 4.93, which is the real behavior of the system, can be achieved by AGM with regard to the initial conditions defined for the system in the following form:

$$u(0) = u_0 \quad , \quad \dot{u}(0) = 0. \tag{4.96}$$

4.1.8.2. Choosing Answer Function

Like in the previous problems, an answer function is chosen as a finite series in the following form:

$$u = e^{-at}\{b\cos(\omega t + \varphi)\} \tag{4.97}$$

Where a , b , ω and φ will be calculated in the next parts.

4.1.8.3. Applying Initial Conditions

Initial conditions are applied on the considered answer function and also on equation 4.93.

a. Apply initial conditions on the answer function:

$$u(0) = u_0 \quad \rightarrow \quad b\cos\varphi = u_0 \tag{4.98}$$

$$\dot{u}(0) = 0 \quad \rightarrow \quad a\cos\varphi + \omega\sin\varphi = 0 \tag{4.99}$$

b. The initial conditions are applied on equation 4.93 and also on its derivative shown by $f(u(IC))$ and $f'(u(IC))$ as follows:

$f(u(t=0))$:

$$b(a^2 - \omega^2)\cos\varphi + 2ab\omega\sin\varphi + \frac{2b}{L\sqrt{L^2 + b^2\cos^2\varphi}}(T_0 L - EAL + EA\sqrt{L^2 + b^2\cos^2\varphi}) = 0 \tag{4.100}$$

$f'(u(t=0))$:

$$mab(3\omega^2 - a^2)\cos\varphi + mb\omega(\omega^2 - 3a^2)\sin\varphi -$$
$$\frac{2EAb^2\cos\varphi}{L(L^2+\cos^2\varphi)}(a\cos\varphi + \omega\sin\varphi) - \frac{2bL}{(L^2+b^2\cos^2)^{3/2}}(T_0 L - EAL +$$
$$EA\sqrt{L^2+b^2\cos^2\varphi})(a\cos\varphi + \omega\sin\varphi) = 0 \quad (4.101)$$

By solving the set of algebraic equations consisting of four equations with four unknowns from equation 4.98 to equation 4.101, the constant coefficients of the answer function can be obtained as

$$a = 0 \quad , \quad b = u_0 \quad , \quad \varphi = 0. \quad (4.102)$$

And the angular frequency can be achieved as follows:

$$\omega = \sqrt{-\frac{-2AEL^2 + 2AE\sqrt{L^2+u_0^{\,2}}L - 2AEuo^2 - 2\sqrt{L^2+u_0^{\,2}}T_0.L}{L^3m + Lmu_0^{\,2}}} \quad . \quad (4.103)$$

By substituting the above values into the answer function, the solution of nonlinear differential equation 4.93 can be acquired by AGM as

$$u(\mathrm{t}) = u_0\cos(\sqrt{\frac{2AEL^2 - 2AE\sqrt{L^2+u_0^{\,2}}L + 2AEu_0^{\,2} + 2\sqrt{L^2+u_0^{\,2}}T_0.L}{L^3m + Lmu_0^{\,2}}}\,t) \quad . \quad (4.104)$$

It is observed that when the system is linear, the angular frequency is obtained according to equation 4.95, which is different from equation 4.103. This means the angular frequency in the nonlinear state is a function of physical properties, including the elasticity module E and the cross-section A, whereas it is not so in the linear manner.

4.1.8.4. Solving the Differential Equation by Numerical Method and Comparing It with AGM

By selecting the physical values for the system, we will have

$$u_0 = 0.2\ m \quad , \quad m = 4\ Kg \quad , \quad E = 60000\ N/m^2$$
$$A = 0.002\ m^2 \quad , \quad L = 1\ m \quad , \quad T_0 = 25\ ^{o}C \quad . \quad (4.105)$$

Therefore, the differential equation 4.93 in the domain $t \in (0\ ,\ 10)$ has been solved numerically in the following table:

Table 4.8. The Results of Achieved Solution by Numerical Method

t	*0.0*	*2*	*4*	*6*	*8*	*10*
$u(t)$	0.2	0.1105	-0.07586	-0.19605	-0.14098	0.038397
$\dot{u}(t)$	0.0	-0.60353	-0.66806	-0.1447	0.51508	0.707233

Here, the obtained solutions have been compared graphically in the following forms:

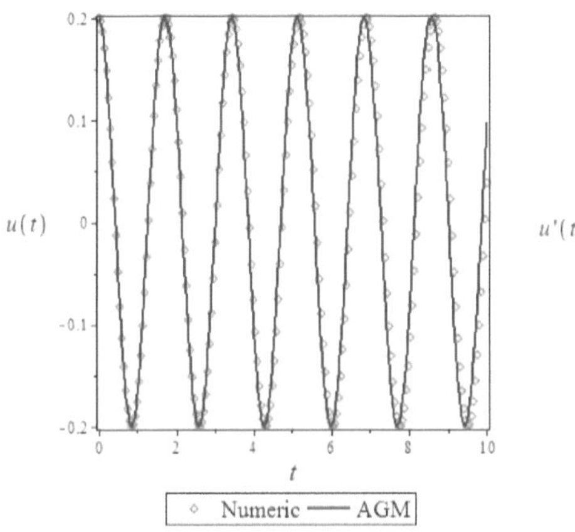

Fig. 4.41. A comparison between the achieved solutions by AGM and numerical method

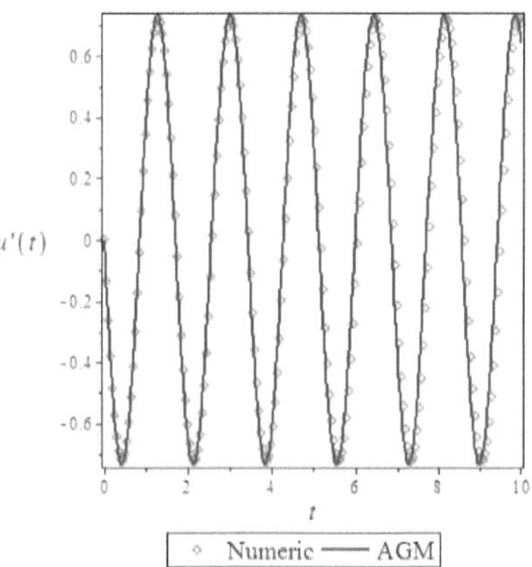

Fig. 4.42. The result of first derivative of the acquired solutions by AGM and numerical method

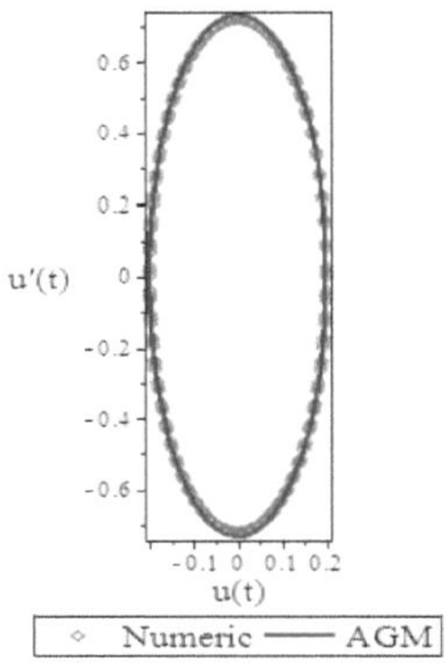

Fig. 4.43. Comparing the resulted phase plans by AGM and RKF45

Finally, the existing error in the achieved solution by AGM has been depicted in the introduced domain as

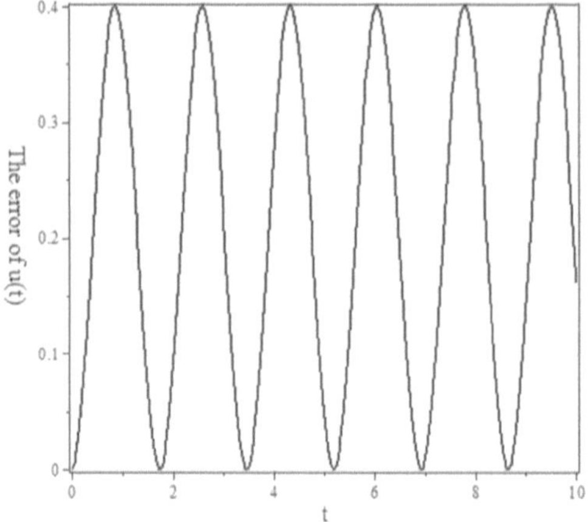

Fig. 4.44. The error chart of the achieved solution by AGM

Example 4.1.9

A pendulum consisting of two masses (m_1 and m_2) has been connected to the end of a horizontal bar and circulates around the point O, according to the following figure. By supposing that the weight of adjoined bars is negligible, obtain the vibrational differential equation governing the introduced system and solve it by AGM.

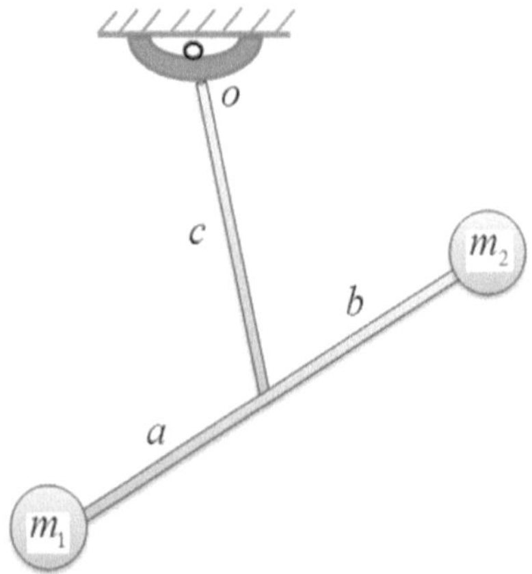

Fig. 4.45. The geometry of the proposed system

Moreover, achieve the value C in the terms of other vibrational parameters of the system when the angular frequency is maximum.

4.1.9.1. Simulation Procedure

Consider the forces diagram as follows:

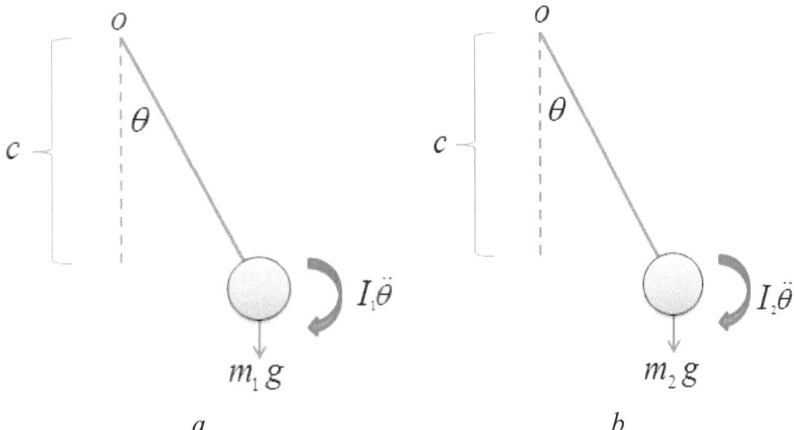

Fig. 4.46. The flow diagrams of the proposed system *a* and *b* are referred to the left and right masses, respectively

By writing the equilibrium equation of existing forces, we will have

$$(\sum I_0)\ddot{\theta}+\sum M_0=0 \;\Rightarrow\; \{m_1(a^2+c^2)+m_2(b^2+c^2)\}\ddot{\theta}+m_1gc\sin\theta+m_2gc\sin\theta=0. \quad (4.106)$$

Then, the above equation can be rewritten in the following form:

$$\ddot{\theta}+\frac{c(m_1+m_2)g}{m_1(a^2+c^2)+m_2(b^2+c^2)}\sin\theta=0 \quad . \quad (4.107)$$

And with the deviation of the system (θ_0), the initial conditions of the system can be achieved as follows:

$$\theta(0)=\theta_0 \qquad , \qquad \dot{\theta}(0)=0. \quad (4.108)$$

4.1.9.2. Solving the Nonlinear Differential Equation by AGM

An answer function is selected in AGM as follows:

$$\theta = e^{-At}\{B\cos(\omega t+\varphi)\}. \tag{4.109}$$

The constant coefficients A , B , ω and φ can be computed by applying the initial conditions.

4.1.9.3. Applying Initial Conditions

Like the previous examples, initial conditions are applied on the answer function and also on differential equation 4.107 in the following forms.

a. Applying the initial conditions on the answer function is done as

$$\theta(t=0)=0 \qquad \rightarrow \qquad B\cos\varphi = \theta_0 \tag{4.110}$$

$$\dot{\theta}(t=0)=0 \qquad \rightarrow \qquad A\cos\varphi + \omega\sin\varphi = 0 \tag{4.111}$$

b. The initial conditions are applied on differential equation 4.107 and also on its derivative shown by $f(\theta(t=0))$ and $f'(\theta(t=0))$ in the following form:

$f(\theta(t=0))$:

$$B(A^2-\omega^2)\cos\varphi + 2AB\omega\sin\varphi + \frac{gc(m_1+m_2)}{m_1(a^2+c^2)+m_2(b^2+c^2)}\sin\{\mathrm{B}\cos\varphi\}=0 \tag{4.112}$$

$f'(\theta(t=0))$:

$$AB(3\omega^2-A^2)\cos\varphi + B\omega(\omega^2-3A^2)\sin\varphi - \frac{bgc(m_1+m_2)(a\cos\varphi+\omega\sin\varphi)}{m_1(a^2+c^2)+m_2(b^2+c^2)}\cos\{B\cos\varphi\}=0 \tag{4.113}$$

By solving the set of algebraic equations consisting of four equations with four unknowns from equation 4.110 to equation 4.113, the constant coefficients of the answer function can be obtained as

$$A = 0 \quad , \quad B = \theta_0 \quad , \quad \varphi = 0. \tag{4.114}$$

And the angular frequency can be acquired as follows:

$$\omega = \sqrt{\frac{gc(m_1 + m_2)\sin(\theta_0)}{[m_1(a^2 + c^2) + m_2(b^2 + c^2)]\theta_0}}. \tag{4.115}$$

When the acquired angular frequency from equation 4.115 becomes maximum, the length of the bar will be computed as follows:

$$\frac{dw}{dc} = 0 \quad \rightarrow \quad c = \frac{\sqrt{(m_1 + m_2)(m_1 a^2 + m_2 b^2)}}{m_1 + m_2}. \tag{4.116}$$

By substituting equation 4.114 and equation 4.115 into equation 4.109, the solution of nonlinear differential equation 4.107 can be acquired by AGM as

$$\theta = \theta_0 \cos\{\sqrt{\frac{gc(m_1 + m_2)\sin(\theta_0)}{[m_1(a^2 + c^2) + m_2(b^2 + c^2)]\theta_0}}\, t\}. \tag{4.117}$$

4.1.9.4. Solving the Differential Equation by Numerical Method and Comparing It with AGM

By selecting the following physical values for the system

$$\theta_0 = 0.3\ Rad \quad , \quad a = 1\ m \quad , \quad b = 1.5\ m \quad , \quad c = 1.2\ m$$
$$m_1 = 4\ Kg \quad , \quad m_2 = 2\ Kg \quad , \quad g = 9.81\ {m}/{s^2} \tag{4.118}$$

Equation 4.107 in the domain $t \in (0\ ,\ 30)$ can be solved numerically in the following table:

Table.4.9. The Results of Numerical Solution

t	0.0	6	12	18	24	30
$\theta(t)$	0.3	0.26959	0.18441	0.06153	-0.073971	-0.194295
$\dot{\theta}(t)$	0.0	0.26534	0.47789	0.59373	0.587354	0.461557

The obtained charts have been compared as follows:

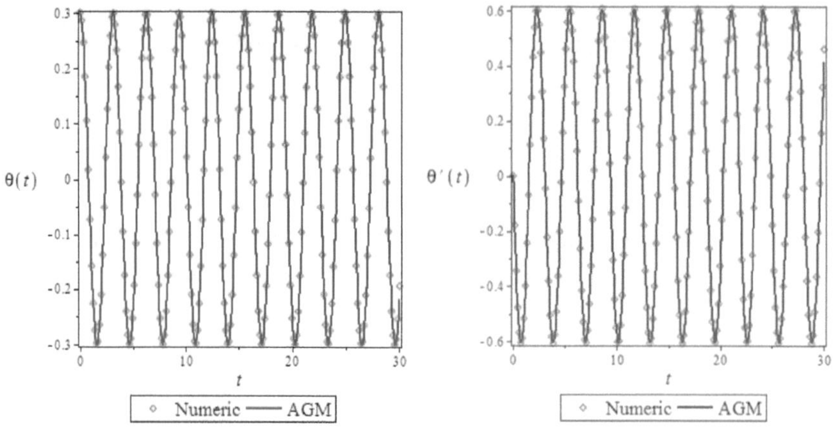

Fig. 4.47. A comparison between the achieved solutions by AGM and numerical method

Fig. 4.48. The result of first derivative of the acquired solutions by AGM and numerical method

After that, the obtained phase plans have been compared graphically in the following figure:

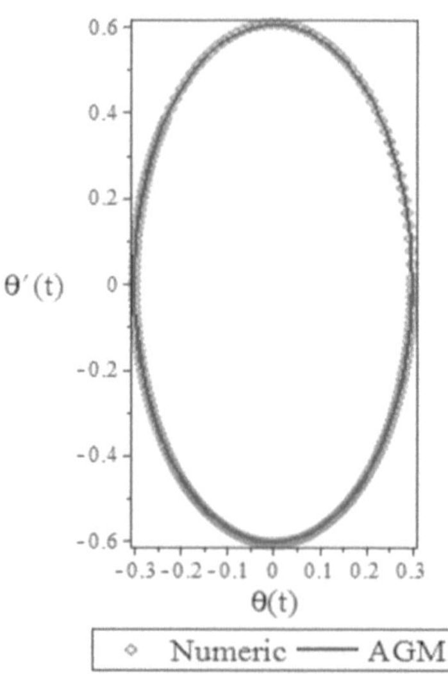

Fig. 4.49. The resulted phase plans by AGM and numerical method

Eventually, on the basis of $\left|\theta(t)_{AGM} - \theta(t)_{Numeric}\right|$, the existing error in the achieved solution by AGM has been depicted in the following form:

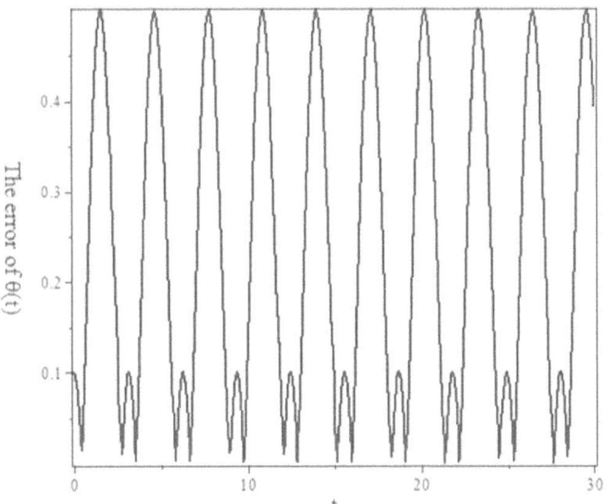

Fig. 4.50. The errors of AGM

Example 4.1.10

Obtain the differential equation governing the vibration of a rectangular plate with the mass m and dimensions $a \times b$ in accordance with the following figure, and then solve it by AGM.

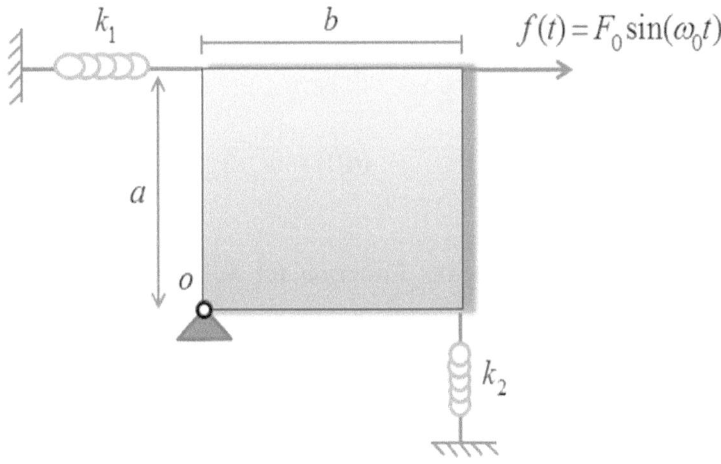

Fig. 4.51. The geometry of the introduced system

4.1.10.1. Simulation Procedure

By applying moment around the point o, the following equation can be written as

$$\sum \vec{M}_0 = 0 \ :$$

$$k_1 a(a\sin\theta) + k_2 b(b\sin\theta) + \mathrm{I}_0\,\ddot{\theta} - a.f(t) = 0. \qquad (4.119)$$

By substituting the inertia moment around the point 0 as $I_0 = \frac{1}{3}(\mathrm{a}^2 + b^2)$ in equation 4.119 and after simplification, we will have

$$\ddot{\theta} + \beta\sin\theta = \eta F_0 \sin(\omega_0 t). \qquad (4.120)$$

Equation 4.120 can be rewritten by utilizing the Taylor series as follows:

$$f(t):\ \ \ddot{\theta} + \beta(\theta - \frac{\theta^3}{3!}) = \eta.F_0.\sin(\omega_0 t). \qquad (4.121)$$

Where the parameters of differential equation 4.121 can be defined as

$$\beta = \frac{3(\mathrm{k}_1 a^2 + k_2 b^2)}{m(a^2 + b^2)} \quad , \quad \eta = \frac{3a}{m(a^2 + b^2)} \qquad (4.122)$$

Then for the initial conditions, we will have

$$\theta(0) = A \quad , \quad \dot{\theta}(0) = 0. \qquad (4.123)$$

4.1.10.2. Choosing an Answer Function for Solving the Differential Equation by AGM

Write the answer function by using a finite series as follows:

$$\theta = b\cos(\omega t + \varphi) + d\cos(\omega_0 t + \phi) \qquad (4.124)$$

Applying the initial conditions

$$\theta(0) = A \qquad \rightarrow \qquad b\cos\varphi + d\cos\varphi = A \tag{4.125}$$

$$\dot{\theta}(0) = 0 \qquad \rightarrow \qquad b\omega\sin\varphi + d\,\omega_0\sin\phi = 0 \tag{4.126}$$

$f(\theta(0))$:

$$-b\omega^2\cos\varphi - d\omega_0^2\cos\phi + \beta\{b\cos\varphi + d\cos\phi - \frac{1}{6}(b\cos\varphi + d\cos\phi)^3\} = 0 \tag{4.127}$$

$f'(\theta(0))$:

$$b\omega^3\sin\varphi + d\omega_0^3\sin\phi + \beta(\omega b\sin\varphi + d\omega_0\sin\phi)\{1 - \frac{1}{2}(d\cos\varphi + d\cos\phi)^2\} = \eta F_0\omega_2 \tag{4.128}$$

$f''(\theta(0))$:

$$b\omega^4\cos\varphi + d\omega_0^4\cos\phi + \beta[\mathrm{b}\,\omega^2\cos\varphi + d\omega_0^2\cos\phi + (b\cos\varphi + d\cos\phi)(b\omega\sin\varphi + d\omega_0\sin\phi) + \frac{1}{2}(b\cos\varphi + d\cos\phi)^2(b\omega^2\cos\varphi + d\omega_0^2\cos\phi)] = 0 \tag{4.129}$$

By solving the set of algebraic equations consisting of five equations with five unknowns from equation 4.125 to equation 4.129, the constant coefficients of the answer function can be obtained very easily.

And for simplicity, the following new variables are introduced as follows:

$$\begin{aligned}
\psi_1 &= \beta(6 - A^2)(\beta A^2 - 2\beta + 2\omega_0^2)\\
\psi_2 &= -\beta(6 - A^2)(\beta A^2 - 6\beta + 6\omega_0^2)(\beta A^2 - 2\beta + 2\omega_0^2)\\
\psi_3 &= \beta^3 A^8 + 8A^6\beta^2\omega_0^2 - 14\beta^2 A^6 + 60\beta^3 A^4 - 72\omega_0^2\beta^2 A^4 + 12A^4\beta\omega_0^4 - 72\beta^3 A^2 -\\
&\quad 72\beta\omega_0^4 A^2 + 144\omega_0^2\beta^2 A^2 - 72\eta^2 F_0^2\omega_0^2\\
\psi_4 &= 1296\beta^2\eta^2 F_0^2 + 432\beta A^2\eta^2 F_0^2\omega_0^2 + 4\beta^4 A^{10} + 1296\omega_0^4\eta^2 F_0^2 - 48\beta^4 A^8 +\\
&\quad 36\beta^2 A^4\eta^2 F_0^2 + 144\beta^4 A^6 - 432\beta^2 A^2\eta^2 F_0^2 - 2592\omega_0^2\beta\eta^2 F_0^2\\
\Delta &= 4\beta\omega_0^2 A^2 - 8\beta^2 A^2 - 24\beta\omega_0^2 + \beta^2 A^4 + 12\beta^2 + 12\omega_0^4\\
\Lambda_1 &= (\beta A^2 + 6\omega_0^2 - 6\beta)^2\\
\Lambda_2 &= (\beta A^2 + 6\omega_0^2 - 6\beta)
\end{aligned} \tag{4.130}$$

Therefore, the coefficients of the answer function can be obtained as follows:

$$b=\frac{\Lambda_1}{3\Delta}\sqrt{\frac{\psi_3}{\psi_2}} \quad , \quad \omega=\sqrt{\frac{\psi_1}{2\Lambda_2}} \quad , \quad \varphi=tg^{-1}(A(\sqrt{\frac{\psi_2}{\psi_3}})$$

$$d=\frac{\sqrt{\psi_4}}{3\Delta} \quad , \quad \phi=tg^{-1}(\frac{6F_0\eta\Lambda_2}{\sqrt{\psi_4}}) \quad . \qquad (4.131)$$

By substituting the above values into the considered answer function, the solution of nonlinear differential equations 4.121 can be acquired by AGM below:

$$\theta(t)=\frac{\Lambda_1}{3\Delta}\sqrt{\frac{\psi_3}{\psi_2}}\cos\{\sqrt{\frac{\psi_1}{2\Lambda_2}}t+tg^{-1}(A(\sqrt{\frac{\psi_2}{\psi_3}}))\}+\frac{\sqrt{\psi_4}}{3\Delta}\cos\{\omega_0 t+tg^{-1}(\frac{6F_0\eta\Lambda_2}{\sqrt{\psi_4}})\}. \qquad (4.132)$$

4.1.10.3. Solving the Differential Equation by Numerical Method and Comparing It with AGM

By selecting the physical values for the system, we will have

$$A=0.1\ Rad \quad , \quad m=25\ Kg \quad , \quad a=25\ cm \quad , \quad b=30\ cm$$

$$k_1=4\ N/m \quad , \quad k_2=5\ N/m \quad , \quad \omega_0=2\ \frac{Rad}{Sec} \quad , \quad F_0=1.5\,N. \qquad (4.133)$$

In this step, differential equation 4.121 in the domain $t \in (0\ ,\ 30)$ has been solved numerically in the following table:

Table 4.10. The Result of Achieved Solution by Numerical Method

t	0.0	6	12	18	24	30
$\theta(t)$	0.1	-0.20254	0.113706	0.314227	-0.099852	-0.110469
$\dot{\theta}(t)$	0.0	-0.120095	-0.255828	0.098578	0.250463	0.0065023

The obtained charts have been compared in the following figures:

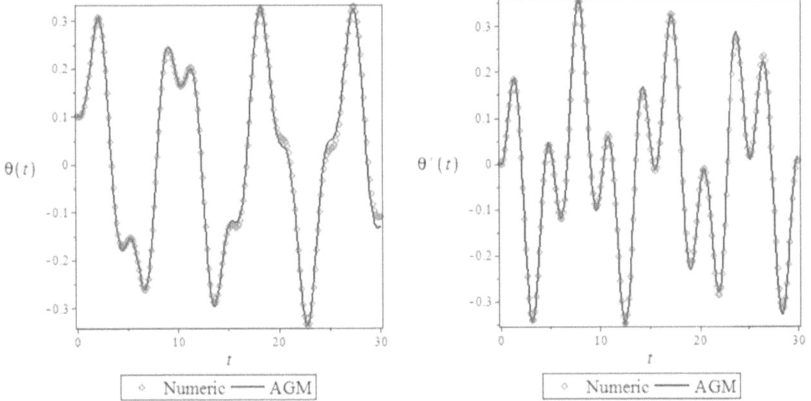

Fig. 4.52. Comparing the achieved solutions by AGM and numerical method

Fig. 4.53. The first derivative of the obtained solutions by AGM and numerical method

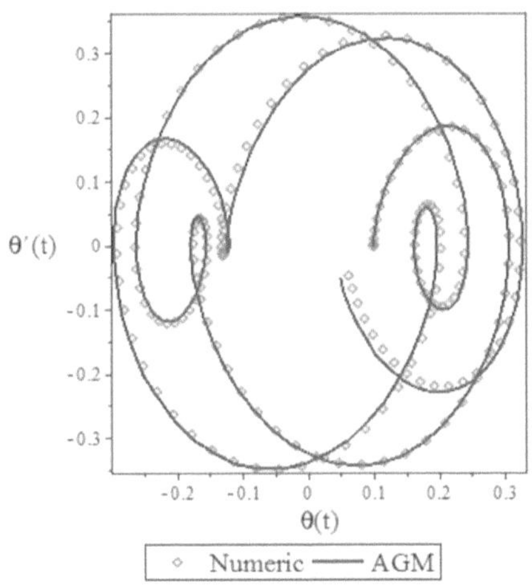

Fig. 4.54. The resulted phase plans by AGM and numerical method

Example 4.1.11

A fin of a helicopter that circulates with the angular velocity Ω has been shown in the following figure:

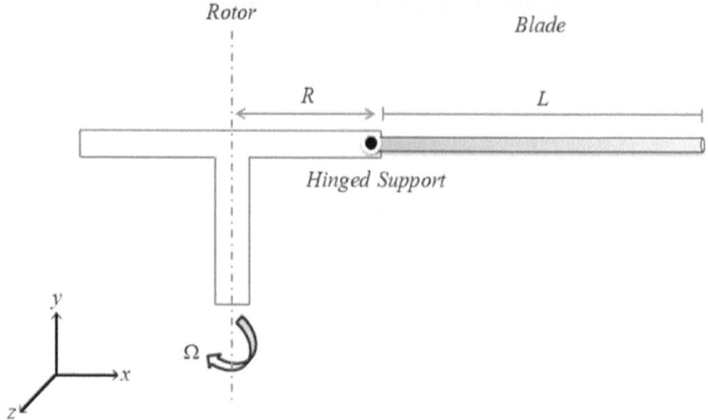

Fig. 4.55. The schematic diagram of the physical model

Determine its natural frequency for the flapping motion, and after obtaining the differential equation governing the proposed system, solve it by AGM. It is necessary to mention that the congruent fin has been attached to the rotor as a hinged end and also has been made of materials that we can neglect its weight.

4.1.11.1. Simulation Procedure

To start up, consider the coordinate system *xyz* to the joint between the rotor and blade. Then we are able to depict the flow diagram as follows:

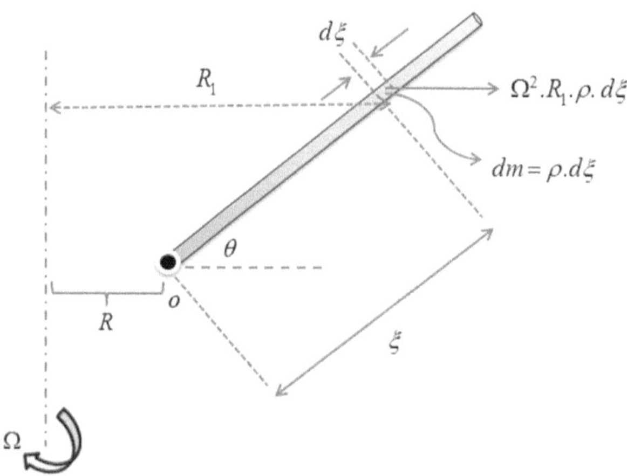

Fig. 4.56. The flowchart of the introduced geometry

On the basis of the dynamic laws, the absolute acceleration of the considered element dm has been shown as

$$a_p = a_0 + \vec{\omega} \times (\vec{\omega} \times \overrightarrow{\rho_{op}}) + \dot{\vec{\omega}} \times \overrightarrow{\rho_{op}}. \tag{4.134}$$

Where in equation 4.134, we will have

$$\vec{\omega} = \Omega j + \dot{\theta} k \quad , \quad \dot{\vec{\omega}} = \ddot{\theta} k + \dot{\theta}\frac{dk}{dt} = \ddot{\theta} k + \dot{\theta}(\Omega\, \mathrm{j} \times \mathrm{k}) = \ddot{\theta} k + \dot{\theta}\Omega i$$
$$a_0 = -R\Omega^2 i \quad , \quad \overrightarrow{\rho_{op}} = (\xi \cos\theta) i + (\xi \sin\theta) j \quad . \tag{4.135}$$

After substituting equation 4.135 into equation 4.134, the result will be

$$a_p = -i(R\Omega^2 + \xi\Omega^2\cos\theta + \xi\dot{\theta}^2\cos\theta - \xi\ddot{\theta}\sin\theta) + j(\xi\dot{\theta}^2\sin\theta + \xi\ddot{\theta}\cos\theta) + k(2\xi\dot{\theta}\Omega\sin\theta) \;. \tag{4.136}$$

Exerting force on the considered element can be obtained as

$$F = \rho . d\xi . a_{p}. \tag{4.137}$$

Now, the moment of the aforementioned force around the hinged joint is computed as follows:

$$M = \int_0^L \vec{\xi} \times F = \int_0^L \vec{\xi} \times \rho . d\xi . a_p = \int_0^L [(\xi^2\ddot{\theta} + R\Omega^2\xi\sin\theta + \xi^2\Omega^2\sin\theta\cos\theta)\rho d\xi\, \vec{k} - (2\xi^2\dot{\theta}\Omega\sin\theta\cos\theta)\rho d\xi\, \vec{j} + (2\xi^2\dot{\theta}\Omega\sin^2\theta)\rho d\xi\, \vec{i}]. \tag{4.138}$$

Since the moment of external forces around the axis z is zero, we can write

$$\int_0^L (\xi^2\ddot{\theta} + R\Omega^2\xi\sin\theta + \xi^2\Omega^2\sin\theta\cos\theta)\rho d\xi = 0 \tag{4.139}$$

$$\int_0^L 2\xi^2\dot{\theta}\Omega\sin\theta\cos\theta)\rho d\xi = M_y \tag{4.140}$$

$$\int_0^L 2\xi^2\dot{\theta}\Omega\sin^2\theta)\rho d\xi = M_{x}. \tag{4.141}$$

The result of equation 4.139 is presented as

$$\frac{1}{3}\rho L^3\ddot{\theta}+\frac{1}{2}\rho R\Omega^2 L^2\sin\theta+\frac{1}{3}\rho a^2 L^3\sin\theta\cos\theta=0. \tag{4.142}$$

Inasmuch as the mass of the rotor is achieved as $\rho L = m$, the simplified form of equation 4.142 is introduced as follows:

$$f(t):\qquad 2\ddot{\theta}+3R\Omega^2\sin(\theta)+L\Omega^2\sin(2\theta)=0. \tag{4.143}$$

Equation 4.143 is nonlinear and can be solved with initial conditions by AGM in the following form:

$$\theta(0)=\theta_0 \qquad , \qquad \dot{\theta}(0)=0. \tag{4.144}$$

It is notable that equation 4.143 has become linear for small θ and can be obtained as

$(\sin\theta\approx\theta)$:

$$2L\ddot{\theta}+(3R\Omega^2+2L\Omega^2)\theta=0. \tag{4.145}$$

And the natural frequency from the linear differential equation 4.145 can be achieved as follows:

$$\omega=\Omega\sqrt{\frac{3R+2L}{2L}}. \tag{4.146}$$

4.1.11.2. Solving the Nonlinear Differential Equation by AGM

Equation 4.143, with the initial conditions from equation 4.144, can be solved by choosing an answer function in the following form:

$$\theta=e^{-at}\{b\cos(\omega t+\varphi). \tag{4.147}$$

4.1.11.3. Applying the Initial Conditions by AGM

$$\theta(0)=\theta_0 \qquad \rightarrow \qquad b\cos\varphi=\theta_0 \tag{4.148}$$

$$\dot{\theta}(0)=0 \qquad \rightarrow \qquad a\cos\varphi-\omega\sin\varphi=0 \tag{4.149}$$

$f(\theta(t=0))$:

$$2a^2b\cos(\varphi)-4ab\sin(\varphi)\omega-2b\cos(\varphi)\omega^2+3R\Omega^2\sin(b\cos(\varphi))+ L\Omega^2\sin(2b\cos(\varphi))=0 \tag{4.150}$$

$$f'(\theta(t=0)):\quad 2a^3bcos(\varphi)-6a^2bsin(\varphi)\omega-6abcos(\varphi)\omega^2+2bsin(\varphi)\omega^3+ 3R\Omega^2cos(bcos(\varphi))(abcos(\varphi)-bsin(\varphi)\omega)+ L\Omega^2cos(2bcos(\varphi))(2abcos(\varphi)-2bsin(\varphi)\omega)=0 \tag{4.151}$$

By solving the set of algebraic equations consisting of four equations with four unknowns from equation 4.148 to equation 4.151, the constant coefficients of the answer function can conveniently be obtained in the following form:

$$a=0 \quad , \quad b=\theta_0 \quad , \quad \varphi=0. \tag{4.152}$$

And the angular frequency can be acquired as follows:

$$\omega=\frac{\sqrt{2}\Omega}{2\theta_0}\sqrt{\theta_0(3\mathrm{R}+\mathrm{L}\sin\theta_0)\sin\theta_0}. \tag{4.153}$$

By substituting the above values into the answer function [equation 4.147], the solution of the nonlinear differential equations 4.143 can be acquired by AGM as follows:

$$\theta(t)=\theta_0\cos\{\frac{\sqrt{2}\Omega}{2\theta_0}\sqrt{\theta_0(3\mathrm{R}+\mathrm{L}\sin\theta_0)\sin\theta_0}\,t\}. \tag{4.154}$$

4.1.11.4. Solving the Differential Equation by Numerical Method and Comparing It with AGM

The physical values of the proposed system are considered as

$$\theta_0 = 0.3\ Rad \quad , \quad L = 4\ m \quad , \quad R = 1\ m \quad , \quad \Omega = 3\ \frac{Rad}{\sec}$$

$$\rho = 1100\ kg/m^3 \quad . \quad (4.155)$$

Then, the differential equation 4.143 in the domain $t \in (0\ ,\ 10)$ has been solved numerically in the following table:

Table 4.11. The Result of Obtained Solution by Numerical Method

t	*0.0*	*2*	*4*	*6*	*8*	*10*
$\theta(t)$	0.3	0.093988	-0.24226	0.244627	0.090106	0.29997
$\dot{\theta}(t)$	0.0	-1.9783	-1.220514	1.19765	1.987184	0.0277228

As a result, the obtained charts have been compared as follows:

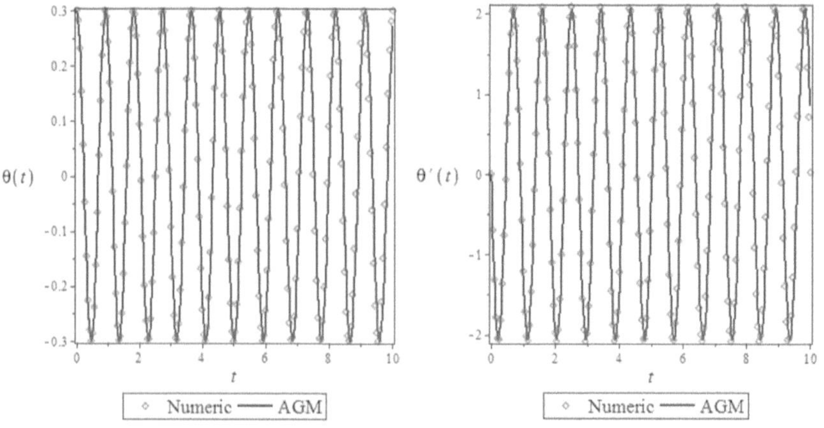

Fig. 4.57. Comparing the achieved solutions by AGM and numerical method

Fig. 4.58. The first derivative of the obtained solutions by AGM and numerical method

Afterward, the obtained phase plans have been compared graphically in the following form:

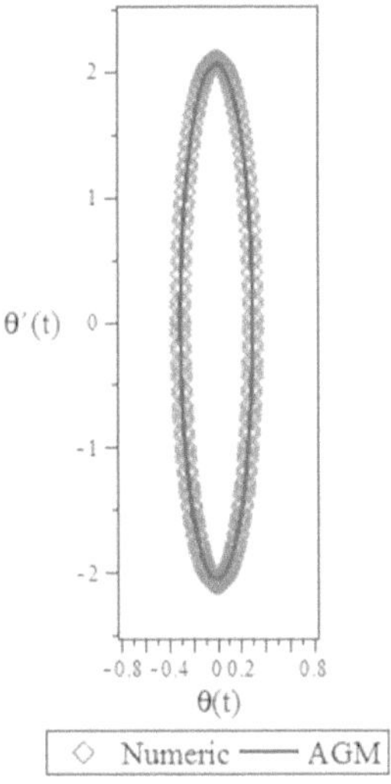

Fig. 4.59. The resulted phase plans by AGM and numerical method

It is noteworthy that after computing $\theta(t)$ by AGM and then utilizing equation 4.140 and equation 4.141, the moment of the forces *M(x), M(y)* can be obtained easily. After substituting and choosing the density value for the considered blade, $\rho = 1100\,kg/m^3$, the moment of the forces M (x) , M (y) can be computed as follows:

$$M_x = -2.901274862\times 10^5 sin(6.868548440t)\times sin(0.3000000000\, cos(6.868548440t))^2 \quad (4.156)$$

$$M_y = -2.901274862\times 10^5 sin(6.868548440t)\times sin(0.3cos(6.868548440t))\times cos(0.3cos(6.868548440t)). \quad (4.157)$$

Eventually, the charts of acquired moments of forces have been shown in the following figures:

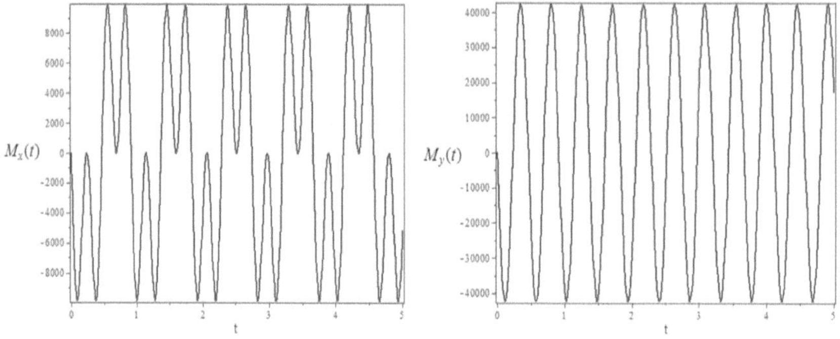

Fig. 4.60. The resulted chart for $M_x(t)$

Fig. 4.61. The resulted chart for $M_y(t)$

4.2. Problems

4.2.1. Consider the following differential equation and solve it by AGM. Then compare the obtained solution by numerical method.

$$f(t):\ \frac{d^2x}{dt^2}+\alpha\frac{dx}{dt}+\beta.x^{\frac{1}{m}}+\lambda.x^{\frac{1}{n}}+\mu.x^{p}=0$$

The related initial conditions are $x(0)=A$, $x'(0)=0$, and the physical parameters are presented as

$$A=0.2\ ,\ \alpha=0.05\ ,\ \beta=0.02\ ,\ \lambda=0.02$$
$$n=3\ ,\ m=5\ ,\ p=4\ ,\ \mu=0.3\ .$$

After depicting the related phase plans by AGM and numerical method, obtain the maximum error existing in the AGM solution.

4.2.2. A congruent disk with the radius R and mass M has been hung from the point A with the distance r from the center of the disk in a vertical plane.

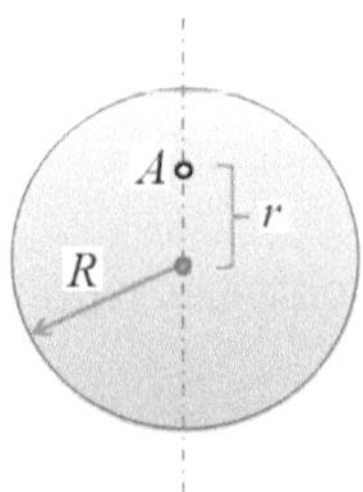

The geometry of the proposed system

Obtain the differential equation governing the vibrational system and solve it by AGM. Then, acquire the amount of r while the frequency of the system is maximum.

HELP: Utilize the equilibrium equation $I_A\ddot{\theta}=\sum M_A$ when the mass inertia moment of the disk can be achieved by $I_A=\frac{1}{2}mR^2+mr^2$.

4.2.3. The Van der Pol's equation is an applied differential equation for systems in the dynamic engineering of vibration, which is mostly utilized in the mechanics of solids, electricity, and the other fields related to the dynamic of vibration. Therefore, solve the following Van der Pol's nonlinear vibrational differential equation by AGM parametrically for positive and negative epsilons.

$$f(t):\ \ddot{x} \mp \varepsilon(1-x^2)\dot{x} + \beta^2 x = 0$$

And the related initial conditions are expressed as $x(0) = A$, $\dot{x}(0) = 0$, where ε and β are the physical constant values for the vibrational systems. In addition, what is the difference between the acquired solutions by positive and negative epsilons ($\varepsilon = \pm 0.1$, $\beta = 1.2$ and $A = 0.05$)?

4.2.4. Solve the following Rayleigh's nonlinear differential equation by AGM parametrically.

$$\ddot{x} \pm (\alpha - \beta\dot{x}^2)\dot{x} + \lambda^2 x = 0$$

α, β and λ are the constant values for the vibrational system in the differential equation, and the related initial conditions are defined as $x(0) = A$, $\dot{x}(0) = 0$. Depict the locus of maximum displacement, velocity, and acceleration points by considering $A = 0.1$, $\alpha = 0.02$, $\beta = 0.1$, $\lambda = 1.2$.

4.2.5. Obtain the solution of nonlinear vibrational differential equation by AGM as follows:

$$f(t):\ \frac{d^2x}{dt^2} + \alpha[x^2\frac{d^2x}{dt^2} + x(\frac{dx}{dt})^2] + \beta x\cos(x) + \lambda(\frac{dx}{dt}) = 0$$

And the related initial conditions are expressed as $x(0) = A$, $\dot{x}(0) = 0$, where $\alpha = 0.2$, $\beta = 0.3$, $\lambda = 0.1$ and $A = 0.1$. Then, evaluate the variation of damping ratio (ξ) and angular frequency (ω) in terms of the initial amplitude of vibration.

4.2.6. Consider the following nonlinear vibrational differential equation.

$$f(t): \ \ddot{x}+c\dot{x}+k_1x+g(x)=0 \quad ; \quad x(0)=A \ , \ x'(0)=0$$

Where the function $g(x)$ is defined as $g(x)=k_2x^2+k_3x^3$. Solve the above equation in the time domain $t\in\{0,\ 80\}$ by AGM when the physical values for the system are defined as $A=0.1$, $c=0.08$, $k=0.1$, $k_2=0.02$, $k_3=0.01$. Moreover, depict the logarithmic reduction factor (δ) and damping initial phase (ϕ) in terms of the initial vibrational amplitude.

4.2.7. Three homogeneous bars with the mass m (for each one) and the length $2L$ have been adjoined from the end and have made the triangle ABC in the following form:

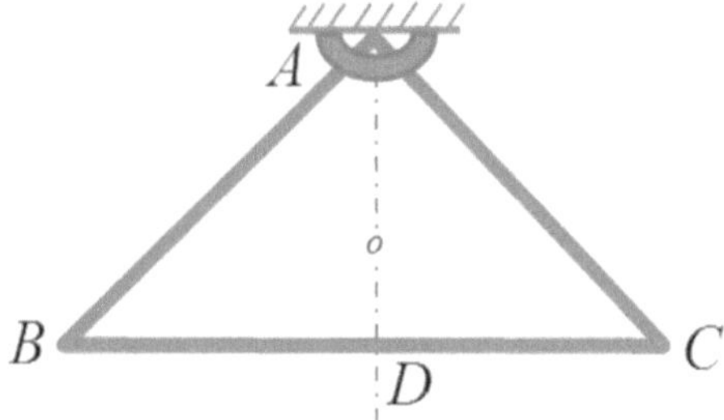

The schematic diagram of the physical model

Achieve the vibrational differential equation and angular frequency by AGM when the triangle can freely circulate around an axis that is vertical on its plane.

Chapter 5

Nonlinear Integrodifferential Equations

5.1. A Short Description of Integrodifferential Equations

Mathematical modeling of real-life phenomena in most of the situations results in applicable equations (take for example ordinary or partial differential equations, integral stochastic equations, and then integrodifferential equations). Many mathematical formulations of physical phenomena contain integrodifferential equations, and these equations arise in a wide variety of fields not only in fluid dynamics but also in biological models and chemical kinetics and so on. Solving integrodifferential equations are usually considered as a difficult task, and most of the researchers have problems while confronting these kinds of equations, especially highly nonlinear ones.

Generally speaking, inasmuch as the derived nonlinear equations from the real-life phenomena in the universe are still difficult to be solved either numerically or theoretically, much attention has been devoted to find the exact solutions,[1, 2, 3] and attempts have been made to search for better and more efficient methods for determining a reliable solution for nonlinear differential equations such as the nonlinear integrodifferential equations.[4] On the basis of the above explanations, many researchers utilized various methods in order to solve nonlinear integrodifferential equations such as Wang and He, who applied VIM to solve integrodifferential equations.[5] Sweilam used VIM to solve both linear and nonlinear boundary value problems for the fourth-order integrodifferential equations.[6]

In recent years, much attention has been devoted to the newly developed manners to gain appropriate solutions for nonlinear equations such as the differential transformation method. This method rapidly gives convergent, successive approximations to the numerical solution and is also applicable for complicated nonlinear differential equations like the research done by A. R. Ahmadi et al.[7] The others are homotopy perturbation method,[8, 9] Adomian decomposition method,[10, 11] perturbation method,[12] EXP-function method.[13–16] To comprehend more, it is better to indicate that a minority of scientific problems have precise and reliable analytic solutions, and most of the aforementioned methods do not have this ability, so these nonlinear equations, particularly complicated nonlinear integrodifferential equations, should be solved by using other methods such as the presented technique in this book.

Eventually, in this chapter, attempts have been made to solve nonlinear integrodifferential equations by using a creative operation. As a matter of fact, two finite series with constant coefficients should be considered as the answer of these categories of nonlinear differential equations; afterward, the obtained equation can be solved by a simple and innovative approach.

5.2. Miscellaneous Examples

Example 5.2.1

The integrodifferential equation, with its boundary conditions, is considered as follows:

$$\frac{d^4u(x)}{dx^4} = 1 + e^x - e^{2x} + 2\int_0^x (u(t)\frac{d^2u(t)}{dt^2})dt. \tag{5.1}$$

And the related boundary conditions are expressed as

$$u(0) = 1 \quad , \quad u'(0) = 1 \quad , \quad u(1) = e \quad , \quad u'(1) = e. \tag{5.2}$$

5.2.1.1. Solving the Differential Equation by AGM

In order to solve integrodifferential equations like the presented problem here, two different answer functions should be considered—one for the integral part, and the other for the rest of the equation in the following form:

$$u(t) = \sum_{i=0}^{5} a_i t^i \quad \rightarrow \quad \text{for the integral part} \tag{5.3}$$

$$u(x) = \sum_{i=0}^{5} a_i x^i \quad \rightarrow \quad \text{for the rest of the equation} \tag{5.4}$$

And the constant coefficients a_0 to a_5 can easily be computed by applying boundary conditions.

The other steps of solution procedure are similar to the previous chapters.

5.2.1.2. Applying Boundary Conditions by AGM

In AGM, the boundary conditions are applied in two ways—on the answer function shown by $u(BC)$ and then on the differential equation shown by $f(u(BC))$ as follows:

a. Applying the boundary conditions on the answer functions of equation 5.1 is done as follows:

$$
\begin{aligned}
u(0) &= 1 &\rightarrow\quad & a_0 = 1\\
u'(0) &= 1 &\rightarrow\quad & a_1 = 1\\
u(1) &= e &\rightarrow\quad & a_5 + a_4 + a_3 + a_2 + a_1 + a_0 = e\\
u'(1) &= e &\rightarrow\quad & a_1 + 2a_2 + 3a_3 + 4a_4 + 5a_5 = e
\end{aligned}
\tag{5.5}
$$

b. And applying the boundary conditions on equation 5.1 after substituting equation 5.3 and equation 5.4 into it, shown by $f(u(\mathrm{x}))$, is done as follows:

$$f(u(x=0)) \quad \rightarrow \quad 24a_4 = 1 \tag{5.6}$$

$$
\begin{aligned}
f(u(\mathrm{x}=1)) \quad \rightarrow \quad & 120a_5 + 24a_4 = 1 + e^1 - e^2 + (\frac{40}{9})a_5^2 + 8a_4a_5 + \frac{52}{7}a_3a_5 +\\
& \frac{24}{7}a_4^2 + \frac{22}{3}a_2a_5 + 6a_3a_4 + 8a_1a_5 + \frac{28}{5}a_2a_4 + \frac{12}{5}a_3^2 + 10a_0a_5 +\\
& 6a_1a_4 + 4a_2a_3 + 8a_0a_4 + 4a_1a_3 + \frac{4}{3}a_2^2 + 6a_0a_3 + 2a_1a_2 + 4a_0a_2
\end{aligned}
\tag{5.7}
$$

By solving the set of algebraic equations composed of six equations with six unknowns from equation 5.5 to equation 5.7, the constant coefficients of the answer functions are computed as

$$
\begin{aligned}
& a_0 = 1 \quad , \quad a_1 = 1 \quad , \quad a_2 = 0.5068 \quad , \quad a_3 = 0.1553\\
& a_4 = 0.0416 \quad , \quad a_5 = 0.0143 \quad .
\end{aligned}
\tag{5.8}
$$

After that, by substituting the constant values of equation 5.8 into equation 5.4, the solution of integrodifferential equation 5.1 is achieved by AGM as follows:

$$u(x) = 0.0143331x^5 + 0.04166666667x^4 + 0.1553854x^3 + 0.5068966x^2 + x + 1. \tag{5.9}$$

5.2.1.3. Numerical Solution of the Differential Equation and Its Comparison with AGM

Equation 5.1, with regard to the boundary conditions in the domain $x \in \{0, 1\}$, can numerically be solved and compared with AGM in table 5.1 as follows:

Table 5.1. The Obtained Numerical Values for $u(x)$ in the Specified Domain

x	*0*	*0.2*	*0.4*	*0.6*	*0.8*	*1*
$u(x)$ *NUM*	1	1.219	1.487	1.817	2.223	2.718
$u(x)$ AGM	1	1.221	1.492	1.822	2.225	2.718

In this step, the obtained charts by numerical method and AGM have been depicted and compared in figure 5.1 and figure 5.2 as follows:

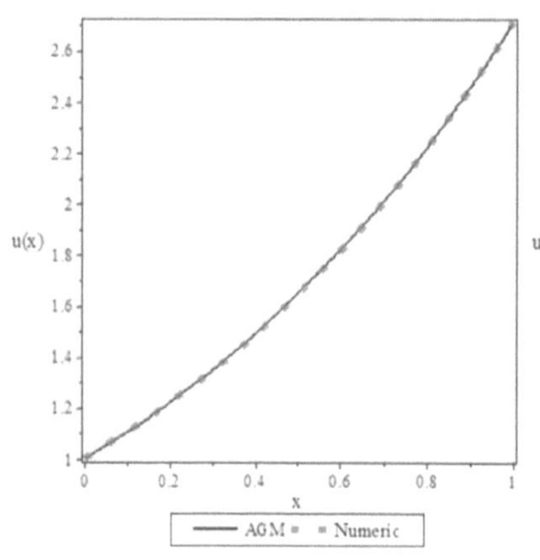

Fig. 5.1. Comparing the obtained solutions, $u(x)$, by AGM and numerical method

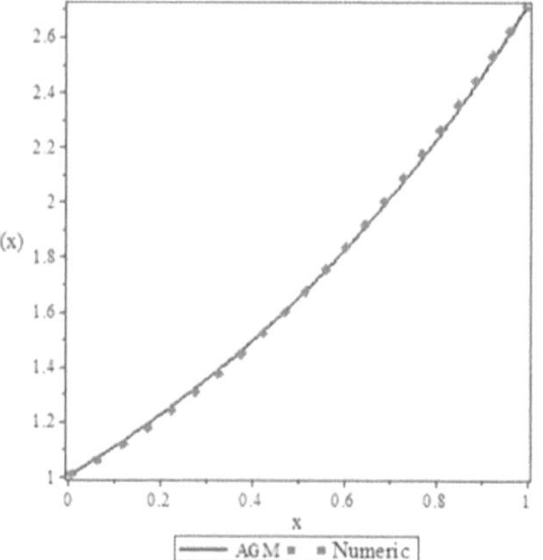

Fig. 5.2. Comparing the first derivative of the obtained solutions, (du / dx), by AGM and numerical method

Example 5.2.2

Consider the following nonlinear integrodifferential equation as

$$\frac{d^2u(x)}{dx^2} = (u(x))^2 + \int_0^x \{ u^2(t) - (\frac{du(t)}{dt})^2 + \frac{du(t)}{dt} \}dt. \tag{5.10}$$

And the related boundary conditions are defined as

$$u(0) = 1 \quad , \quad u(1) = 2. \tag{5.11}$$

5.2.2.1. Solving the Differential Equation by AGM

On the basis of the given explanation in the previous example, two different answer functions are considered in the following form:

$$u(t) = \sum_{i=0}^{7} a_i t^i \quad \rightarrow \quad \text{for the integral part} \tag{5.12}$$

$$u(x) = \sum_{i=0}^{7} a_i x^i \quad \rightarrow \quad \text{for the rest of the equation} \tag{5.13}$$

After applying boundary conditions, the constant coefficients a_0 to a_7 can easily be computed similar to the previous example.

5.2.2.2. Applying Boundary Conditions by AGM

The boundary conditions are applied in two ways.

a. Applying the boundary conditions on the answer function of equation 5.10 is done as follows:

$$u(0) = 1 \quad \rightarrow \quad a_0 = 1$$
$$u(1) = 2 \quad \rightarrow \quad a_7 + a_6 + a_5 + a_4 + a_3 + a_2 + a_1 + a_0 = 2. \tag{5.14}$$

b. Applying the boundary conditions on equation 5.10 is done after substituting equation 5.12 and equation 5.13 into it in the following form:

$$f(u(x=0)) \quad \rightarrow \quad 2a_2 = (a_0)^2 \tag{5.15}$$

$$f(u(\mathrm{x}=1)) \quad \rightarrow \quad ....^{*} \tag{5.16}$$

Also, applying the boundary conditions on the first derivative of the differential equation shown by $f'(u(BC))$ is done as follows:

$$f'(u(x=0)) \quad \rightarrow \quad 6a_3 = a_0^{\ 2} + 2a_0a_1 - a_1^{\ 2} + a_1 \tag{5.17}$$

$$f'(u(x=1)) \quad \rightarrow \quad ....^{*} \tag{5.18}$$

Moreover, for the second derivative of $f(u(BC))$, we will have

$$f''(u(x=0)) \quad \rightarrow \quad 24a_4 = 2a_0a_1 + 4a_0a_2 + 2a_1^{\ 2} - 4a_1a_2 + 2a_2 \tag{5.19}$$

$$f''(u(x=1)) \quad \rightarrow \quad ....^{*} \tag{5.20}$$

Finally, by solving the set of algebraic equations consisting of eight equations with eight unknowns from equation 5.14 to equation 5.20, the constant coefficients of the answer function are computed as

$$a_0 = 1 \quad , \quad a_1 = 0.0998 \quad , \quad a_2 = 0.5 \quad , \quad a_3 = 0.2149 \quad , \quad a_4 = 0.1258$$
$$a_5 = 0.0586 \quad , \quad a_6 = -0.0252 \quad , \quad a_7 = 0.0259 \quad . \tag{5.21}$$

By substituting the constant values of equation 5.21 into equation 5.13, the solution of integrodifferential equation is achieved by AGM as follows:

$$u(x) = 1 + 0.0998x + 0.5x^2 + 0.2149x^3 + 0.1258x^4 + 0.0586x^5 - 0.0252x^6 + 0.0259x^7. \tag{5.22}$$

5.2.2.3. Numerical Solution of the Differential Equation and Its Comparison with AGM

In regard to the boundary conditions in the domain $x \in \{0,\ 1\}$, equation 5.10 can be solved numerically and compared with AGM in the following table:

---

* For simplification, the mathematical calculations of the marked items have been omitted, and for more details, please see the solution of integrodifferential equation in the appendices.

Table 5.2. The Obtained Numerical Values for $u(x)$ in the Specified Domain

x	*0*	*0.2*	*0.4*	*0.6*	*0.8*	*1*
$u(x)$ *NUM*	1	1.0358	1.1239	1.2839	1.5521	2
$u(x)$ AGM	1	1.0419	1.1374	1.3067	1.5795	2

In this part, the obtained charts by numerical method and AGM have been compared graphically in figure 5.3 as follows:

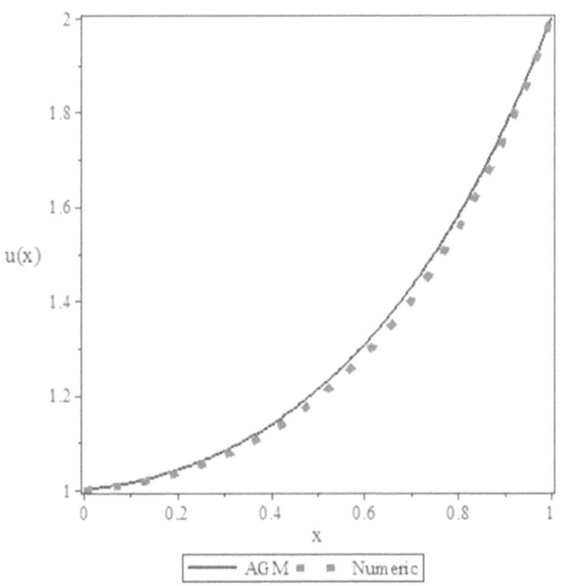

Fig. 5.3. Comparing the obtained solutions, $u(x)$, by AGM and numerical method

Example 5.2.3

The second-order nonlinear differential equation, with the following boundary conditions, is considered as

$$\frac{d^2u(x)}{dx^2} - u(x)\frac{du(x)}{dx} = (u(x))^2 + \int_0^x \{ u^2(t) - (\frac{du(t)}{dt})^2 + \frac{du(t)}{dt} \} dt \tag{5.23}$$

$$BC: \quad u(0) = 0 \quad , \quad u(1) = 1. \tag{5.24}$$

5.2.3.1. Solving the Differential Equation by AGM

Based on the given explanations, the following two answer functions are considered as follows:

$$u(t) = \sum_{i=0}^{7} a_i t^i \quad \rightarrow \quad \text{for the integral part} \tag{5.25}$$

$$u(x) = \sum_{i=0}^{7} a_i x^i \quad \rightarrow \quad \text{for the rest of the equation} \tag{5.26}$$

It is notable that the constant coefficients (a_0 to a_7) will easily be obtained by applying boundary conditions.

5.2.3.2. Applying Boundary Conditions in AGM

Like the previous examples, boundary conditions are applied on equation 5.23 and also on the obtained equation from substituting equation 5.25 and equation 5.26 into equation 5.23 as

$$u(0) = 0 \quad \rightarrow \quad a_0 = 0 \tag{5.27}$$

$$u(1) = 1 \quad \rightarrow \quad a_7 + a_6 + a_5 + a_4 + a_3 + a_2 + a_1 + a_0 = 1 \tag{5.28}$$

then

$$f(u(x=0)) \quad \rightarrow \quad -a_0 a_1 + 2a_2 = a_0^{\,2} \tag{5.29}$$

$$f(u(x=1)) \quad \rightarrow \quad ....^{*} \tag{5.30}$$

Then, boundary conditions are applied on the derivatives of the achieved equation in the following form:

---

* For simplification, the mathematical calculations of the marked items have been omitted, and for more details, please see the solution of integrodifferential equation in the appendices.

For the first derivative, we will have

$$f'(u(x=0)) \quad \rightarrow \quad -2a_0a_2 - a_1^2 + 6a_3 = a_0^2 + 2a_0a_1 - a_1^2 + a_1 \tag{5.31}$$

and

$$f'(u(x=1)) \quad \rightarrow \quad ....^* \tag{5.32}$$

After that, the application of boundary conditions on the second derivative of the obtained equation is done as follows:

$$f''(u(x=0)) \quad \rightarrow \quad -6a_0a_3 - 6a_1a_2 + 24a_4 = 2a_0a_1 + 4a_0a_2 + 2a_1^2 - 4a_1a_2 + 2a_2. \tag{5.33}$$

Afterward,

$$f''(u(x=1)) \quad \rightarrow \quad ....^* \tag{5.34}$$

By solving a set of algebraic equations consisting of eight equations with eight unknowns from equation 5.27 to equation 5.34, the constant coefficients of equation 5.26 are gained in the form of

$$a_0 = 0 \;,\; a_1 = 0.76425 \;,\; a_2 = 0 \;,\; a_3 = 0.12737 \;,\; a_4 = 0.04867,$$
$$a_5 = 0.12139 \;,\; a_6 = -0.13965 \;,\; a_7 = 0.07795 \;. \tag{5.35}$$

After substituting the constant coefficients of equation 5.35 into equation 5.26, the solution of nonlinear differential equation 5.23 is acquired as

$$u(x) = 0.7642x + 0.1273x^3 + 0.0486x^4 + 0.1213x^5 - 0.1396x^6 + 0.0779x^7. \tag{5.36}$$

---

* For simplification, the mathematical calculations of the marked items have been omitted, and for more details, please see the solution of integrodifferential equation in the appendices.

Then the charts of the obtained solution and its derivative are depicted in figure 5.4 and figure 5.5 as

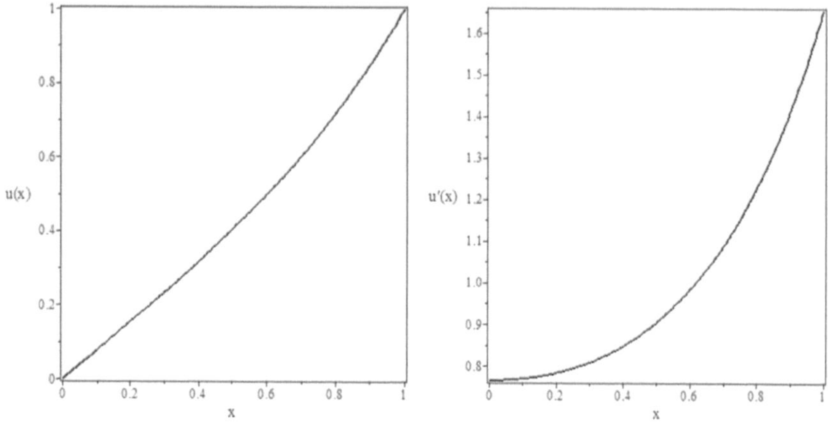

Fig. 5.4. The result of the obtained solutions by AGM

Fig. 5.5. The first derivative of the achieved solution by AGM

5.2.3.3 Numerical Solution

In table 5.3, equation 5.23 is solved numerically (RKF45) in regard to the boundary conditions in the given domain of $x \in \{0,1\}$ as follows:

Table 5.3. The Obtained Values for $u(x)$ and du / dx on the Basis of Numerical Solution of Equation 5.23

x	*0*	*0.2*	*0.4*	*0.6*	*0.8*	*1*
$u(x)$	0	0.1548	0.3173	0.4989	0.7172	1
$u'(x)$	0.7686	0.7859	0.8485	0.9812	1.2246	1.6368

5.2.3.4. Comparing the Obtained Solutions by AGM and Numerical Method

On the basis of the obtained results in table 5.3 and the acquired solution by AGM in equation 5.36, we will have the following comparisons in figure 5.6 and figure 5.7 as

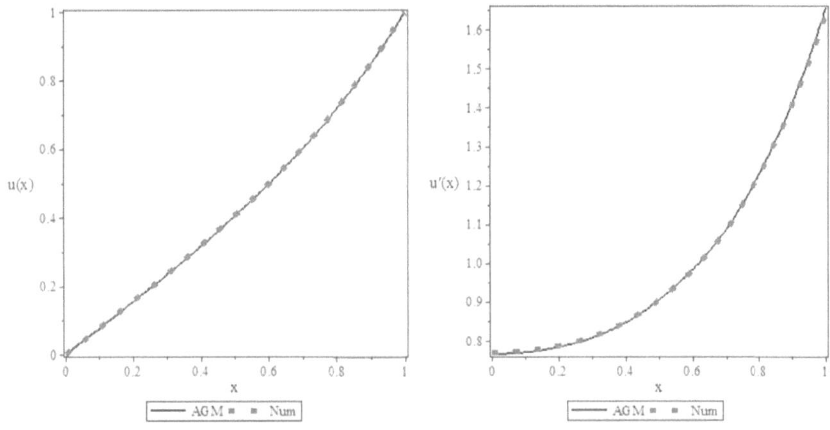

Fig. 5.6. The result of obtained solutions by AGM and numerical method

Fig. 5.7. The first derivative of the achieved solution by AGM and numerical method

Example 5.2.4

Consider the following nonlinear differential equation with the boundary conditions as

$$\frac{\frac{d^2u(x)}{dx^2}}{[1+(\frac{du(x)}{dx})^2]^{\frac{3}{2}}} = \frac{w.x}{2EI_o}\int_0^x [\,1+(\frac{du(t)}{dt})^2\,]^{0.5}\,dt \tag{5.37}$$

$$u(0) = 0 \quad , \quad u'(0) = 0. \tag{5.38}$$

5.2.4.1. Solving the Differential Equation by AGM

Due to the given explanations, two finite series with constant coefficients are considered in order to solve the second-order nonlinear integrodifferential equation in the following form:

$$u(t) = \sum_{i=0}^{5} a_i\, t^i \tag{5.39}$$

$$u(x) = \sum_{i=0}^{5} a_i \, x^i \tag{5.40}$$

It is citable that the constant coefficients (a_0 to a_5) will easily be computed by applying boundary conditions.

5.2.4.2. Applying Boundary Conditions in AGM

Based on the given explanations, boundary conditions are applied on equation 5.40 and on the obtained equation from substituting the considered answer functions into equation 5.37 as follows:

$$u(0) = 0 \quad \rightarrow \quad a_0 = 0 \tag{5.41}$$

$$u'(0) = 0 \quad \rightarrow \quad a_1 = 0. \tag{5.42}$$

Then, boundary conditions are applied on the achieved equation and also on its derivatives as the following:

Applying boundary conditions on the yielded equation, we have

$$f(u(x=0)) \quad \rightarrow \quad \frac{2a_2}{(a_1^2+1)^{\frac{3}{2}}} = 0. \tag{5.43}$$

The application of boundary conditions on the derivatives of the obtained equation is done as follows:

$$f'(u(x=0)) \quad \rightarrow \quad \frac{6a_3}{(a_1^2+1)^{\frac{3}{2}}} - \frac{12a_2^2 a_1}{(a_1^2+1)^{\frac{5}{2}}} = 0. \tag{5.44}$$

Afterward,

$$f''(u(x=0)) \quad \rightarrow \quad \frac{24a_4}{(a_1^2+1)^{\frac{3}{2}}} - \frac{108a_1a_2a_3}{(a_1^2+1)^{\frac{5}{2}}} + \frac{120a_2^3a_1^2}{(a_1^2+1)^{\frac{7}{2}}} - \frac{24a_2^3}{(a_1^2+1)^{\frac{5}{2}}} = \frac{w(a_1^2+1)^{0.5}}{(EI_o)}. \tag{5.45}$$

And finally, we have

$$f'''(u(x=0)) \qquad \rightarrow \qquad \ldots\ldots.^{*} \tag{5.46}$$

By solving a set of algebraic equations consisting of six equations with six unknowns from equation 5.41 to equation 5.46, the constant coefficients of equation 5.40 are gained as

$$a_0 = 0 \quad , \quad a_1 = 0 \quad , \quad a_2 = 0 \quad , \quad a_3 = 0 \quad , \quad a_4 = \frac{0.04166\mathrm{w}}{\mathrm{EI}_o} \quad , \quad a_5 = 0. \tag{5.47}$$

After substituting the constant coefficients of equation 5.47 into equation 5.40, the solution of the nonlinear integrodifferential equation 5.37 is acquired as

$$u(x) = \frac{0.04166w}{EI_o} x^4. \tag{5.48}$$

By considering the physical values below

$$w = -20 \quad , \quad E = 2000000 \quad , \quad I_o = 0.00005 \tag{5.49}$$

* For simplification, the mathematical calculations of the marked items have been omitted, and for more details, please see the solution of integrodifferential equation in the appendices.

The charts of the obtained solution and its derivative are illustrated in figure 5.8 and figure 5.9 as

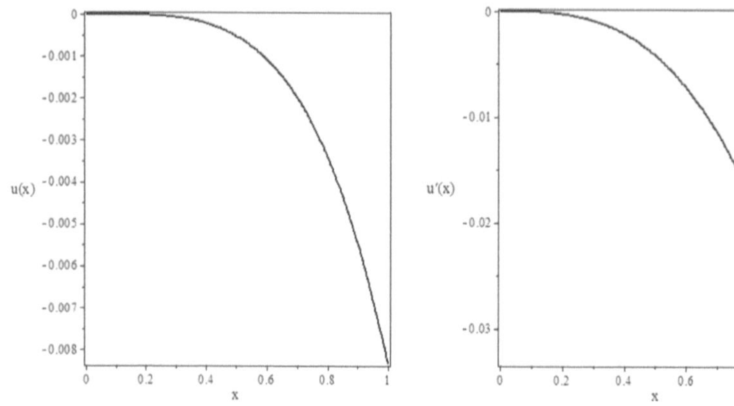

Fig. 5.8. The result of obtained solution by AGM

Fig. 5.9. The first derivative of the yielded solution by AGM

5.2.4.3. Numerical Solution

In table 5.4, equation 5.37 is solved numerically (RKF45) in regard to the boundary conditions in the given domain of $x \in \{0,1\}$ as follows:

Table 5.4. The Obtained Numerical Values for $u(x)$ and $u'(x)$ in the Specified Domain

x	*0*	*0.2*	*0.4*	*0.6*	*0.8*	*1*
$u(x)$	0	-0.000013	-0.00021	-0.00108	-0.0034	-0.0083
$u'(x)$	0	-0.00026	-0.0021	-0.00720	-0.01706	-0.0333

5.2.4.4. Comparing the Obtained Solutions by AGM and Numerical Method

As regards the obtained results in table 5.4 and the acquired solution by AGM, we will have the following comparisons in figure 5.10 and figure 5.11 as

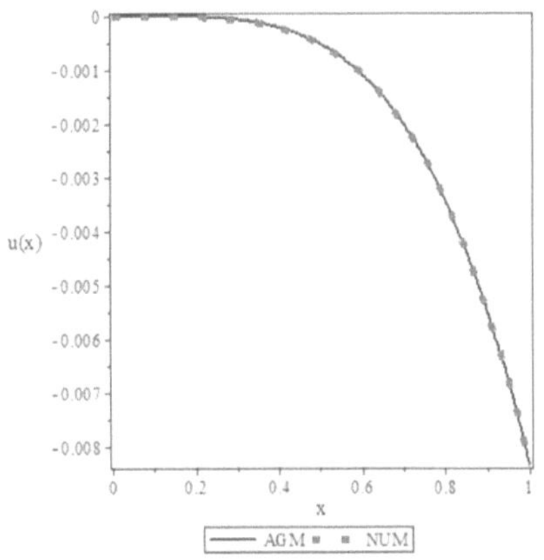

Fig. 5.10. Comparing the obtained solutions, $u(x)$, by AGM and numerical method

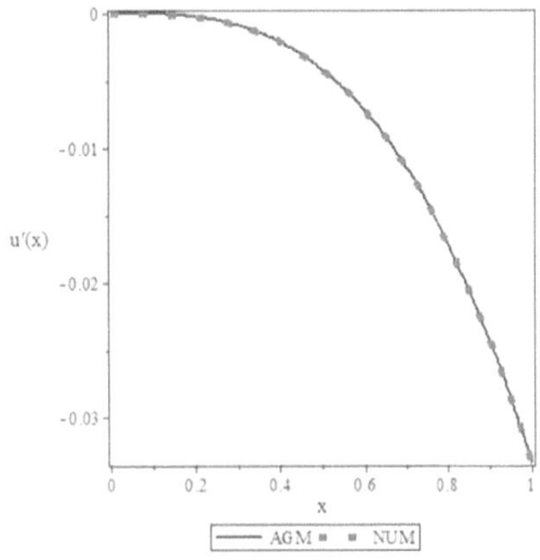

Fig. 5.11. Comparing the first derivative of the obtained solutions, *(du/dx)*, by AGM and numerical method

5.3. Problems

5.3.1. The following integrodifferential equation is considered as

$$u'+u^2-x=\int_1^x[u^2(t)-u^3(t)-u(t)]dt,$$

and the related boundary condition is expressed in the form of

$$u(0)=0.$$

Solve the above equation by AGM in the domain $x\in\{0,\,1\}$ and obtain the maximum error of the achieved solution when the considered answer function consists of eight sentences.

5.3.2. Suppose the following set of nonlinear integrodifferential equation and solve it by AGM.

$$\begin{cases} u'(x)=\int_0^x\{u^3(s)\text{-}v(s)u'(s)e^{\text{-}x+s}\}ds \\ v(s)=B+\int_0^x v^2(s)\text{-}u(s)v(s)\cos(s\text{-}x)\}ds \end{cases}$$

The related boundary condition is $u(0)=A$. Then, compare the achieved solution by AGM and numerical method graphically in $x\in\{0,\,1\}$ when $A=0.1$ and $B=0.5$.

5.3.3. The second-order nonlinear differential equation with the following boundary conditions is introduced as

$$u''(x)=u(x).u'(x)+\int_0^x[u(t)+u'(t).u^2(t)-3u^2(t)]dt$$

$$u(0)=0 \quad , \quad u'(1)=1.$$

On the basis of the given explanations in the fifth chapter, solve the aforementioned equation by AGM and compare it by numerical method in $x \in \{0,1\}$.

5.3.4. Consider the following nonlinear differential equation with the boundary conditions as

$$u''(x) = e^{2x} + \int_0^x \{e^{2(x-t)}\ [\frac{du(t)}{dt} u^2(t) - \frac{du(t)}{dt} + u(t) + 1\ ]\ \} dt$$

$$u(0) = u'(0) = 0.$$

Without utilizing Taylor series expansion for converting exponential term, solve the presented problem by AGM.

5.4. References

1. X. Y. Wang, "Exact and Explicit Solitary Wave Solutions for the Generalized Fisher Equation," *Physics Letters A* 131 (1988): 277–279.

2. A. Jeffrey and M. N. B. Mohamad, "Exact Solutions to the KdV-Burgers' Equation," *Wave Motion*, vol. 14, no. 4 (December 1991): 369–375.

3. He Bin et al., "New Exact Solutions of the Double Sine-Gordon Equation Using Symbolic Computations Original Research Article," *Applied Mathematics and Computation*, vol. 186, no. 2 (15 March 2007): 1334–1346.

4. N. H. Sweilam, "Fourth Order Integro-Differential Equations Using Variational Iteration Method," *Computers & Mathematics with Applications* (2006), doi:10.1016/j.camwa.12.055.

5. S. Q. Wang and J. H. He, "Variational Iteration Method for Solving Integro-Differential Equations," *Physics Letters A*, vol. 367, no. 3 (23 July 2007): 188–191.

6. N. H. Sweilam, "Fourth Order Integro-Differential Equations Using Variational Iteration Method," *Computers and Mathematics with Applications*, vol. 54, nos. 7–8 (October 2007): 1086–1091.

7. A. R. Ahmadi et al., "A comprehensive analysis of the flow and heat transfer for a nanofluid over an unsteady stretching flat plate," *Powder Technology* 258 (2014): 125–133.

8. D. D. Ganji, "The Application of He's Homotopy Perturbation Method to Nonlinear Equations Arising in Heat Transfer," *Physics Letters A*, vol. 355, nos. 4–5 (10 July 2006): 337–341.

9. Ganji, "The Application of He's Homotopy Perturbation Method to Nonlinear Equations Arising in Heat Transfer," 337–341.

10. M. Sheikholeslami et al., "Analytical Investigation of Jeffery-Hamel Flow with High Magnetic Field and Nanoparticle by Adomian Decomposition Method," *Applied Mathematics and Mechanics* 3, no. 1 (2012): 1553–1564.

11. M. Sheikholeslami, D. D. Ganji, and H. R. Ashorynejad, "Investigation of

Squeezing Unsteady Nanofluid Flow Using ADM," *Powder Technology* 239 (2013): 259–265.

12. D. D. Ganji et al., "A Comparative Comparison of He's Method with Perturbation and Numerical Methods for Nonlinear Vibrations Equations," International Journal of Nonlinear Dynamics in Engineering and Sciences 1, no. 1 (2009): 1–20.

13. Ji-Huan He and Xu-Hong Wu, *Chaos, Solitons and Fractals*, vol. 30, no. 3 (November 2006): 700–708.

14. Xu-Hong Wu and Ji-Huan He, *Computers and Mathematics with Applications*, vol. 54, nos. 7–8 (2007): 966–986.

15. Z. Z. Ganji, D. D. Ganji, and H. Bararnia, "Approximate General and Explicit Solutions of Nonlinear BBMB Equations Exp-Function Method," *Applied Mathematical Modelling*, vol. 33, no. 4 (2009): 1836–1841.

16. Z. Z. Ganji, D. D. Ganji, and A. Asgari, "Finding General and Explicit Solutions of High Nonlinear Equations by the Exp-Function Method," *Computers and Mathematics with Applications*, vol. 58, nos. 11–12 (December 2009): 2124–2130.

Chapter 6

Application of AGM in Partial Differential Equations

6.1. Partial Differential Equations

Partial differential equations that arise in real-world physical problems are often too complicated to be solved exactly. And even if an exact solution is obtainable, the required calculations may be too complicated to be practical, or it might be difficult to interpret the outcome. In fact, solving nonlinear equations can guide researchers to comprehend the physical phenomena deeply and sometimes leads them to investigate some facts that are not easily understood through common observations. On the basis of the above explanations, here we aim to discuss and investigate fluids and the vibrational behavior of the systems analytically with a suitable and short technique.

6.2. A Broad Range of Applied Examples

Example 6.2.1

The present example is concerned with calculating the two-dimensional velocity profile of a viscous flow for an incompressible fluid along the leading edge of a flat plate by using the continuity and motion equations on the basis of the following figure:

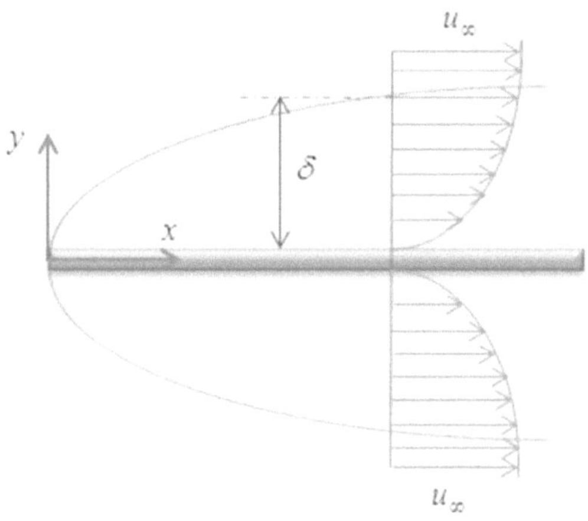

Fig. 6.1. The schematic diagram of the physical model

It is noteworthy that μ and ρ are dependent on the thickness of velocity profile, along with the direction of fluid flow motion, as follows:[1]

$$\mu = \mu_0 e^{-\alpha(\frac{y}{\delta})} \quad ; \quad \rho = \rho_0 e^{-\beta(\frac{xy}{\delta^2})}. \tag{6.1}$$

The partial differential equations governing the system in a steady state, according to figure 6.1, are as follows:

continuity equation [2]

$$\frac{\partial}{\partial x}(\rho u) + \frac{\partial}{\partial y}(\rho V) = 0 \tag{6.2}$$

the equation of motion at x-component

$$\rho(u\frac{\partial u}{\partial x} + V\frac{\partial u}{\partial y}) = \frac{\partial}{\partial x}(\mu\frac{\partial u}{\partial x}) + \frac{\partial}{\partial y}(\mu\frac{\partial u}{\partial y}). \tag{6.3}$$

The boundary conditions are the following:

$$\begin{aligned} u(0) &= u_\infty \quad \text{at} \quad x = 0 \\ v(0) &= 0 \quad \text{at} \quad y = 0 \\ v(\delta) &= u_\infty \quad \text{at} \quad y = \delta. \end{aligned} \tag{6.4}$$

After the Taylor series expansion of equation 6.1 and choosing the first two sentences of it, we have

$$\mu = \mu_0(1 - \alpha\frac{y}{\delta}) \quad ; \quad \rho = \rho_0(1 - \beta\frac{xy}{\delta^2}). \tag{6.5}$$

6.2.1.1. Solving the Partial Differential Equations with AGM

Fluid displacement in x and y directions, which means u and v, respectively, can be acquired by the method of separation of variables in the forms of

$$u = u(x).v(y) \quad , \quad V = u(x).v(y). \tag{6.6}$$

By substituting the above obtained equations into equations 6.2 and 6.3, we will have

$$g(x,y):\ \rho.v(y)\frac{du(x)}{dx} + u(x)\frac{d}{dy}(\rho.v(y)) = 0 \tag{6.7}$$

and

$$f(x,y):\ \rho\{u(x).v^2(y)\frac{du(x)}{dx} + v(y)u^2(x)\frac{dv(y)}{dy}\} = u(x)(\mu\frac{d^2\,v(y)}{dy^2}$$

$$\mu.v(y)\frac{d^2u(x)}{dx^2}) \quad . \tag{6.8}$$

In AGM, the answers of equation 6.7 and equation 6.8 are considered as finite series of polynomials with constant coefficients as

$$u(x) = \sum_{n=0}^{2} c_n x^n = c_0 + c_1 x + c_2 x^2 \tag{6.9}$$

$$v(y) = \sum_{n=0}^{5} d_n y^n = d_0 + d_1 y + d_2 y^2 + d_3 y^3 + d_4 y^4 + d_5 y^5. \tag{6.10}$$

It is notable that the constant coefficients c_0 to c_2 and d_0 to d_5 are obtained by applying boundary conditions from equation 6.4.

6.2.1.2. Applying the Boundary Conditions on the Set of Partial Differential Equations

Like the previous chapters, applying the boundary conditions is done in order to compute the constant coefficients of the solution of partial differential equations, which, in this case, are the velocity profile in x and y directions, u and v, in two ways.

a. Applying the boundary conditions on equation 6.9 and equation 6.10 is done as follows:

$$\begin{cases} u(x) = u(BC) \\ v(y) = v(BC) \end{cases} \tag{6.11}$$

It is notable that BC is the abbreviation of "boundary conditions."

b. The boundary conditions are applied on equation 6.7 and equation 6.8 and their derivatives after substituting the considered answer functions into equation 6.7 and equation 6.8 in the forms of

$$\begin{cases} g\{u(BC), v(BC)\} = 0 \quad , \quad g'\{u(BC), v(BC)\} = 0 \\ f\{u(BC), v(BC)\} = 0 \quad , \quad f'\{u(BC), v(BC)\} = 0 \end{cases} \tag{6.12}$$

Applying boundary conditions on the answer functions in regard to equation 6.4 is done as

$$\begin{cases} x = 0 \rightarrow & u(0) = u_\infty & \text{so} & c_0 = u_\infty \\ y = 0 \rightarrow & v(0) = 0 & \text{therefore} & d_0 = 0 \\ y = \delta \rightarrow & v(\delta) = u_\infty & \text{eventually} & d_0 + d_1\delta + d_2\delta^2 + d_3\delta^3 + d_4\delta^4 + d_5\delta^5 = u_\infty . \end{cases} \tag{6.13}$$

In this step, the application of boundary conditions on $g(x, y)$ and f(x, y), with regard to the values $x = 0, \; y = 0$ to δ, is expressed as

$$f(0,0): \quad \rho_0 c_0 d_0 (c_1 d_0 + c_0 d_1) = \frac{-\mu_0}{\delta}(\alpha c_0 d_1 - 2c_0 d_2 \delta - 2c_2 d_0 \delta). \tag{6.14}$$

And

$$f(0,\delta):$$

$$\begin{aligned} &\rho_0 c_0 (d_0 + d_1\delta + d_2\delta^2 + d_3\delta^3 + d_4\delta^4 + d_5\delta^5)\{c_1(d_0 + d_1\delta + d_2\delta^2 + \\ &d_3\delta^3 + d_4\delta^4 + d_5\delta^5) + c_0(d_1 + 2d_2\delta + 3d_3\delta^2 + 4d_4\delta^3 + 5d_5\delta^5)\} = \\ &-\frac{\mu_0 \alpha c_0}{\delta}(d_1 + 2d_2\delta + 3d_3\delta^2 + 4d_4\delta^3 + 5d_5\delta^4) + \mu_0(1-\alpha)\{c_0(2d_2 + \\ &6d_3\delta + 12d_4\delta^2 + 20d_5\delta^3) + 2c_2(d_0 + d_1\delta + d_2\delta^2 + d_3\delta^3 + d_4\delta^4 + d_5\delta^5)\}. \end{aligned} \tag{6.15}$$

Then

$$g(0,0): \quad c_1 d_0 + c_0 d_1 = 0. \tag{6.16}$$

Afterward, we have

$$\begin{aligned} g(0,\delta): \quad & (d_0 + d_1\delta + d_2\delta^2 + d_3\delta^3 + d_4\delta^4 + d_5\delta^5)(c_1 - \frac{\beta c_0}{\delta}) + \\ & c_0(d_1 + 2d_2\delta + 3d_3\delta^2 + 4d_4\delta^3 + 5d_5\delta^4) = 0 \end{aligned} \quad . \tag{6.17}$$

For the first derivative of the main differential equation, we will have

$$\begin{aligned} f'(0,0): \quad & {}_0(c_1^2 d_0^2 + 2c_0 d_0^2 c_2 + 4c_0 d_0 d_1 c_1 + c_0^2 d_1^2 + 2c_0^2 d_0 d_2 = \\ & -\frac{\mu\,\alpha}{\ }(c_1 d_1 + 4c_0 d_2 + 2c_2 d_0) + 2\ {}_0(c_1 d_2 + 3c_0 d_3 + c_2 d_1) \end{aligned} . \tag{6.18}$$

After that,

$$f'(0,\delta): \ f'(u(0), v(\delta)) = 0. \tag{6.19}$$

Constant coefficients c_0 to c_2 and d_0 to d_5 are gained by solving a set of equations consisting of nine equations with nine unknowns from equation 6.13 to equation 6.19.

To simplify, the following new variables are considered:

$$\psi = \mu_0 \rho_0 \delta u_\infty^2 (\alpha\beta - 17\alpha - \beta + 16) - \mu_0^2 (61\alpha^2 - 8\beta) \tag{6.20}$$

and

$$\begin{aligned} \Omega = \{ & \mu_0^4 \alpha\beta (54\alpha^3 - 102\alpha^2 - 24\alpha + 433\alpha\ \beta - 306\alpha^2\beta + 81\alpha\beta - \\ & 24\alpha + 136) + \mu_0^2 \rho_0^2 \delta^2 u_\infty^4 (146\alpha\beta - 82\alpha^2\beta + \alpha^2\beta^2 - 2\alpha\beta^2 - 64\beta + \\ & \beta^2 + \alpha^2) + 2\mu_0^3 \rho_0 \delta u_\infty^2 (82\alpha^2\beta - 42\alpha^2\beta - 44\alpha\beta + 4\beta - 26\alpha^2\beta^2 + \\ & 9\alpha^3\beta^3 + 25\alpha\beta^2 - 643\alpha^3 - 1596\alpha + 1760\alpha^2 - 8\beta^2) + 8\mu_0 \rho_0^3 \delta^3 u_\infty^6 \beta(1-\alpha) - \\ & \mu_0^4 (3191\alpha^4 + 14744\alpha^2 - 11200\alpha^3 - 8640\alpha + 64\beta - 64\beta^2 + 1904)\}^{0.5} \end{aligned} \tag{6.21}$$

then

$$\xi_1 = \mu_0^2 (112\alpha - 17\alpha\beta + 9\beta\alpha^2 - 52) \tag{6.22}$$

After that

$$\xi_2 = -\rho_0 \delta u_\infty^2 (1-\alpha) + \mu_0 (8 - 17\alpha + 9\alpha^2) \tag{6.23}$$

$$\lambda = \frac{1}{\xi_2}(\xi_1 + \psi + \Omega) \tag{6.24}$$

$$\Delta = \rho_0 \beta u_\infty^2 \delta - 2\lambda - 12\mu_0 + \frac{9\alpha\lambda}{4} + 16\mu_0\alpha \tag{6.25}$$

Due to the aforementioned new variables, the constant coefficients of the answer functions are obtained as follows:

$$c_0 = u_\infty \quad , \quad c_1 = \frac{-u_\infty}{\delta}(4-\beta+\frac{\lambda}{4\mu_0}) \quad , \quad c_2 = \frac{u_\infty \Delta}{2\mu_0\delta^2(1-\alpha)}$$
$$d_0 = 0 \quad , \quad d_1 = 0 \quad , \quad d_2 = 0 \quad , \quad d_3 = 0$$
$$d_4 = \frac{u_\infty}{\delta^4}(1-\frac{\lambda}{4\mu_0}) \quad , \quad d_5 = \frac{u_\infty \lambda}{4\mu_0\delta^5} \quad . \tag{6.26}$$

After substituting the obtained parameters from equation 6.26 into equation 6.9 and equation 6.10, the solution of the presented problem, which is the velocity profile, along with x and y directions, is achieved in the form of

$$u(x) = u_\infty\{1-(4-\beta+\frac{\lambda}{4\mu_0})\}(\frac{x}{\delta})+\frac{\Delta}{2\mu_0(1-\alpha)}(\frac{x}{\delta})^2 \tag{6.27}$$

and

$$\text{v}(y) = u_\infty\{(1-\frac{\lambda}{4\mu_0})(\frac{y}{\delta})^4+\frac{\Delta}{4\mu_0}(\frac{y}{\delta})^5\}. \tag{6.28}$$

Select the following physical values:

$$u_\infty = 0.00425 \ (\frac{m}{s}) \quad , \quad \delta = 25 \ (cm) \quad , \quad L = 0.7 \ (m) \quad , \quad \rho_0 = 980 \ (\frac{kg}{m^3})$$
$$\alpha = 0.18 \quad , \quad \beta = 0.11 \quad . \tag{6.29}$$

According to equation 6.29, the velocity profiles in x and y directions are rewritten as

$$u(x) = 0.00425+0.0522x+1.5825x^2 \tag{6.30}$$

and

$$v(y) = 8.6638y^4-30.3034y^5. \tag{6.31}$$

Eventually, in this step, the charts of velocity profiles are illustrated graphically as.

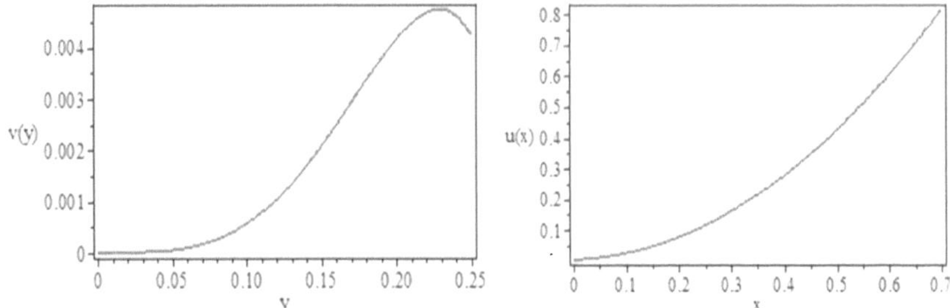

Fig. 6.2. The acquired velocity profile in y-direction by AGM

Fig. 6.3. The acquired velocity profile in x-direction by AGM

With regard to equation 6.6, the velocity profiles are obtained as follows:

$u = u(x).v(x)$ therefore:

$$u(x,y)=u_\infty^2\{(1-(4-\beta+\frac{\lambda}{4\mu_0}))(\frac{x}{\delta})+\frac{\Delta}{2\mu_0(1-\alpha)}(\frac{x}{\delta})^2\}\{(1-\frac{\lambda}{4\mu_0})(\frac{y}{\delta})^4+\frac{\Delta}{4\mu_0}(\frac{y}{\delta})^5\}. \quad (6.32)$$

Based on the introduced physical values, we will have

$$u(x,y)=(0.00425+0.0522x+1.5825x^2)(8.6638y^4-30.3034y^5). \quad (6.33)$$

The 3-D velocity profile $u(x,y)$ in Cartesian coordinates is depicted as follows:

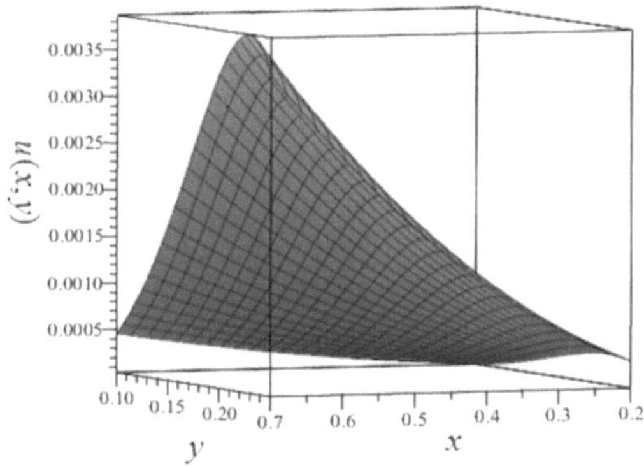

Fig. 6.4. The resulted velocity profile in x- and y-directions by AGM

Afterward, the contour of velocity profile is charted in Cartesian coordinates as below.

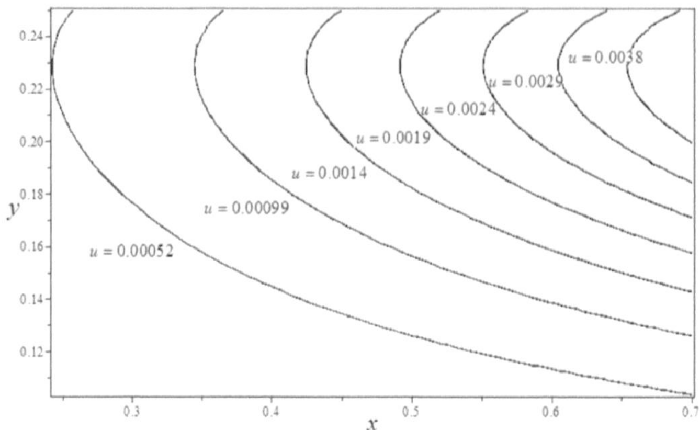

Fig. 6.5. The obtained contour from velocity profile according to equation 6.33 by AGM

Example 6.2.2

According to figure 6.6, two cylinders with fixed and hinged ends are considered. To start the solution procedure, we should determine the differential equation of bending vibration governing the cylindrical structure in regard to the driving forces exerting on it.[3]

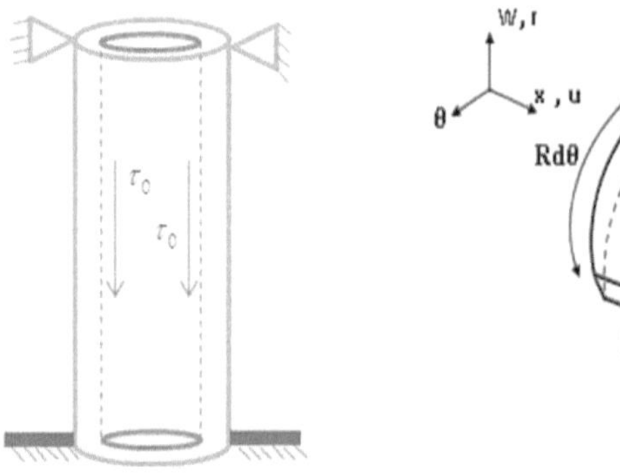

Fig. 6.6. The physical model of the problem

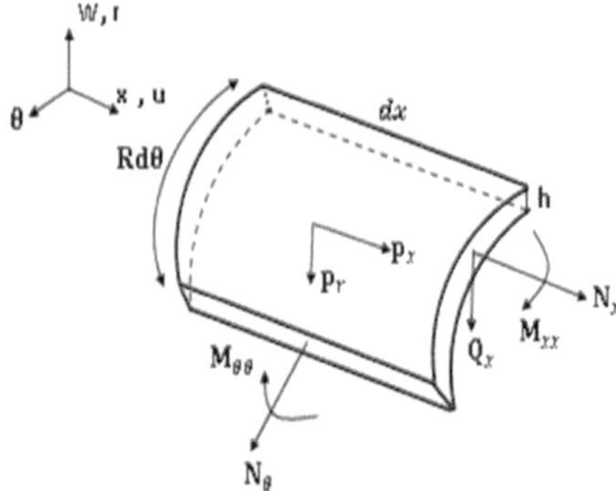

Fig. 6.7. The surface of the cylindrical shell

Due to the above explanations, we will have

$$RD\frac{\partial^4 W(x,t)}{\partial x^4}+\frac{Eh}{R}W(x,t)-\nu Nx+P_rR+\rho A\frac{\partial^2 W(x,t)}{\partial t^2}=0 \quad (6.34)$$

$$A=2\pi R.h. \quad (6.35)$$

The parameters P_r and N_x in equation 6.34 are driving forces that resulted from the vibration of the cylindrical structure, and $W(x,t)$ is the instantaneous displacement of the cylindrical structure along the radius. And also, A is the shell cross-section area, which is computed with respect to the shell thickness h and its radius R .

So equation 6.34 is solved with the method of separation of variables as follows:

$$W(x,t)=W(x).z(t). \quad (6.36)$$

After substituting equation 6.35 and equation 6.36 into equation 6.34, the result is gained in the form of

$$\frac{D}{2\pi\rho h}\frac{1}{W(x)}\frac{d^4W(x)}{dx^4}+\frac{E}{2\pi\rho R^2}-\frac{\upsilon N_x-P_rR}{2\pi\rho hRW(x)\mathrm{z(t)}}+\frac{1}{z(t)}\frac{d^2z(t)}{dt^2}=0. \quad (6.37)$$

In equation 6.37, the parameter D is the bending stiffness of the shell, which is defined as below:

$$D=\frac{Eh^3}{12(1-\upsilon^2)}. \quad (6.38)$$

After substituting the square of modal frequency $({\omega^2}_n)$ in the first two sentences and the second two sentences of equation 6.37, we will have

$$\frac{D}{2\pi\rho hW(x)}\frac{d^4W(x)}{dx^2}+\frac{E}{2\pi\rho R^2}=\omega_n^2 \quad (6.39)$$

$$-\frac{\upsilon N_x - P_r R}{2\pi\rho h R W(x).z(t)} + \frac{1}{z(t)}\frac{d^2 z(t)}{dt^2} = -\omega_n^2. \tag{6.40}$$

Equation 6.40 can be explained in terms of the modal mass M_n and the modal force $f_n(t)$ as

$$\frac{\partial^2 z(t)}{\partial t^2} + \omega_n^2 z(t) = \frac{f_n(t)}{M_n}. \tag{6.41}$$

In equation 6.41, M_n and $f_n(t)$ are computed in the forms of

$$M_n = \int_0^L (2\pi\rho h R W^2(x))dx \tag{6.42}$$

and

$$f_n(t) = \int_0^L (\upsilon N_x - P_r.R)W(x)dx. \tag{6.43}$$

6.2.2.1. Solution

The solution of equation 6.39 is yielded in the form of

$$W(x) = ae^{-i\lambda_n x} + be^{-\lambda_n x} + ce^{i\lambda_n x} + de^{\lambda_n x}. \tag{6.44}$$

Here, we define the following new variable λ_n as follows:

$$\lambda_n = \left(\frac{h(2\pi\rho R^2\omega_n^2 - E)}{RD}\right)^{0.25} \quad , \quad n = 1,2,3,\ldots. \tag{6.45}$$

The appearance of equation 6.44 can be changed by rewriting it in the hyperbolic form as below:

$$W(x) = c_1\cosh(\lambda_n x) + c_2\sinh(\lambda_n x) + c_3\cos(\lambda_n x) + c_4\sin(\lambda_n x). \tag{6.46}$$

The solution of equation 6.41 is obtained in the form of

$$Z(t) = A\sin(\omega_n t) + B\cos(\omega_n t) + \frac{f_n(t)}{\omega_n^2 M_n}(\cos(\omega_n t) - 1). \tag{6.47}$$

By supposing there is no displacement and initial velocity (initial conditions) for cylinders, the coefficients A and B are considered zero. Therefore, the time indicator equation, which is equation 6.47, is acquired as follows:

$$Z(t) = \frac{f_n(t)}{\omega_n^2 M_n}(\cos(\omega_n t) - 1). \tag{6.48}$$

6.2.2.2. Applying Boundary Conditions

The boundary conditions for the cylinders, which have been shown as hinged and fixed ends in figure 6.6, are applied on equation 6.46 as follows:

$$\text{Fixed end} \begin{cases} W(0) = 0 & \rightarrow \quad c_1 + c_3 = 0 \\ W'(0) = 0 & \rightarrow \quad c_2 + c_4 = 0 \end{cases} \tag{6.49}$$

$$\text{Hinged end} \begin{cases} W(L) = 0 & \rightarrow c_1\cosh(\lambda_n L) + c_2\sinh(\lambda_n L) + c_3\cos(\lambda_n L) + c_4\sin(\lambda_n L) = 0 \\ W''(L) = 0 & \rightarrow c_1\cosh(\lambda_n L) + c_2\sinh(\lambda_n L) - c_3\cos(\lambda_n L) - c_4\sin(\lambda_n L) = 0 \end{cases}. \tag{6.50}$$

To obtain the nonzero answer from the set of equation 6.49 and equation 6.50, the determinant of the coefficients matrix of the set of equations should obtain zero as

$$\det = \begin{vmatrix} 1 & 0 & 1 & 0 \\ 0 & 1 & 0 & 1 \\ \cosh(\lambda_n L) & \sinh(\lambda_n L) & \cos(\lambda_n L) & \sin(\lambda_n L) \\ \cosh(\lambda_n L) & \sinh(\lambda_n L) & -\cos(\lambda_n L) & -\sin(\lambda_n L) \end{vmatrix} = 0 \; . \tag{6.51}$$

By defining a new variable, $\beta_n = \lambda_n . L$, the answer of the aforementioned determinant is acquired in the form of

$$\det = \cosh(\beta_n)\sin(\beta_n) - \sinh(\beta_n)\cos(\beta_n) = 0. \tag{6.52}$$

By solving equation 6.52, the parameter β_n and also λ_n from equation 6.45 can be achieved easily. Finally, from this equation, the modal frequencies ω_n can be computed. It is citable that whenever equation 6.52 is drawn as the following determinant function, the values of β_n with ω_n can be observed in the charts below:

Fig. 6.8. Depicting the roots of determinant function

Eventually, by solving equation 6.52, the values of β_n are acquired as follows:

Table 6.1. The Results of Computing β_n for Different Modal Frequencies

n	1	2	3	4	5	6	7	8	9	...
β_n	3.92	7.07	10.21	13.35	16.49	19.63	22.77	25.92	29.06	...

According to the values of table 6.1 and the defined new variable $\beta_n = \lambda_n .L$, we can get the parameter λ_n as follows:

$$\lambda_n = (2n + 0.5)(\frac{\pi}{2L}) . \tag{6.53}$$

By substituting equation 6.53 into equation 6.45, the vibrational modal frequency is gained as below:

$$\omega_n = \frac{\sqrt{2}}{2R}\sqrt{\frac{hE + DR^2\lambda_n^4}{\pi\rho h}} \quad , \quad n = 1, 2, 3, \ldots . \tag{6.54}$$

6.2.2.3. Computing Modal Shapes of Vibration

In this step, equation 6.49 and the first part of equation 6.50 are simultaneously solved in terms of one of the constant coefficients c_1, c_2, c_3, c_4, which are in terms of C_4. According to the above explanations, we will have

$$c_1 = \frac{\sinh(\lambda_n L)-\sin(\lambda_n L)}{\cosh(\lambda_n L)-\cos(\lambda_n L)}c_4 \quad , \quad c_2 = -c_4$$
$$c_3 = -\frac{\sinh(\lambda_n L)-\sin(\lambda_n L)}{\cos(\lambda_n L)-\cos(\lambda_n L)}c_4 \quad . \tag{6.55}$$

Substituting equation 6.55 into equation 6.46, and with respect to the variable $\beta_n = \lambda_n .L$, the function of vibrational modal shape is achieved as below:

$$W_n(x) = c_4\{\frac{\sinh(\beta_n)-\sin(\beta_n)}{\cosh(\beta_n)-\cos(\beta_n)}(\cos(\frac{\beta_n}{L}x)-\cosh(\frac{\beta_n}{L}x))+\sinh(\frac{\beta_n}{L}x)-\sin(\frac{\beta_n}{L}x)\}. \tag{6.56}$$

With regards to the values of β_n from table 6.1, the charts of the modal shapes of the structure element, along with the length of the cylinder, can be observed as follows:

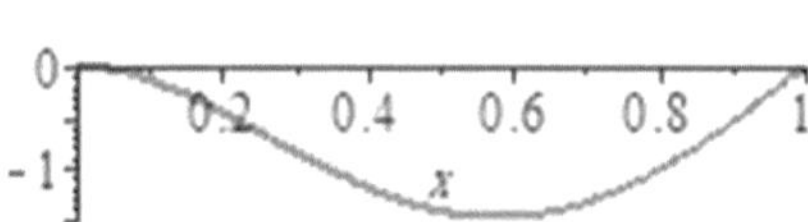

Fig. 6.9. The first mode of vibration $(n = 1)$

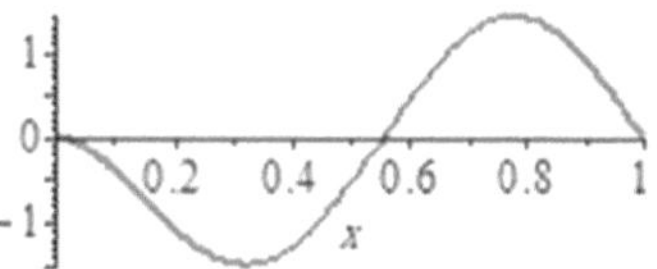

Fig. 6.10. The second mode of vibration $(n = 2)$

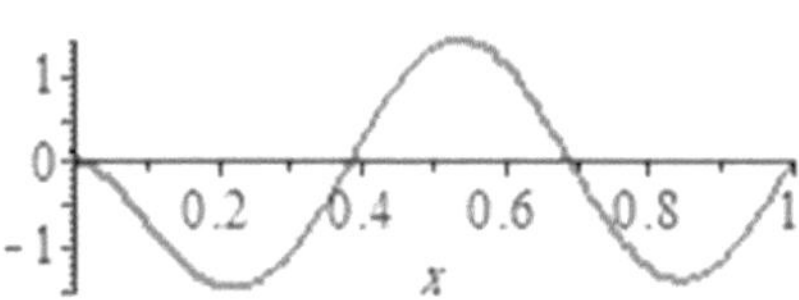

Fig. 6.11. The third mode of vibration $(n = 3)$

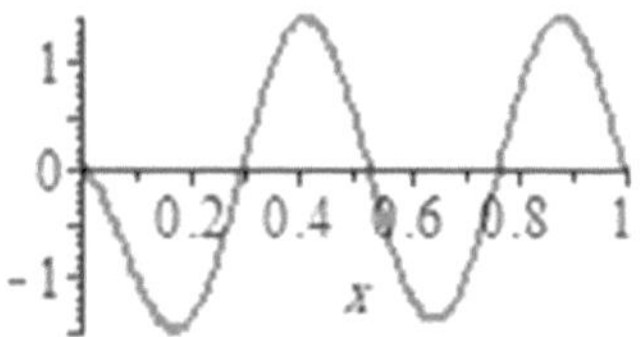

Fig. 6.12. The fourth mode of vibration $(n = 4)$

By acquiring the function of modal forms from equation 6.56 and the time indicator function from equation 6.48, we can achieve the displacement of structure elements in different times as regards equation 6.36, and then for each mode of vibration, we will have

$$W_n(x,t) = W_n(x).Z_n(t) \qquad ; \qquad n = 1,2,3,.... \tag{6.57}$$

And for all the moods, the displacement of the cylindrical structure element is gained as follows:

$$W(x,t) = \sum_{n=1}^{\infty} W_n(t).Z_n(t). \tag{6.58}$$

The following physical values for the structure are introduced as

$$L = 1\ (m)\ ,\ E = 8\times10^{10}\ (N/m^2)\ ,\ \upsilon = 0.3\ ,\ h = 0.04\ (m)\ ,\ \rho = 2800\ (Kg/m^3). \tag{6.59}$$

In this part, from two cylinders carrying non-Newtonian fluid flow, we just analyze the vibration of the outer cylinders. So the analyses of vibrations of the inner cylinder and outer cylinder are the same. It is citable that the radius of the outer cylinder is 0.8(m).

Driving forces (P_r and N_x) exerting on the cylinder can be yielded as follows:

$$N_x = \tau_0.L. \tag{6.60}$$

The parameter τ_0 in equation 6.60, which is shear stress on the walls of the outer cylinder, and the parameter P_r, which is the differential pressure exerting on the walls of the cylinders along the radius and the length of the cylinder, has been computed on the basis of Ostwald-de Waele model as [4]

$$\tau_{r\theta} = -m\{r\frac{\partial}{\partial r}(\frac{u_\theta}{r})\}^n. \tag{6.61}$$

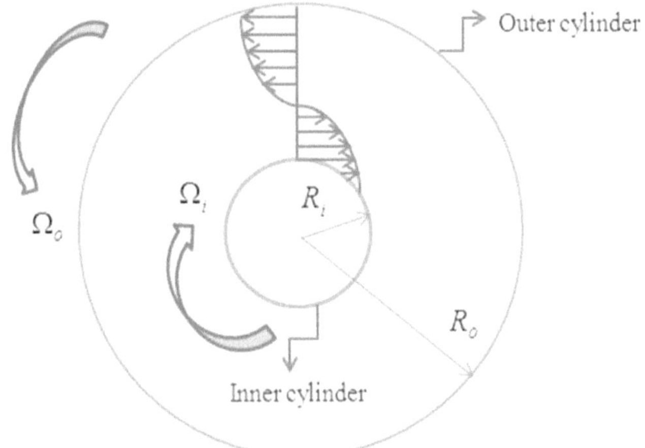

Fig. 6.13. The schematic diagram of the presented system from the top side

On the basis of figure 6.13, $u_\theta(r)$ can be achieved from the differential equation governing the non-Newtonian fluid between the two cylinders with the radii R_o and R_i and the length L, wherein each of the cylinders rotates with angular velocities Ω_o and Ω_i for the outer and inner cylinders, respectively as

$$f(r):\frac{\partial}{\partial r}\{-mr^2(r\frac{\partial}{\partial r}\left(\frac{u_\theta}{r}\right))^n\}=0. \tag{6.62}$$

In the above equation, n and m are constant values and are dependent on the fluid properties. Afterward, boundary conditions governing this system are introduced as

$$\begin{cases} u_\theta(R_i)=-\Omega_i R_i \\ u_\theta(R_0)=\Omega_0 R_0. \end{cases} \tag{6.63}$$

Then, with the following physical values

$$\begin{aligned} &R_0=80(cm) \quad , \quad R_i=50(cm) \quad , \quad m=0.3 \quad , \quad n=1.3 \\ &\Omega_0=10\left(Rad/_{Sec}\right) \quad , \quad \Omega_i=15\ (Rad/_{Sec}) \end{aligned} \tag{6.64}$$

The fluid flow between two cylinders is achieved as follows:

$$u_\theta(r) = -52.28 + 126.3r - 89.4r^2 + 32.32r^3. \tag{6.65}$$

Based on the aforementioned achievements, the shearing stress on the walls of cylinders from equation 6.61 is acquired as

$$\tau_{r\theta} = 0.3(-89.54r + 64.78r^2 + \frac{52.4}{r})^{1.3}. \tag{6.66}$$

In regard to equation 6.61, the shearing stress exerting on the walls of the outer cylinder is acquired in the form of

$$\tau_o = -\tau_{r\theta}\Big|_{r = R_o.} \tag{6.67}$$

It is noteworthy that the motion equation in r and z directions for the presented system can be written as

$$\begin{aligned} &r-component \quad : \frac{\partial p}{\partial r} = \frac{\rho u_\theta^2}{r} \\ &z-component \quad : \frac{\partial p}{\partial z} = \rho g_z. \end{aligned} \tag{6.68}$$

And from the above equation, the pressure of fluid (P) is achieved as below:

$$p = p(r,z) \rightarrow \ dp = \left(\frac{\partial p}{\partial r}\right)dr + \left(\frac{\partial p}{\partial z}\right)dz. \tag{6.69}$$

Then after substituting equation 6.68 into equation 6.69, the result is

$$dp = \frac{\rho u_\theta^2}{r}dr + \rho g dz. \tag{6.70}$$

By integrating from equation 6.70, the differential pressure is yielded as follows:

$$\Delta P = \int_{R_i}^{R_0} \frac{\rho u_\theta^2(r)}{r} dr + \int_0^L \rho g dz. \tag{6.71}$$

Based on the given physical values from equation 6.59, the shearing stress on the walls of outer cylinder and the differential pressure of the fluid along the length and radius of the cylinders have been computed as

$$P_r = \Delta P = 17070 \quad \left(\frac{N}{m^2}\right) \quad , \quad N_x = 30,8,1 \quad (N.m). \tag{6.72}$$

With regard to the above physical values, we can observe the instantaneous displacement and the structural vibration in different modes, in different times, and in a time period $T = \dfrac{2\pi}{\omega_n}$ from equation 6.57 as follows:

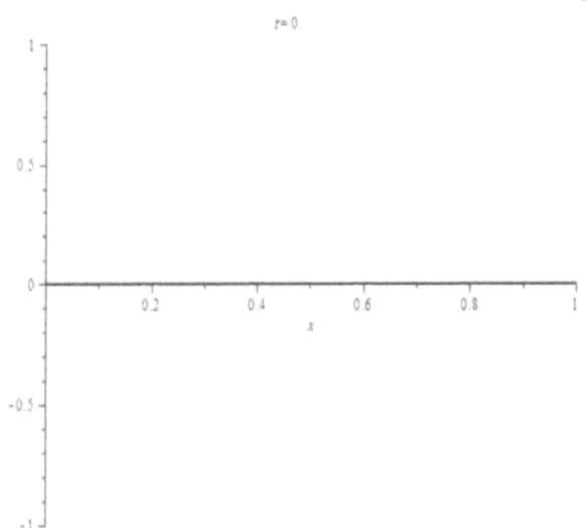

Fig. 6.14. Instantaneous displacement of the structure in the first mode of vibration in $t = 0$

Fig. 6.15. Instantaneous displacement of the structure in the first mode of vibration in $t = 0.00044$

Then for the second mode of vibration, we will have

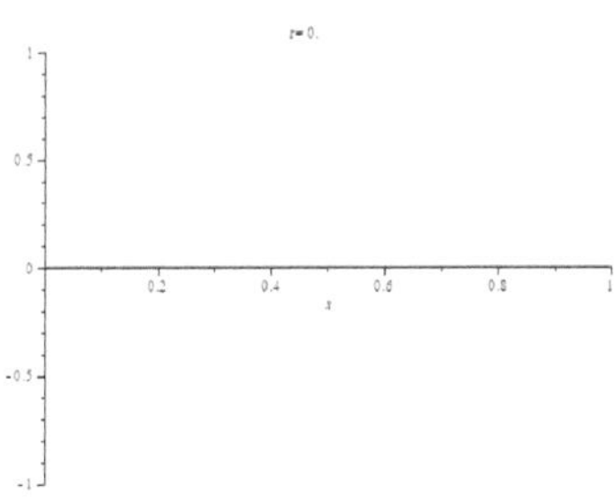

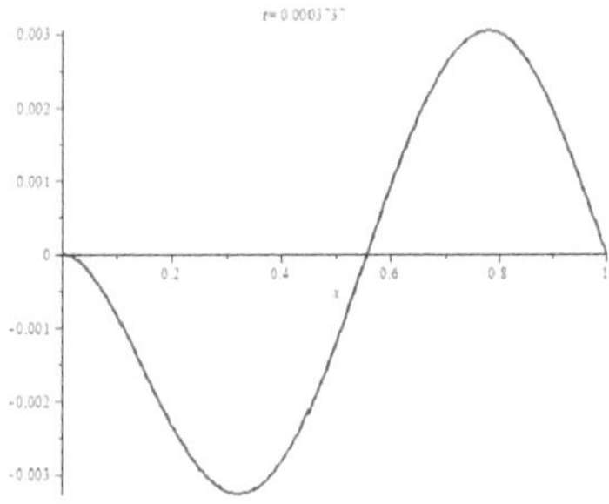

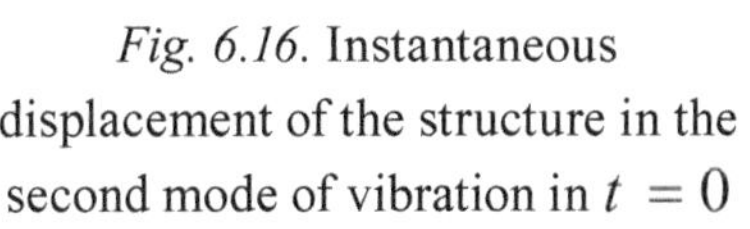

Fig. 6.16. Instantaneous displacement of the structure in the second mode of vibration in $t = 0$

Fig. 6.17. Instantaneous displacement of the structure in the second mode of vibration in $t = 0.00037$

6.2.2.4. The Bending Moment of Vibration in Various Modes

By determining the modal displacement of elements from equation 6.57, the bending moment along the cylinder length $M_{xx}(x,t)$ and along the angular cross-sectional area of the shell of the cylinder $M_{\theta\theta}(x,t)$ are gained as follows:

$$M_{xx} = -D\frac{\partial^2 W_n(x,t)}{\partial x^2} \quad , \quad M_{\theta\theta} = -\upsilon D\frac{\partial^2 W_n(x,t)}{\partial x^2}. \tag{6.73}$$

As a result, the charts for the first two modes of the bending moment of the vibration are depicted in the following forms:

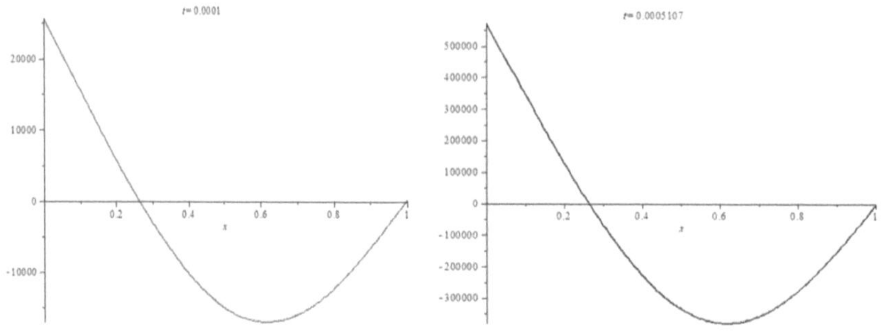

Fig. 6.18. $M_{xx}(x,t)$ for the first mode of vibration in $t = 0.0001$

Fig. 6.19. $M_{xx}(x,t)$ for the first mode of vibration in $t = 0.00051$

After that, for the second mode of vibration, we will have

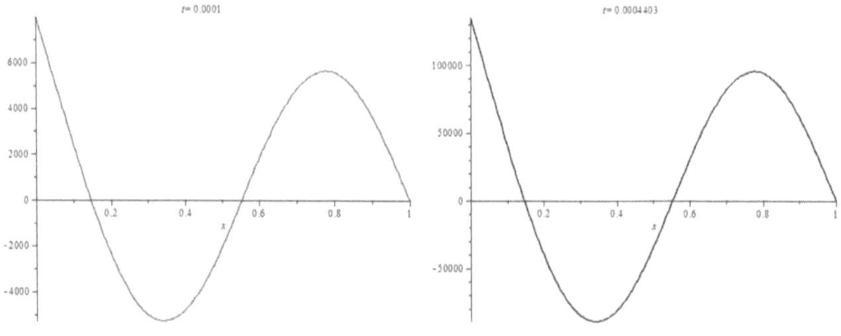

Fig. 6.20. $M_{xx}(x,t)$ for the second mode of vibration in $t = 0.0001$

Fig. 6.21. $M_{xx}(x,t)$ for the second mode of vibration in $t = 0.00044$

Afterward, the bending moment along the angular cross-sectional area of the cylindrical shell is depicted in the first and second modes of vibration as

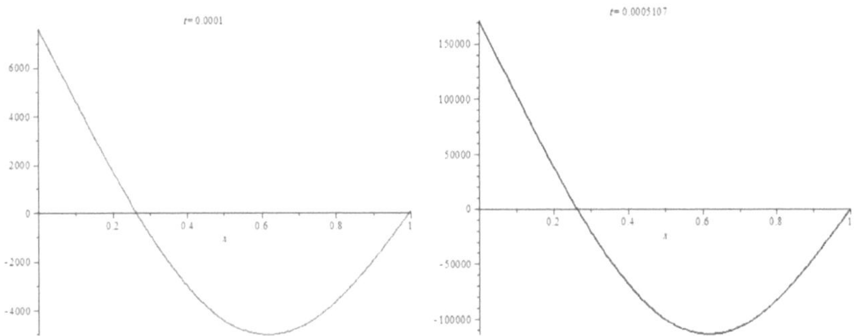

Fig. 6.22. $M_{\theta\theta}(x,t)$ for the first mode of vibration in $t = 0.0001$

Fig. 6.23. $M_{\theta\theta}(x,t)$ for the first mode of vibration in $t = 0.00051$

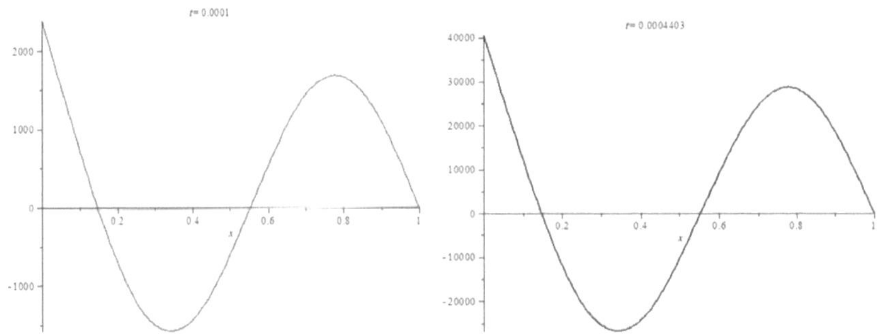

Fig. 6.24. $M_{\theta\theta}(x,t)$ for the second mode of vibration in $t = 0.0001$

Fig. 6.25. $M_{\theta\theta}(x,t)$ for the second mode of vibration in $t = 0.00044$

In addition, the shearing force of vibration along the cylinder length on its wall is obtained as follows:

$$Q_x = -D\frac{\partial^3 W_n(x,t)}{\partial x^3} \tag{6.74}$$

The first two modes of the shearing force are charted in the form of

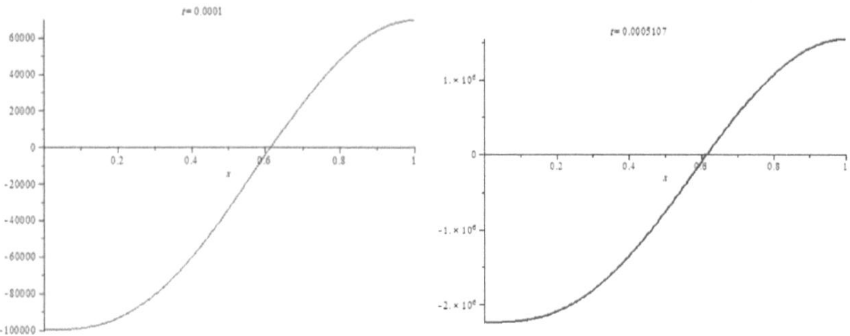

Fig. 6.26. Q_x for the first mode of vibration in $t = 0.0001$

Fig. 6.27. Q_x for the first mode of vibration in $t = 0.00051$

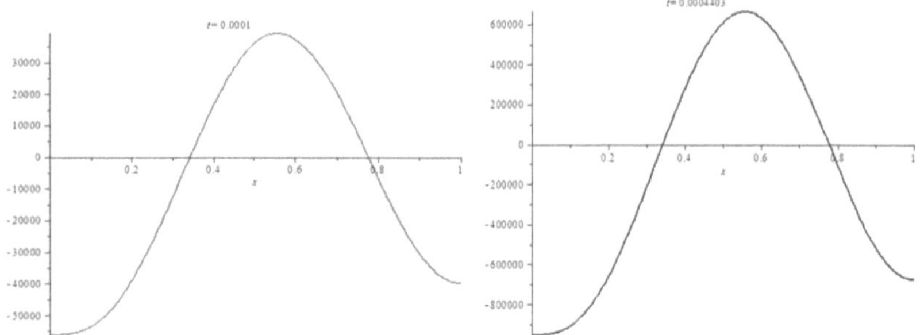

Fig. 6.28. Q_x for the second mode of vibration in $t = 0.0001$

Fig. 6.29. Q_x for the second mode of vibration in $t = 0.00044$

Then, the vibrational axial force along the length of the cylinder N_{xx} and along the surface of the cylinder shell $N_{\theta\theta}$ is computed as below:

$$N_{xx} = \frac{-Eh}{1-\upsilon^2}(\upsilon \frac{W_n(x,t)}{R}) \quad , \quad N_{\theta\theta} = \upsilon N_x - \frac{Eh}{R} W_n(x,t). \qquad (6.75)$$

The charts below have been shown for the axial force in the first mode as follows:

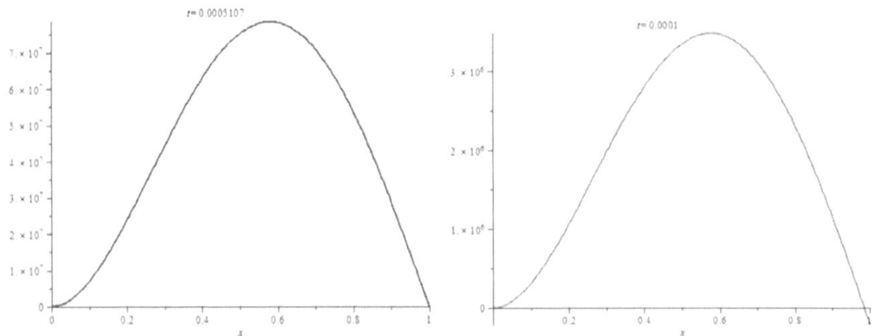

Fig. 6.30. N_{xx} for the first mode of vibration in $t = 0.0001$

Fig. 6.31. N_{xx} for the first mode of vibration in $t = 0.00051$

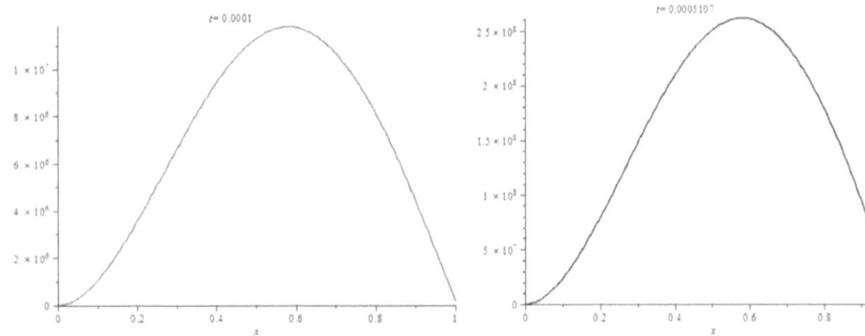

Fig. 6.32. $N_{\theta\theta}$ for the first mode of vibration in $t = 0.0001$

Fig. 6.33. $N_{\theta\theta}$ for the first mode of vibration in $t = 0.00051$

Example 6.2.3

Most of the chemical reactions that are with mass transfer are produced by catalysts to increase or decrease the rate of the reactions.[5] The form and physic of the catalyst is one of the vital and important parameters for determining the kind of differential equation governing the chemical system. In the present case study, we are going to solve the differential equation governing a chemical reaction on the surface of a catalyst, whose shape is like a cylinder.[6, 7] The main aim of solving the aforementioned equation is to obtain the concentration of reactive substances, along with the radius and length of the mentioned cylindrical catalyst. In order to achieve the aforementioned items, solving the obtained differential equation is necessary; but in most cases, classic analytic solutions cannot be used. So the equation should be solved by utilizing particular methods.

In the introduced problem, attempts have been made to solve partial differential equations governing on the system of chemical reactions by compounding AGM and Laplace transform theorem. To understand more, it is better to indicate that in the first step, partial differential equations will be converted into ordinary differential equations by applying Laplace transform theorem, and then it is the best time to utilize AGM in order to solve the obtained ODEs. And eventually, by calculating the inverse Laplace transform from the obtained equation, the desired solution is acquired. Additionally, it is necessary to mention that, basically, the Laplace transform theorem is applied in differential equations in which the time parameter is considered as one of their independent variables. But in this case study, we are going to solve a partial differential equation with two independent variables (x and r) in the following procedure: First, we consider the biggest independent variable of the presented problem, which is x and is obtainable in accordance with the physic of the problem. Then, we calculate the Laplace transform of the mentioned equation in terms of x instead of time, which is one of the important parts of this case study, and then the obtained ordinary differential equation will be solved by AGM.

Basically, the differential equations governing the chemical reactions systems with catalysts are complicated and mostly are obtained by writing mass and energy balance laws.[8]

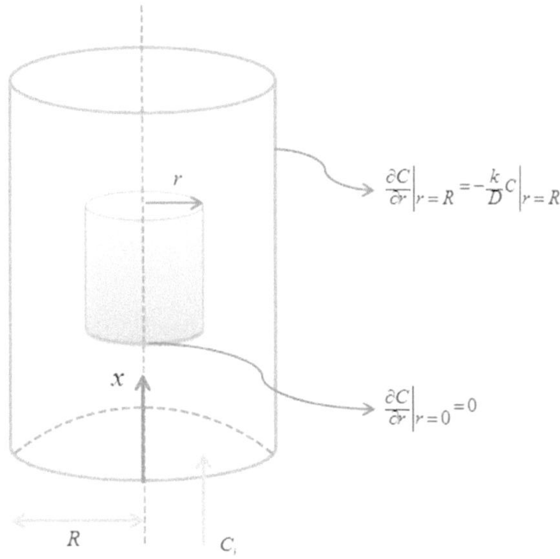

Fig. 6.34. The schematic diagram of the physical model

In the present paper according to figure 6.34, we assume that reactant (A), in which C_i is its concentration, enters a cylindrical catalyst surface and makes the following reaction:

$$A \rightarrow B. \tag{6.76}$$

Then, the rate of chemical reaction is obtained as follows:

$$r_A = \frac{-dN_A}{dt} = KC_A. \tag{6.77}$$

It is notable that along with the radius (r), the mass transfer phenomenon has happened and the concentration will be changed, along with the length of the cylinder (x-direction). In regard to the above explanations, the differential equation governing the mentioned system (mass transfer on the surface of the cylindrical catalyst) in the cylindrical coordinates is defined as

$$V\frac{\partial C(r,x)}{\partial x} = D\frac{1}{r}\frac{\partial}{\partial r}(r\frac{\partial C(r,x)}{\partial x}). \tag{6.78}$$

After that, the boundary conditions for the differential equation governing the mass transfer system are expressed as

$$C(r,0) = c_i \tag{6.79}$$

$$\frac{\partial C(0,x)}{\partial r} = 0 \quad , \quad \frac{\partial C(R,x)}{\partial r} = \frac{-K}{D} C \bigg|_{r=R}. \tag{6.80}$$

Inasmuch as the length of the cylinder is big and significant $(L >> r)$ in comparison with its radius, it is logical to consider the boundary conditions of the main differential equation, along with x-direction, as initial conditions. Now calculate the Laplace transform of equation 6.78, which is in terms of x. As a result, we will have [9]

$$Vr\{C(r,s) - C(r,0)\} = D\frac{d}{dr}\{r\frac{dC(r,s)}{dr}\} \tag{6.81}$$

In the aforementioned equation, (s) is defined as the Laplace parameter, and we are able to rewrite equation 6.81 in the following form:

$$f(r): Vr(\bar{C} - C_i) = D\frac{d}{dr}(r\frac{d\bar{C}}{dr}). \tag{6.82}$$

And consequently, the introduced boundary conditions are converted as follows:

$$\bar{C}'(0) = 0 \quad , \quad \bar{C}'(R) = \frac{-K}{D}\bar{C}\bigg|_{r=R}. \tag{6.83}$$

It is noteworthy that in equation 6.83, the parameter $\bar{C}$ is defined as $\bar{C}(r, s)$, so the introduced partial differential equation is converted into an ODE in terms of Laplace parameter (s), which is solvable by AGM.

6.2.3.1. Solving the Obtained Differential Equation by AGM

In the previous section, the presented partial differential equation has been converted into an ordinary differential equation by utilizing the Laplace transform theorem. In AGM, it is possible to choose a finite series of polynomials with constant coefficients as the answer of the achieved ordinary differential equation in the following form:

$$\bar{C}=\sum_{n=0}^{5} a_n r^n = a_0 + a_1 r + a_2 r^2 + a_3 r^3 + a_4 r^4 + a_5 r^5. \tag{6.84}$$

In the above equation, the constant coefficients (a_0 to a_5) will be obtained by applying boundary conditions.

6.2.3.2. Applying Boundary Conditions

The boundary conditions are applied in two ways in AGM:

a. Applying the boundary conditions on the answer of the main differential equation, which is equation 6.84, is expressed in the form of

$$\bar{C}=\bar{C}(BC). \tag{6.85}$$

It is notable that BC is the abbreviation of "boundary conditions."

In this step, the above procedure is also applied as

$$\bar{C}'(0)=0 \quad \rightarrow \quad a_1 = 0. \tag{6.86}$$

And then

$$\bar{C}'(R)=\frac{-K\bar{C}(R)}{D} \quad \rightarrow \quad a_1 + 2a_2 R + 3a_3 R^2 + 4a_4 R^3 + 5a_5 R^4 = -\frac{K}{D}(a_0 + a_1 R + a_2 R^2 + a_3 R^3 + a_4 R^4 + a_5 R^5). \tag{6.87}$$

b. Applying the boundary conditions on the main differential equation, which is equation 6.82 and is named $f(r)$, and also on its derivatives is done after substituting equation 6.84 into differential equation 6.82 in the following form:

$$f(\bar{C}(r)) \quad \rightarrow \quad f(\bar{C}(\mathrm{B.C}))=0 \quad , \quad f'(\bar{C}(\mathrm{B.C}))=0. \tag{6.88}$$

In this step in regard to the above explanations, the application of boundary conditions on the obtained equation is done as follows:

$$f(\bar{C}(R)):$$
$$D\{a_1+2a_2R+3a_3R^2+4a_4R^3+5a_5R^4+R(2a_2+6a_3R+12a_4R^2+20a_5R^3)\}= RV\{s(a_0+a_1R+a_2R^2+a_3R^3+a_4R^4+a_5R^5)-C_i\} \tag{6.89}$$

Now, we are allowed to apply boundary conditions on the derivatives of the yielded equation as

$$f'(\bar{C}(0)): \quad \rightarrow \quad 4Da_2=V(a_0s-c_i). \tag{6.90}$$

Then

$$f'(\bar{C}(R)):$$
$$D\{4a_2+12a_3R+24a_4R^2+40a_5R^3+R(6a_3+24a_4R+60a_5R^2)\}= V\{s(a_0+a_1R+a_2R^2+a_3R^3+a_4R^4+a_5R^5)-C_i\}+RV(a_1+2a_2R+ 3a_3R^2+4a_4R^3+5a_5R^4)S \quad . \tag{6.91}$$

For the second derivative of the obtained equation, we will have

$$f''(\bar{C}(R)): \quad \rightarrow \quad 18Da_3=2Va_1S. \tag{6.92}$$

By solving a set of algebraic equations consisting of six equations with six unknowns from equation 6.86 to equation 6.92, except equation 6.88, the constant coefficients of equation 6.84 can be obtained easily.

To simplify, the following new variables are considered as

$$\xi=27DV^2R^3S^2+154DVKR^2S+400D^2RVS+3V^2KR^4S^2+800KD^2$$

$$\psi_1 = 200DKR + KR^3VS + 27DVR^2S + 400D^2$$
$$\psi_2 = R^4V^2S^2 - 23DR^2S + 400D^2$$
$$\psi_3 = 25D - 3VR^2S \quad . \tag{6.93}$$

By substituting the obtained constant coefficients into equation 6.84, the solution of equation 6.82 will be achieved in terms of (r) and Laplace parameter (s) as follows:

$$a_0 = \frac{VR\psi_1C_i}{\xi}\ ,\ a_1 = 0\ ,\ a_2 = -\frac{VK\psi_2C_i}{2D\xi}\ ,\ a_3 = 0\ ,\ a_4 = -\frac{V^2K\psi_3C_iS}{2D\xi}\ ,\ a_5 = -\frac{V^3RKC_iS^2}{D\xi}. \tag{6.94}$$

And after substituting the obtained constant coefficients of the above equation into equation 6.84, we will have

$$\bar{C}(r,s) = \frac{VR\psi_1C_i}{\xi} - \frac{VK\psi_2C_i}{2D\xi}r^2 - \frac{V^2K\psi_3C_iS}{2D\xi}r^4 - \frac{V^3RKC_iS^2}{D\xi}r^5. \tag{6.95}$$

It is noteworthy that the above equation has been written in terms of the variable (r) and the parameter of Laplace transform (s), which is calculated by its inverse Laplace transform { L^{-1} }; the function of changing the concentration for the reactant (A) will be acquired in terms of $C(r, x)$.

For simplicity, the following new variables are considered as

$$\Omega = 3529K^2R^2 + 9200DKR + 40000D^2$$
$$\lambda = \frac{D(77KR + 200D)}{VR^2(KR + 9D)}\ ,\ \phi = \frac{D\sqrt{\Omega}}{R^2V(KR + 9D)}\ ,\ \eta = R^2VD(KR + 9D)^2. \tag{6.96}$$

Then in accordance with the given variables, the solution of the presented problem is expressed in the following form:

$$\begin{aligned}C(r,x) = {} & \frac{C_i}{18\Omega}\{-3V^2\Omega KR^3r^2(KR+9D)(R+2r)(R-r)^2 Dirac(x)+\\ & e^{-\frac{1}{3}\lambda x}[\{4KRD\Omega V(77KR+200D)r^5 - 3KR(179KR+625D)r^4 +\\ & KR^3(223KR+1021D)r^2 + 6R^4(k^2R^2+36KRD+243D^2)\}\cosh(\frac{\phi}{3}x)+\\ & DRV\sqrt{\Omega}\{-4K(4729K^2R^2+20000KDR+40000D^2)r^5 +\\ & 3KR(1025\times 10^5 D^2+62325KRD+1138K^2R^2) - KR^3(166417KRD+\\ & 495800D^2+18371K^2R^2)+6R^4(9D+KR)(5400D^2+4321KRD+\\ & 523K^2R^2)\sinh(\frac{\phi}{3}x)\}]\}\end{aligned} \quad . \quad (6.97)$$

It is necessary to mention that in the above equation, $(Dirac)$ is Dirac delta function.

By choosing the following physical values, we will have

$$c_i = 0.6 \;,\; V=1 \;,\; D=0.015 \;,\; R=0.3 \;,\; k=0.02 \;,\; L=5 . \quad (6.98)$$

The 3-D chart of the achieved solution, $C(r,x)$, along with the radius (r) and the length (x), after substituting the above physical values into equation 6.97, is illustrated as follows:

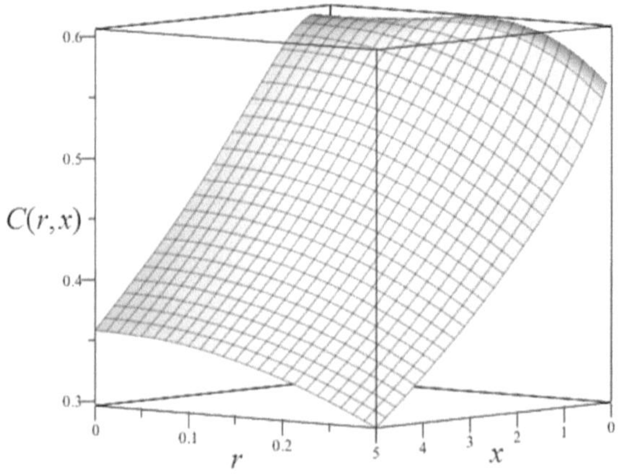

Fig. 6.35. The 3-D concentration chart of the reactant, along with the radius and length of the cylinder

Then, a 2-D contour of concentration can be displayed as

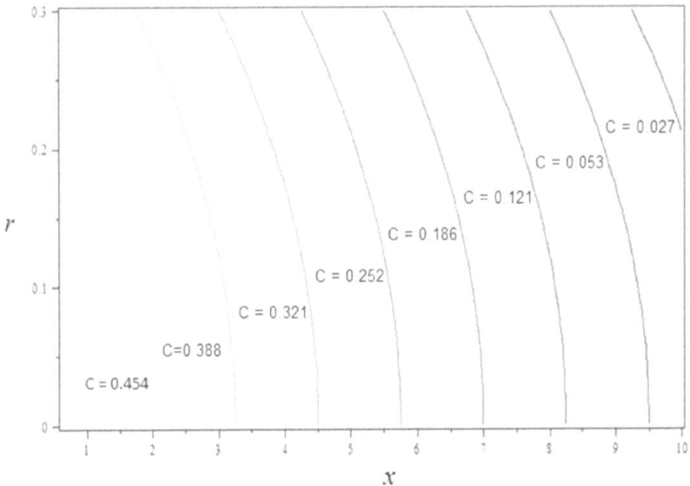

Fig. 3.36. The 2-D concentration contour of the mentioned problem

6.2.3.3. Comparing the Achieved Solutions by AGM and Numerical Method

In regard to the given physical values, the numerical solution of the presented problem in the specified domain of $x \in \{1 , 5\}$ is achieved as follows:

Table 6.2. The Obtained Numerical Results of Concentration along with the Radius in Five Different Spots of the Length of the Cylinder

x \ r	*0 m*	*0.05 m*	*0.1 m*	*0.15 m*	*0.2 m*	*0.25 m*	*0.3 m*
1 (m)	0.583	0.573	0.566	0.635	0.537	0.511	0.482
2 (m)	0.515	0.511	0.502	0.491	0.474	0.451	0.425
3 (m)	0.457	0.453	0.445	0.435	0.420	0.400	0.377
4 (m)	0.405	0.401	0.396	0.386	0.372	0.355	0.333
5 (m)	0.360	0.355	0.349	0.342	0.329	0.316	0.295

The following charts are depicted in five different spots from the length of the cylinder, which is $x = 1,2,3,4,5\,(m)$, by AGM and numerical method in order to compare the achieved solutions together in the following form:

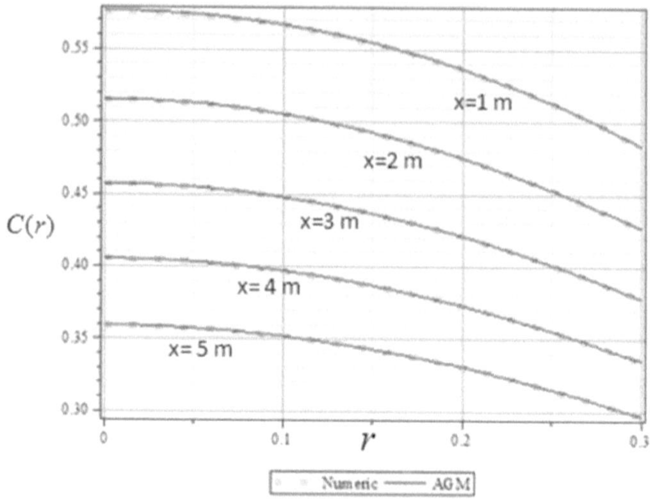

Fig. 6.37. Comparing the obtained solutions by AGM and numerical method (RKF45) in five different spots of the length of the cylinder

As regards figure 6.37, which compares the obtained solutions by AGM and numerical method, it is concluded that AGM is a reliable and precise method for solving a broad range of differential equations, and particularly compounding AGM with Laplace transform theorem will be a suitable approach for solving partial differential equations.

Example 6.2.4

Carbon nanotubes (CNTs) have attracted much industrial and academic attention because of their exciting potential applications in various fields such as sensors,[10] nanocomposites,[11] molecular devices,[12] or advanced materials with electronic properties.[13] New research shows that they may also hold the potential for revolutionizing medical research with magnetic resonance imaging of individual molecules. Furthermore, carbon nanotubes are more resistant than steel and have been considered as one of the strongest and hardest materials up to now. Their impressive electric and thermal properties make them an extremely versatile material. Hollow on the inside and only one atom thick, they lend themselves to a large variety of potential uses from tennis rackets and bulletproof vests to electronic components and energy-storage devices. It is necessary to mention that CNTs were discovered by Sumio Iijima in 1991 while studying the surfaces of graphite electrodes, which are used in an electric arc discharge. His observation and analysis of the nanotube structures started a new direction in carbon research that complemented the excitement and activities in this field. The structure of nanotubes remains completely different from traditional carbon fibers, which have been used industrially for several decades—for instance, in reinforcements in airplane frame parts, batteries, and so on. For the first time, this innovation (CNTs) represents the most perfect structure of which is entirely known at the atomic level. It is related to the predictable property of CNTs that mainly distinguishes them from other carbon fibers and puts them, along with molecular fullerene species, in a special category of prototype materials. Among the nanotubes, two varieties that differ in the arrangement of their grapheme cylinders have attracted much attention from researchers. Multi-walled nanotubes (MWNTs) are collections of several concentric grapheme cylinders and are larger structures compared to single-walled nanotubes (SWNTs), which are individual cylinders of 1–2 nm diameter. The research undertaken in this problem is devoted to solve the differential equations governing the nonlinear vibration of single-walled carbon nanotubes, which are embedded in an elastic medium by utilizing a new idea. This is related to the fact that classic analytic methods are not able to solve a wide variety of nonlinear differential equations such as the presented problem here.

Consider a CNT with the length l, Young's modulus E, density ρ, cross-sectional area A, and cross-sectional moment I, which is embedded in an elastic medium, as illustrated in the following picture:

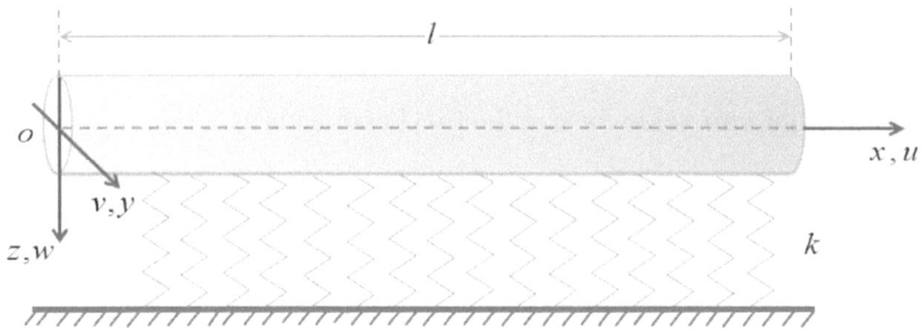

Fig. 6.38. The schematic diagram of the physical model

Suppose that $u(x,t)$, $v(x,t)$ and $w(x,t)$ are displacements corresponding to the axial, tangential, and radial directions, respectively, in terms of the special coordinate x and the time variable t. The free vibration equations of embedded nanotubes, considering the geometric nonlinearity of the structure, are[14]

$$\rho A\frac{\partial^2 u}{\partial t^2}-EA\frac{\partial^2 u}{\partial x^2}=\frac{1}{2}(EA-N)\frac{\partial}{\partial x}\left[(\frac{\partial v}{\partial x})^2+(\frac{\partial w}{\partial x})^2-2\frac{\partial u}{\partial x}[(\frac{\partial v}{\partial x})^2+(\frac{\partial w}{\partial x})^2]\right]+G \quad (6.99)$$

$$\rho A\frac{\partial^2 w}{\partial t^2}-N\frac{\partial^2 w}{\partial x^2}=EI_1\frac{\partial^4 w}{\partial x^4}(EA-N)\frac{\partial}{\partial x}(e\frac{\partial w}{\partial x})+F_1 \quad (6.100)$$

$$e=\frac{\partial u}{\partial x}-(\frac{\partial u}{\partial x})^2+\frac{1}{2}\left[(\frac{\partial v}{\partial x})^2+(\frac{\partial w}{\partial x})^2\right]. \quad (6.101)$$

In addition to the case in which the longitudinal excitation and the axial force are zero, the longitudinal inertia is also neglected. These assumptions simplify equation 6.99 to the following relation:

$$\frac{\partial^2 u}{\partial x^2}=-\frac{1}{2}\frac{\partial}{\partial x}\left[(\frac{\partial v}{\partial x})^2+(\frac{\partial w}{\partial x})^2\right]. \quad (6.102)$$

After integrating the above equation with respect to x, it is obtained that $\frac{\partial u}{\partial x}=e(t)-\frac{1}{2}\left[(\frac{\partial v}{\partial x})^2+(\frac{\partial w}{\partial x})^2\right]$; therefore, $u(x,t)$ can be achieved as

$$u(x,t)=c(t)+x\,e(t)-\frac{1}{2}\int_0^1\left[(\frac{\partial v}{\partial x})^2+(\frac{\partial w}{\partial x})^2\right]dx. \quad (6.103)$$

Assuming the zero value for u at the two beam ends as

$$u(0,t)=0 \quad \text{and} \quad u(l,t)=0. \quad (6.104)$$

Then, utilizing the above equations into equation 6.103, we will have

$$e(t)=\frac{1}{2l}\int_0^1\left[(\frac{\partial v}{\partial x})^2+(\frac{\partial w}{\partial x})^2\right]dx. \quad (6.105)$$

As a result, equation 6.99 and equation 6.100 can be transformed to the following equations:

$$\rho A\frac{\partial^2 w}{\partial t^2}+EI\frac{\partial^4 w}{\partial x^4}=\left[\frac{EI}{2l}\int_0^1\left[(\frac{\partial v}{\partial x})^2+(\frac{\partial w}{\partial x})^2\right]dx\right]\frac{\partial^2 w}{\partial x^2}+F_1 \tag{6.106}$$

$$\rho A\frac{\partial^2 v}{\partial t^2}+EI\frac{\partial^4 w}{\partial x^4}=\left[\frac{EI}{2l}\int_0^1\left[(\frac{\partial v}{\partial x})^2+(\frac{\partial w}{\partial x})^2\right]dx\right]\frac{\partial^2 w}{\partial x^2}+F_2. \tag{6.107}$$

In the above equations, F_1 and F_2 are prescribed transverse loads that are considered as the interaction pressure per unit axial length between the outmost tube and the surrounding medium, which can be described by the Winkler model as

$$F_1(x,t)=-kw \quad \text{and then} \quad F_2(x,t)=-kv. \tag{6.108}$$

For a nanotube with simply supported ends, the displacement functions $v(x,t)$ and $w(x,t)$ can be expressed as

$$w(x,t)=W(t)\sin\frac{m\pi x}{l} \qquad \text{After that} \qquad v(x,t)=V(t)\sin\frac{m\pi x}{l}. \tag{6.109}$$

The above expressions satisfy the end conditions $w=0$ and $M=0$ at both $x=0$ and $x=l$, where M is the bending moment. After substituting equation 6.108 and equation 6.109 into equation 6.106 and equation 6.107, the nonlinear differential equations in terms of time functions $v(t)$ and $w(t)$ for a SWNT can be derived as

$$\begin{cases}\dfrac{d^2W}{dt^2}+(\dfrac{\pi^4 EI}{\rho AL^4})W+(\dfrac{\pi^4 E}{4\rho L^4})W^3+(\dfrac{\pi^4 E}{4\rho L^4})V^3W=0 & (6.110)\\[2ex] \dfrac{d^2V}{dt^2}+(\dfrac{\pi^4 EI}{\rho AL^4})V+(\dfrac{\pi^4 E}{4\rho L^4})V^3+(\dfrac{\pi^4 E}{4\rho L^4})VW^2=0 & (6.111)\end{cases}$$

It is notable that W and V in the above formulae are defined as functions of time, $W(t)$ and $V(t)$.

Eventually, the final form of the differential equation governing the mentioned system is expressed as

$$\begin{cases} f(t): \ddot{w} + \alpha w + \beta w^3 + \lambda v^3 w = 0 & (6.112) \\ g(t): \ddot{v} + \alpha v + \beta v^3 + \lambda v w^2 = 0 & (6.113) \end{cases}$$

In the aforementioned formulas, the parameters α, β and λ are defined as follows:

$$\alpha = \frac{\pi^4 EI}{\rho A L^4} \quad , \quad \beta = \lambda = \frac{\pi^4 E}{4\rho L^4}. \tag{6.114}$$

In this step, the initial conditions are expressed for the nonlinear vibrational differential equations in the forms of

$$W(0) = X_0 \quad , \quad \dot{W}(0) = 0 \quad , \quad V(0) = A \quad , \quad \dot{V}(0) = 0. \tag{6.115}$$

6.2.4.1. Solving the Differential Equations by AGM

The solution of the given set of differential equations in this approach is considered as two polynomials of Fourier series with constant coefficients as follows:

$$W = e^{-at}\{b\cos(\omega_1 t + \phi_1)\} \tag{6.116}$$

$$V = e^{-ct}\{d\cos(\omega_2 t + \phi_2)\}. \tag{6.117}$$

The constant coefficients of the above equations—which are angular frequencies (ω_1, ω_2), a, b, c, d and initial vibrational phases (φ_1, φ_2)—can easily be gained by applying the initial and boundary conditions. It is citable that in vibrational problems, there are always initial conditions, but boundary conditions can exist or not in these kinds of problems.

6.2.4.2. Applying Initial or Boundary Conditions

The constant coefficients of the considered answers can be gained by applying the initial or boundary conditions in this new approach.

Due to the proposed physical model, there are no boundary conditions, so the constant coefficients of equation 6.116 and equation 6.117 are obtained just with respect to the given initial conditions that have been presented in equation 6.115.

It is noteworthy that in AGM, initial or boundary conditions are applied in two ways as follows:

a. In general, the initial conditions are applied on the considered answer functions in the form of

$$\mathrm{w} = \mathrm{w}(IC) \quad , \quad \mathrm{v} = v(IC). \tag{6.118}$$

For simplicity, IC is considered as the abbreviation of the initial conditions.

As a result, applying the initial conditions on equation 6.116 and equation 6.117 is done as

$$\mathrm{W}(0) = X_0 \quad \rightarrow \quad b\cos\phi_1 = X_0 \tag{6.119}$$

and

$$W'(0) = 0 \quad \rightarrow \quad b(a\cos\phi_1 + \omega_1 \sin\phi_1) = 0. \tag{6.120}$$

After that,

$$\mathrm{V}(0) = A \quad \rightarrow \quad d\cos\phi_2 = A. \tag{6.121}$$

Then, we have

$$V'(0) = 0 \quad \rightarrow \quad d(\mathrm{c}\cos\phi_2 + \omega_2 \sin\phi_2) = 0. \tag{6.122}$$

b. The application of initial or boundary conditions on the main set of differential equations—which, in this case, is defined by $f(t)$ and $g(t)$, and also on its derivatives—is done in the following general form:

$$f(w(t)) \quad \rightarrow \quad f(w(IC))=0 \quad , \quad f'(w(IC))=0 \quad , \quad \cdots \tag{6.123}$$

$$g(v(t)) \quad \rightarrow \quad g(v(IC))=0 \quad , \quad f'(v(IC))=0 \quad , \quad \cdots. \tag{6.124}$$

It is concluded that after substituting equation 6.116 and equation 6.117, which have been considered as the answer of the main set of differential equations, into equation 6.112 and equation 6.113, the initial conditions are applied on the obtained equations and also on their derivatives.

Therefore, the initial conditions are applied on the obtained set of equations as follows:

$f(w(0))$:

$$b(\alpha-\omega_1^2+a^2)\cos\phi_1+2ab\omega_1\sin\phi_1+b(\beta b^2\cos^2\phi_1+\lambda d^3\cos^3\phi_2)\cos\phi_1=0 \tag{6.125}$$

$f'(w(0))$:

$$\begin{aligned}&ab(3\omega_1^2-a^2-\alpha)\cos\phi_1+b\omega_1(\omega_1^2-3a^2-\alpha)\sin\phi_1-3\beta b^3(a+\omega_1\sin\phi_1)\cos^2\phi_1-\\&3\lambda d^3b(c+\omega_2\sin\phi_2)\cos\phi_1\cos^2\phi_2-\lambda d^3b(a\cos\phi_1+\omega_1\sin\phi_1)\cos^3\phi_2=0\end{aligned} \quad . \tag{6.126}$$

Then for the first derivative of the obtained set of equations, we will have

$f(v(0))$:

$$d(c^2-\omega_2^2+\alpha)\cos\phi_2+2cd\omega_2\sin\phi_2+d(d^2\cos^2\varphi_2+\lambda b^2\cos^2\phi_1)\cos\phi_2=0 \tag{6.127}$$

$f'(v(0))$:

$$\begin{aligned}&cd(3\omega_2^2-c^2-\alpha)\cos\phi_2+d\omega_2(3c^2+\alpha-\omega_2^2)\sin\phi_2-3\beta d^3(c.\cos\phi_2+\omega_2\sin\phi_2)\cos^2\phi_2-\\&\lambda d b^2(c.\cos\phi_2+\omega_2\sin\phi_2)\cos^2\phi_1-2\lambda db^2(a\cos\phi_1+\omega_1\sin\phi_1)\cos\phi_1\cos\phi_2=0\end{aligned} \quad . \tag{6.128}$$

By solving a set of algebraic equations consisting of eight equations with eight unknowns from equation 6.119 to equation 6.128, with the exception of equation 6.123 and equation 6.124, the constant coefficients a, b, c, d, φ_1, φ_2, ω_1 and ω_2 can be yielded easily.

Therefore, the constant coefficients of the considered answers are achieved as follows:

$$a=0 \quad , \quad b=X_0 \quad , \quad c=0 \quad , \quad d=-1 \quad , \quad \phi_1=0 \quad , \quad \phi_2=\pi . \tag{6.129}$$

And then, the angular frequencies (ω_1, ω_2) can be computed in the forms of

$$\omega_1=\sqrt{\alpha+\beta X_0^2+\lambda A^3} \quad , \quad \omega_2=\sqrt{\alpha+\beta A^2+\lambda X_0^2} . \tag{6.130}$$

After substituting the obtained values from equation 6.129 and equation 6.130 into equation 6.116 and equation 6.117, the solution of the mentioned problem will be obtained as

$$W(t)=X_0\cos\{\sqrt{\alpha+\beta X_0^2+\lambda A^3}t\} \tag{6.131}$$

$$V(t)=A\cos\{\sqrt{\alpha+\beta A^2+\lambda A X_0^2}t\} . \tag{6.132}$$

By selecting the physical values below

$$X_0=0.2 \quad , \quad A=0.1 \quad , \quad \alpha=0.3 \quad , \quad \beta=0.1 \quad , \quad \lambda=0.2 \tag{6.133}$$

$$\omega_1=0.5515\left(\frac{Rad}{Sec}\right) \quad , \quad \omega_2=0.5558\left(\frac{Rad}{Sec}\right) \tag{6.134}$$

the solution of the presented problem and the related angular frequencies are rewritten as follows:

$$W(t)=0.2\cos(0.5515t) \quad , \quad V(t)=0.1\cos(0.5558t) . \tag{6.135}$$

Then the charts of the obtained solutions and their phase planes are drawn in the following:

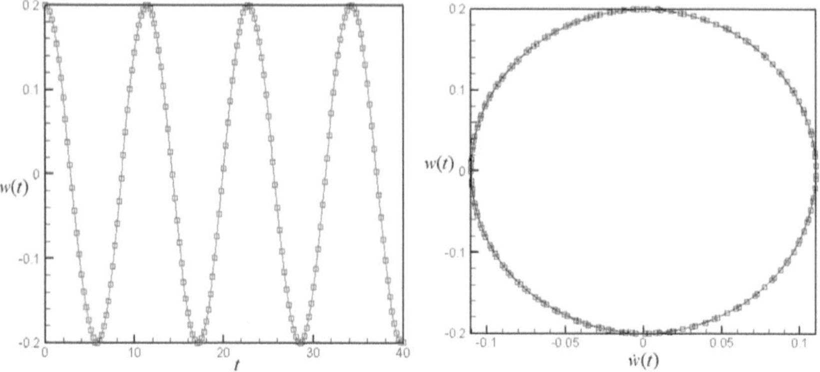

Fig. 6.39. The diagram of the obtained solution, $W(t)$, by AGM

Fig. 6.40. The resulted phase plane for the obtained $W(t)$ by AGM

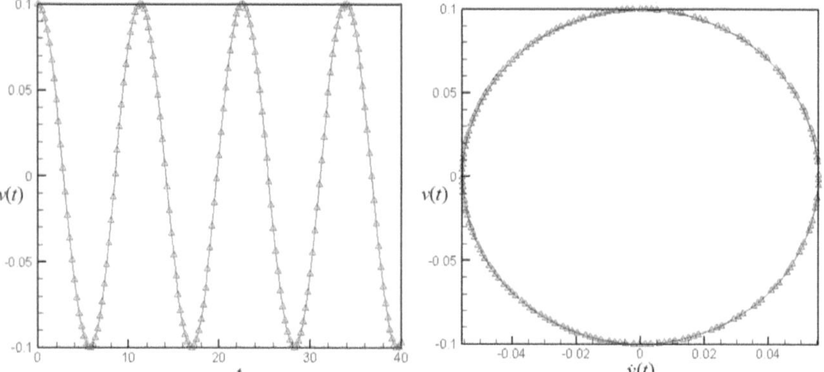

Fig. 6.41. The chart of the obtained solution, $V(t)$, by AGM

Fig. 6.42. The resulted phase plane for the obtained $V(t)$ by AGM

6.2.4.3. Numerical Solution

With regard to the given physical values from equation 6.133 and the determined domain of $t \in \{0\,,\,40\}$, defined in terms of second (sec), the solution of the proposed set of differential equations is expressed numerically in the following table:

Table 6.3. The Obtained Numerical Solution for $W(t)$, $V(t)$ and Their Derivatives according to the Given Physical Values

t	0	5	10	15	20
$W(t)$	0.2	-0.185	0.1422	-0.07837	0.002627
$W'(t)$	0	-0.04188	0.0774	-0.1012	0.1099
$V(t)$	0.1	-0.09311	0.7345	-0.04378	0.008147
$V'(t)$	0	-0.02027	0.0376	-0.04976	0.055139

Continuation of Table 6.3

t	25	30	35	40
$W(t)$	0.07324	-0.1385	0.18285	-0.19998
$W'(t)$	-0.10233	0.07945	-0.04466	0.002975
$V(t)$	0.02869	-0.06177	0.0866	-0.09965
$V'(t)$	-0.05312	0.04391	-0.02854	0.009033

Due to the obtained solutions and the obtained numerical solution, whose results have been presented in table 6.3, we will have the following comparisons:

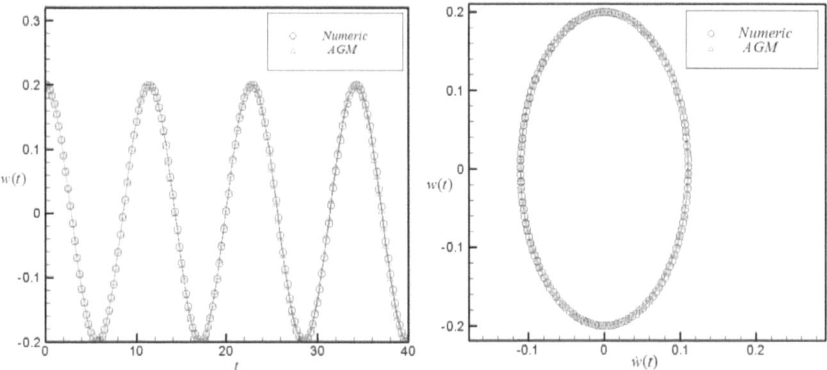

Fig. 6.43. A comparison between the achieved solution, $W(t)$,by AGM and numerical method

Fig. 6.44. Comparing the resulted phase planes for $W(t)$ by AGM and numerical method

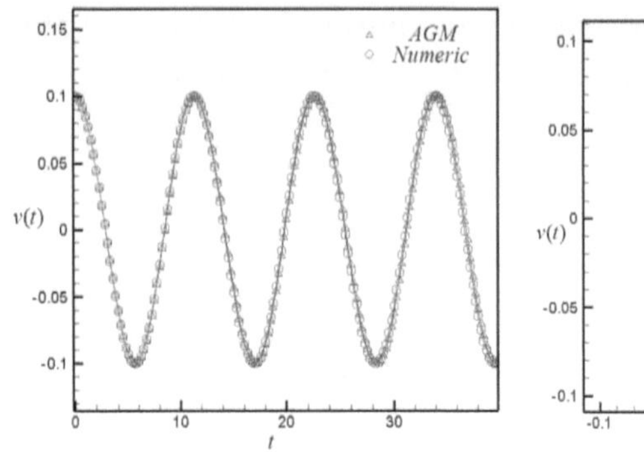

Fig. 6.45. A comparison between the achieved solution, $V(t)$,by AGM and numerical method

Fig. 6.46. Comparing the resulted phase planes for $V(t)$ by AGM and numerical method

In accordance with the displacement functions of the mentioned beam

$$w(x,t) = W(t)\sin\frac{m\pi x}{L} \quad , \quad v(x,t) = V(t)\sin\frac{m\pi x}{L}. \tag{6.136}$$

After substituting the obtained solutions by AGM from equation 6.135 into equation 6.136, the displacement charts for various modes of vibration (take for example in the second mode of vibration, (ω_2) can be achieved by selecting $m = 2$ and $L = 1$) as follows:

It is notable that by considering $m = 1, 2, 3, \ldots$, we will be able to draw the displacement functions in the other modes of vibrations.

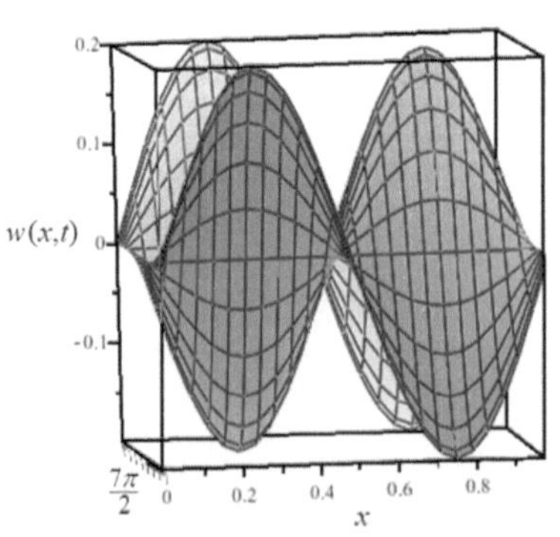

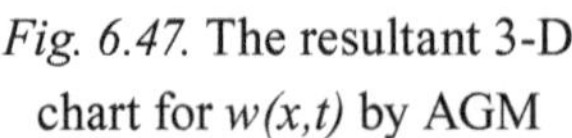

Fig. 6.47. The resultant 3-D chart for *w(x,t)* by AGM

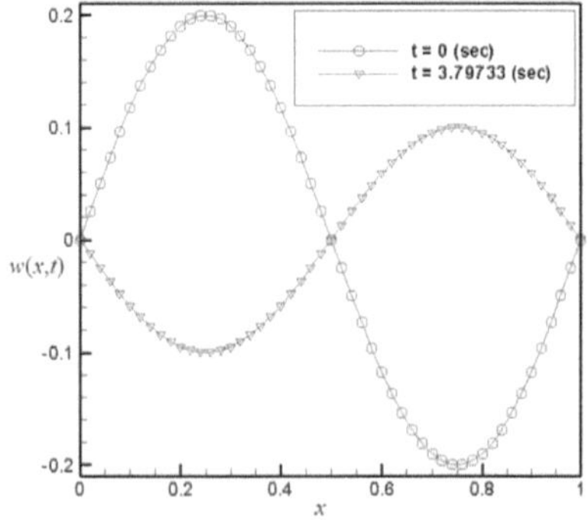

Fig. 6.48. The obtained displacement chart for *w(x,t)* by AGM

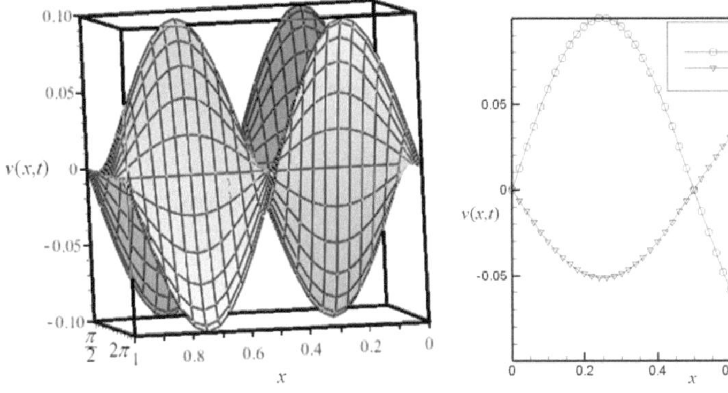

Fig. 6.49. The resultant 3-D chart for *v(x,t)* by AGM

Fig. 6.50. The obtained displacement chart for *v(x,t)* by AGM

In the introduced example, attempts have been made to analyze the nonlinear vibrations of single-walled carbon nanotubes, and as a result, the vibrational displacements in different directions and the related phase planes have been achieved and illustrated graphically. Moreover, the relevant angular frequencies have been obtained, and also, the displacement charts for the second mode of vibration have been depicted in two and three dimensions.

Example 6.2.5

Generally speaking, all unknowns of the problems of the theory of elasticity in dynamic state consist of three components of displacement, six stress components, and six strain components, along with time; and then the mentioned problems will be solved by creating three equilibrium equations, six compatibility equations of strains, and finally, six equations that are related to the structural behavior. But a wide variety of problems, due to their geometric forms and types of loading on them, can easily be changed into simpler forms and solved. In the discussion of 2-D problems, it is considered that if the plate thickness is small in comparison with the plate dimensions and also if the loading of the problem is limited to the internal forces of the plate, the problem can be solved in the state of 2-D stress with acceptable approximation. Investigating dynamically distributed loading on structural plates and reaction of plate structures as vibration is of great importance for users in civil engineering.[15] Plate vibrations in the presence of dynamic forces and loads exerted on them are, in fact, the investigation of the maximum instantaneous strength of the structural elements, which are dependent on some parameters such as stresses, bending, shearing forces, instantaneous displacement due to the vibration of the structures.

In this example, the vibration of circular plates in the presence of dynamic loads has been assessed and analyzed by the elasticity theory in dynamic state, and also, bending moments and instantaneous shearing forces have been discussed completely. Then, attempts have been made to obtain modal frequencies by presenting a general formula for circular plates, which can be considered as a novelty in the field of vibrations. A couple of differential equations for vibrational plates by the separation equation have been solved, and consequently, modal mass and shearing force have been studied completely. Moreover, modal frequencies (ω_n) and modal shapes have generally been achieved and depicted graphically.

According to figure 6.51, in order to gain the motion equations in terms of displacement function, it is essential to write Newton's first law for an optional volume element in a coordinate. As regards Newton's second law for equilibrium of forces, in accordance with figure 6.51, which is a circular plate element, the vibrational differential equation governing a circular plate with dynamic normal load in bending state is achieved as follows:[16]

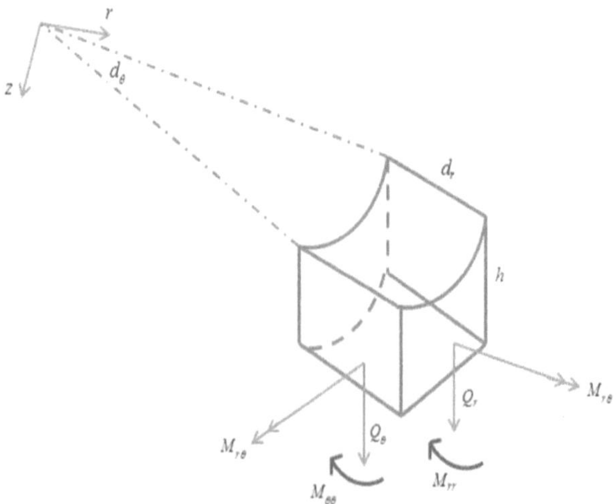

Fig. 6.51. The arbitrary circular plate element of the mentioned physical model with all the forces and moments exerted on it

$$D.\nabla^4 + \bar{m}\frac{\partial^2 u_z}{\partial t^2} = q(r,\theta,t) \tag{6.137}$$

In partial differential equation 6.137, the parameter $\bar{m}$ is defined as the unit mass per area and is expressed as $\bar{m} = \rho.h$, where h and ρ are thickness (m) and density (kg/m^3) of the circular plate, and also, D is bending rigidity (N/m^2) of the structure that is written as

$$D = \frac{Eh^3}{12(1-\upsilon^2)}. \tag{6.138}$$

Where υ is Poisson's ratio. Therefore, equation 6.137 can be rewritten as follows:[17]

$$D.\nabla^2(\frac{\partial^2 u_z}{\partial r^2} + \frac{1}{r}\frac{\partial u_z}{\partial r} + \frac{1}{r^2}\frac{\partial^2 u_z}{\partial \theta^2}) + \rho h\frac{\partial^2 u_z}{\partial t^2} = q. \tag{6.139}$$

The operator ∇^2 in polar coordinates is defined as follows:

$$\nabla^2 = \frac{\partial^2}{\partial r^2} + \frac{1}{r}\frac{\partial}{\partial r} + \frac{1}{r^2}\frac{\partial^2}{\partial \theta^2}. \tag{6.140}$$

Here, as regards figure 6.52 in the following, a circular plate is considered, whose surrounding is fixed, and the distributed load of vibration $q(r,t)$ is exerted on it. Since loading in different points of the plate is applied homogeneously, the forces exerted on the plate are axisymmetric. In axisymmetric state, all the terms of differential equation 6.139 are independent from the angular θ. So all the derivatives in relevance to θ are equal to zero; and as a result, in axisymmetric state, the governing differential equation 6.139 will be changed as

$$\frac{D}{r}\frac{\partial}{\partial r}\{r\frac{\partial}{\partial r}(\frac{1}{r}\frac{\partial}{\partial r}(r\frac{\partial u_z}{\partial r}))\} + \rho h\frac{\partial^2 u_z}{\partial t^2} = q(r,t). \tag{6.141}$$

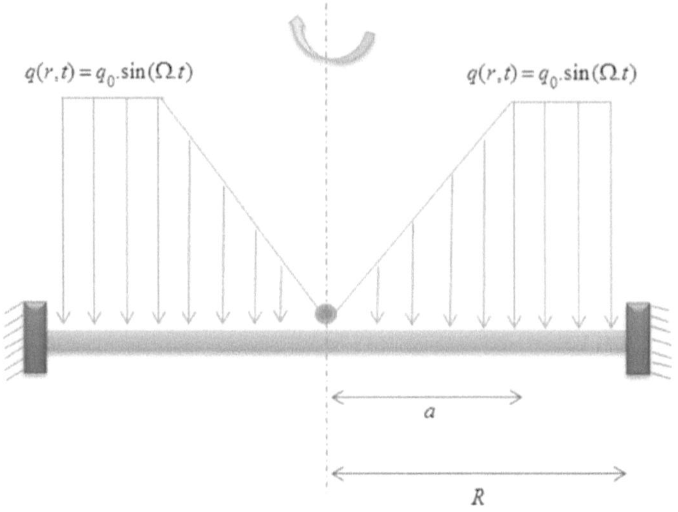

Fig. 6.52. The schematic diagram of the symmetrical circular plate

Therefore, in regard to figure 6.52, $q(r,t)$ is written as follows:

$$q(r,t)=\sin(\Omega t).\begin{cases}(\frac{r}{a})q_0 & 0\le r\le a\\ q_0 & a\le r\le R\end{cases}. \tag{6.142}$$

6.2.5.1. Solving the Differential Equation

Differential equation 6.141 is solved by the separation equation as follows:[18–20]

$$u_z = V(r).z(t). \tag{6.143}$$

After substituting equation 6.143 into equation 6.141 and after arranging and substituting square of modal frequency, two differential equations have been yielded in the following forms:

$$\frac{D}{\rho h V(r)}\{\frac{d^4V(r)}{dr^4}+\frac{2}{r}\frac{d^3V(r)}{dr^3}-\frac{1}{r^2}\frac{d^2V(r)}{dr^2}+\frac{1}{r^3}\frac{dV(r)}{dr}\}=\omega_n^2 \tag{6.144}$$

$$\frac{1}{z(t)}\frac{d^2z(t)}{\partial t^2}-\frac{\sin(\Omega.t)}{\rho hV(v)z(t)}\begin{cases} q_0(\frac{r}{a}) & 0\le r\le a \\ q_0 & a\le r\le R \end{cases}=-\omega_n^2. \tag{6.145}$$

And then, equation 6.145 can be rewritten as

$$\frac{d^2z(t)}{dt^2}+\omega_n^2z(t)=\frac{1}{M_n}f_n(\tau). \tag{6.146}$$

In equation 6.146, the parameters $f_n(\tau)$ and M_n are force function and modal mass of the plate vibration, respectively, which are defined as equations below:

$$M_n=\int_0^{2\pi}\int_0^{R}\rho hV_n^2(r).rdrd\theta \tag{6.147}$$

$$f_n(\tau)=\int_0^{2\pi}\{(\int_0^{a}q_0(\frac{r}{a})+\int_a^{R}q_0)V_n(v).rdr\}\sin(\Omega t)d\theta. \tag{6.148}$$

In equation 6.148, $V_n(r)$ is the function of modal shapes, which, by applying boundary conditions, will be computed in the next part of the example.

6.2.5.1.1. Computing Modal Frequencies and Modal Shapes

By solving equation 6.144 with regard to the boundary conditions of clamping ends, modal frequencies and shapes can be calculated. The answer of differential equation 6.144 is achieved as the Bessel function of the first and second kinds (J and Y) from zero order as follows:

$$V(r)=c_1J_0(\lambda r)+c_2Y_0(\lambda r)+c_3J_0(i\lambda r)+c_4Y_0(i\lambda r). \tag{6.149}$$

It is notable that in the above equation, the parameter i is an imaginary part of complex number, and λ is defined as follows:

$$\lambda = \sqrt{\omega_n \sqrt{\frac{\rho h}{D}}}. \tag{6.150}$$

6.2.5.2. Applying Boundary Conditions

In regard to the circular plate with fixed surrounding and also with axisymmetric loading, boundary conditions can be defined as the following:

The plane slope at the circle center $(r = 0)$ is zero because of the axisymmetric state in the circular plane:

$$\frac{dV(r)}{dr} = 0 \quad at \quad r = 0. \tag{6.151}$$

Due to the characteristics of Bessel function from the second kind, (Y), in $r = 0$, we have

$$Y_0 = \infty. \tag{6.152}$$

According to equation 6.152 and the answer of differential equation 6.149, coefficients (c_2 and c_4) will be zero in order to keep the stability of the system.

$$c_2 = 0 \quad , \quad c_4 = 0 \tag{6.153}$$

As a result, the answer of differential equation 6.149 can be rewritten in accordance with equation 6.153 as follows:

$$\mathrm{V}(r) = c_1 J_0(\lambda r) + c_3 J_0(i\lambda r). \tag{6.154}$$

Due to the fixed end of the circular plate, displacement and plane slope in points, which are joined to the body, will obtain zero as follows:

$$B.C: \begin{cases} V(R) = 0 & \text{therefore} \quad c_1 J_0(\lambda R) + c_3 J_0(i\lambda R) = 0 \quad (6.155) \\ \dfrac{dV(R)}{dr} = 0 & \text{and then} \quad -c_1 \lambda J_1(\lambda R) + i\lambda c_3 J_1(i\lambda R) = 0 \quad (6.156) \end{cases}$$

In order to compute a nonzero answer from equation 6.155 and equation 6.156, the matrix determinant of the aforementioned set of equations should be zero as [21]

$$\det = \begin{vmatrix} J_0(\lambda R) & J_0(i\lambda R) \\ J_1(\lambda R) & iJ_1(i\lambda R) \end{vmatrix} = 0. \tag{6.157}$$

The answer of equation 6.157 is yielded by introducing new variable $(\beta_n = \lambda . R)$ as

$$\det = J_0(i\beta_n).J_1(\beta_n) - iJ_0(\beta_n).J_1(i\beta_n) = 0. \tag{6.158}$$

If equation 6.158 is depicted as a function in Cartesian coordinates, the roots of the equation can be observed in the following chart:

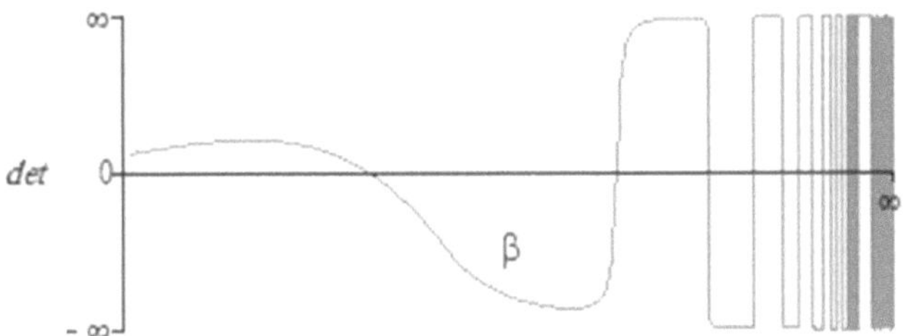

Fig. 6.53. The roots of the determinant function

Finally, the values of (β_n) from equation 6.158 are acquired as follows:

Table 6.4. The Results of Computing β_n for Different Modal Frequencies

(n)	*1*	2	3	4	5	6	7	8	9	*10*
β_n	3.1962	6.306	9.439	12.577	15.716	18.856	21.997	25.137	28.278	31.420

On the basis of the obtained values from table 6.4 and changing variable $(\beta_n = \lambda . R)$, parameter (λ) can be gained as follows:

$$\lambda = \frac{\beta_n}{R} = (1.9974957n + 0.0226)(\frac{\pi}{2R}). \tag{6.159}$$

By substituting equation 6.159 and changing variable $(\beta_n = \lambda.R)$ into equation 6.150, modal frequencies can be achieved as

$$\omega_n = \sqrt{\frac{D}{\rho h}}(\frac{\beta_n}{R})^2. \tag{6.160}$$

6.2.5.3. Computing the Equation of Modal Shapes

By computing the coefficient C_3 in terms of the coefficient C_1 from equation 6.155 or equation 6.156 and substituting into equation 6.154, the equation of modal shapes can be calculated as follows:

$$V_n(r) = \frac{c_1}{J_1(i\beta_n)}\{J_0(\beta_n \frac{r}{R}).J_1(i\beta_n) + iJ_1(\beta_n).J_0(i\beta_n \frac{r}{R})\}. \tag{6.161}$$

In regards to the values of β_n from table 6.4 or equation 6.159, the modal shapes can be depicted in different modes for the circular plate as below.

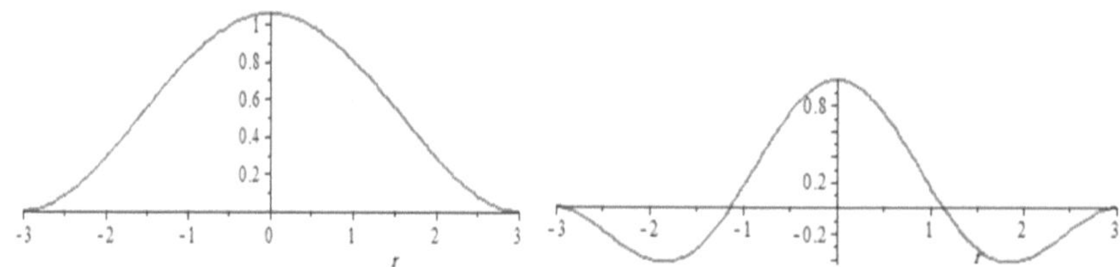

Fig. 6.54. The first mode of vibration (n=1)

Fig. 6.55. The second mode of vibration (n=2)

6.2.5.4. Solving the Differential Equation of Time Index

By determining the function of modal shape from equation 6.161, the functions of modal mass M_n and modal force $f_n(\tau)$ from integrated equations 6.147 and 6.148 can easily be achieved. And also, by determining modal mass and force, the differential equation of time index, which is equation 6.146, can simply be computed. It is notable that in order to compute the function of time index, initial conditions (initial displacement and initial velocity) are needed—in which, by supposing the lack of initial conditions, the answer of the time index differential equation can be yielded in the form of

$$z_n(t) = \frac{1}{\omega_n M_n} \int_0^t f_n(\tau) \sin\{\omega_n (t-\tau)\} d\tau . \tag{6.162}$$

By determining the displacement of time index $z_n(t)$ from equation 6.162 and the modal shape function $V_n(r)$ from equation 6.161, the vibrational displacement of structure element of the circular plate for every mode (n) can be acquired as follows:

$$U_z(r,t) = u_n(r).z_n(t). \tag{6.163}$$

Afterward, the displacement of the structure can be gained for all the modes according to the following equation:

$$U_z(r,t) = \sum_{n=1}^{\infty} u_n(r).z_n(t). \tag{6.164}$$

6.2.5.5. Calculating Bending Moments, Stresses, and Shearing Forces in Every Mode of Vibration

Bending moments for each mode of vibration (n) are achieved according to the following equations:

$$\begin{cases} M_{rr} = -D\left(\dfrac{\partial^2 u_z}{\partial r^2} + \dfrac{\upsilon}{r}\dfrac{\partial u_z}{\partial r}\right) \\ M_{\theta\theta} = -D\left(\dfrac{1}{r}\dfrac{\partial u_z}{\partial r} + \upsilon \dfrac{\partial^2 u_z}{\partial r^2}\right). \end{cases} \tag{6.165}$$

And shearing forces for each mode of vibration are computed in regard to the equations below:

$$\begin{cases} Q_r = -D\dfrac{\partial}{\partial r}(\dfrac{\partial^2 u_z}{\partial r^2} + \dfrac{1}{r}\dfrac{\partial u_z}{\partial r}) \\ Q_\theta = 0 \end{cases} \quad . \tag{6.166}$$

In this step, the stresses exerted on the plate for each mode of vibration are acquired in the forms of

$$\begin{cases} \sigma_{rr} = -\dfrac{E.h}{1-\upsilon^2}(\dfrac{\partial^2 u_z}{\partial r^2} + \dfrac{\upsilon}{r}\dfrac{\partial u_z}{\partial r} \\ \sigma_{\theta\theta} = -\dfrac{E.h}{1-\upsilon^2}(\dfrac{1}{r}\dfrac{\partial u_z}{\partial r} + \upsilon\dfrac{\partial^2 u_z}{\partial r^2}) \\ \sigma_{r\theta} = 0 \end{cases} \quad . \tag{6.167}$$

By choosing the following physical values for the structure such as

$$q_0 = 50(N/m^2) \quad , \quad E = 3\times 10^7 (N/m^2) \quad , \quad \upsilon = 0.3 \quad , \quad h = 3(cm)$$
$$R = 3(m) \quad , \quad a = 2(m) \quad , \quad \Omega = 2(Rad/\sec) \quad , \tag{6.168}$$

the displacement of the circular plate, bending moment, shearing force, and stress can be observed in the following charts for each mode of vibration. It is noteworthy that each time period, $T = \dfrac{2\pi}{\omega_n}$, is divided into four time domains as follows:

Table 6.5. The Obtained Modal Frequencies in Different Modes of Vibrations

n	1	2	3	4	5	6	7	8	9	10
ω_n	1.067	4.152	9.303	16.516	25.790	37.125	50.522	65.979	83.497	103.077

In regard to the aforementioned table in the first mode of vibration, the following charts are depicted on the basis of $\omega_1 = 1.067$:

The displacement chart of the circular element

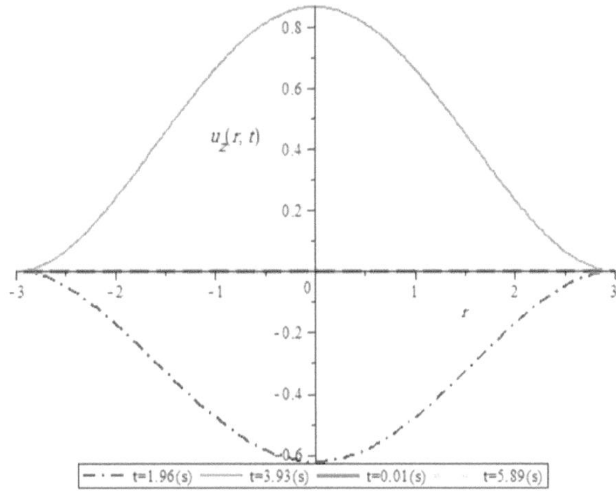

Fig. 6.56. A comparison among various instantaneous displacements, $u_z(r,t)$, of the mentioned structure in the first mode of vibration for $t = 0.01\ ,1.96\ ,3.93\ ,\ 5.89(s)$

The chart of bending moment

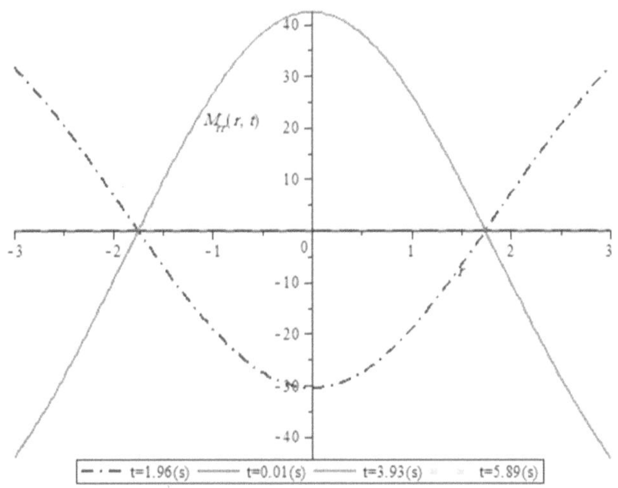

Fig. 6.57. A comparison among various $M_{rr}(r,t)$ of the mentioned structure in the first mode of vibration for $t = 0.01\ ,1.96\ ,3.93\ ,\ 5.89(s)$

Fig. 6.58. A comparison among various $M_{\theta\theta}(r,t)$ of the mentioned structure in the first mode of vibration for $t = 0.01\ ,1.96\ ,3.93\ ,\ 5.89(s)$

Then, the chart of shearing force is depicted as follows:

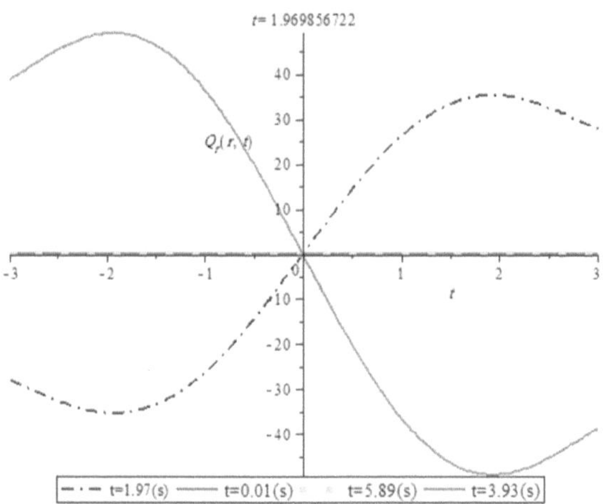

Fig. 6.59. A comparison among various $Q_r(r,t)$ of the mentioned structure in the first mode of vibration for $t = 0.01\ ,1.97\ ,3.93\ ,\ 5.89(s)$

Afterward, the stress charts are illustrated as

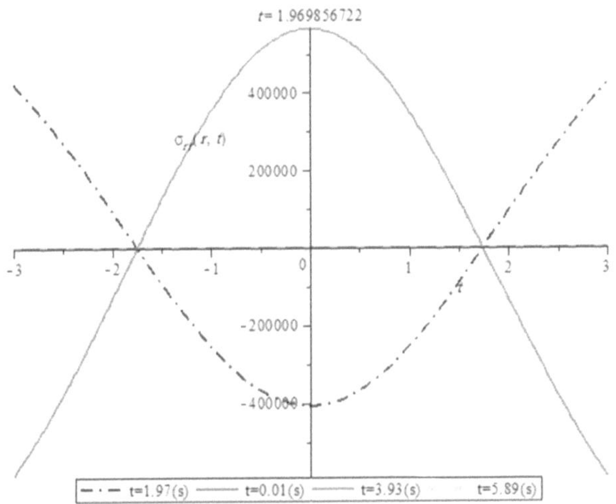

Fig. 6.60. A comparison among various $\sigma_{rr}(r,t)$ of the mentioned structure in the first mode of vibration for $t = 0.01\ ,1.97\ ,3.93\ ,\ 5.89(s)$

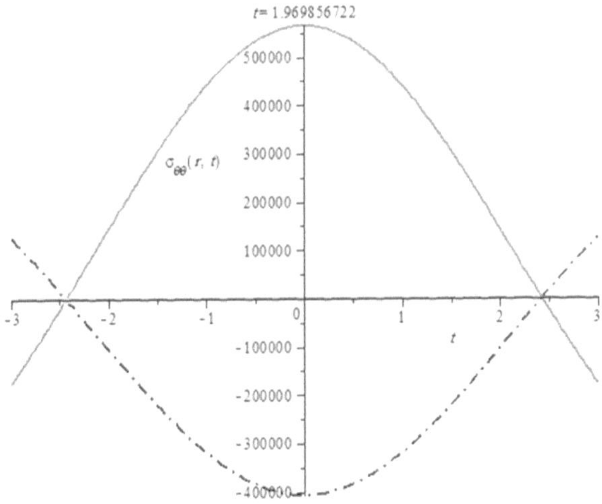

Fig. 6.61. A comparison among various $\sigma_{\theta\theta}(r,t)$ of the mentioned structure in the first mode of vibration for $t = 0.01\ ,1.96\ ,3.93\ ,\ 5.89(s)$

Eventually, in the second mode of vibration, in which $\omega_2 = 4.152$, the following charts are illustrated:

The displacement chart of the circular element

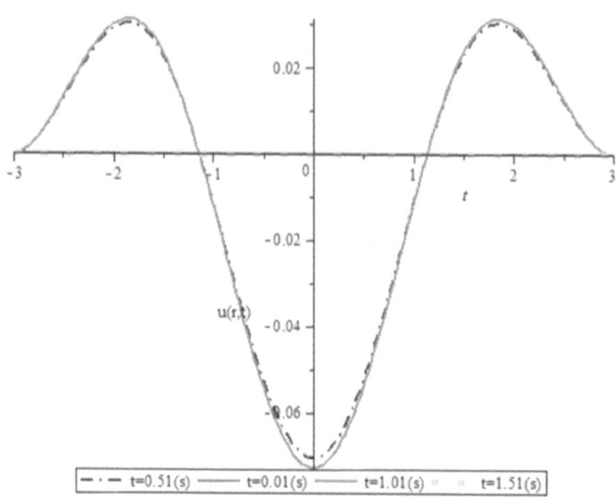

Fig. 6.62. A comparison among various instantaneous displacements, $u_z(r,t)$, of the mentioned structure in the second mode of vibration for $t = 0.01\ ,0.51\ ,1.01\ ,\ 1.51(s)$

In accordance with the obtained displacement charts, it is clear to say that displacement in the first mode of vibration is more than the next ones; and also, the same result is obtained for bending moment, shearing and stress forces. Furthermore, inasmuch as the mentioned circular plate is axisymmetric, the charts of modal shapes have been acquired symmetrically.

It is notable that the other charts of the second mode and upper modes of vibration, such as stress, shearing force, and bending moment, will easily be computed like the above procedure.

6.2.5.6. Conclusion

To summarize the whole procedure of the presented example, reading the following lines is recommended.

In order to perform the aforementioned procedure, Newton's first law for an arbitrary circular plate element and Newton's second law for equilibrium of forces have been applied to achieve the vibrational differential equation governing a circular plate with dynamic normal load in bending state. Since loading in different points of the mentioned plate is applied homogeneously, the forces exerted on the plate are axisymmetric; and as everyone knows, in axisymmetric state, all the terms of the obtained differential equation 6.139 are independent from the angular θ. So all the derivatives in relevance to θ are equal to zero, and as a result, the equation governing the present system is simplified significantly. After that, by utilizing separation law, modal frequencies and modal shapes have been acquired in regard to the boundary conditions. Consequently, the functions of modal mass and modal frequencies have been achieved. In this step, the time index function has been yielded in order to compute the vibrational displacement of the circular plate for each mode of vibration on the basis of the most general solution, which is the sum of all solutions (separation law). Eventually, the values of bending moment, shearing force, and shearing stress for the first and second mode of vibrations have been acquired and depicted completely.

6.3. Problems

6.3.1. Analyze the heat transfer along with the radius of a cylinder in accordance with the following figure:

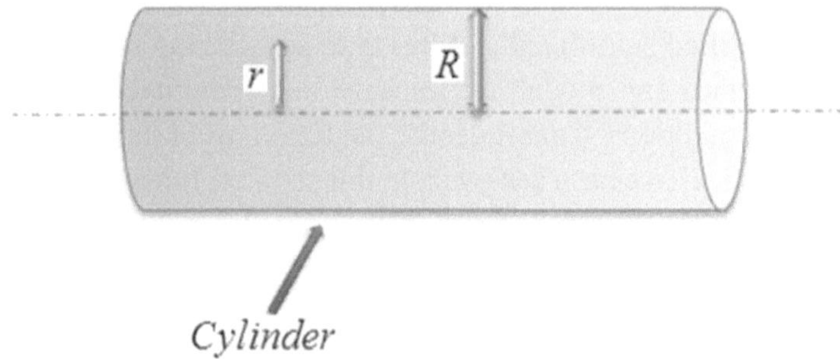

The schematic diagram of the physical model

Then, compare the achieved solution by numerical method (RKF45) to check the precision of acquired solution by AGM.

The partial differential equation governing the cylinder is introduced as follows:

$$\frac{1}{r}\frac{\partial}{\partial r}\left(r\frac{\partial T(r,t)}{\partial r}\right)=\frac{1}{\alpha}\frac{\partial T(r,t)}{\partial t}.$$

And the boundary and initial conditions of the mentioned cylinder are expressed in the forms of

$$B.C:\begin{cases}T(R,t)=T_1\\ \dfrac{\partial T}{\partial r}(0,t)=0\end{cases}\quad,\qquad IC:T(r,0)=T_0.$$

After that, obtain the 3-D chart of temperature distribution in the Cartesian coordinates and depict the 2-D contour for the temperature distribution in the cylinder in terms of centigrade.

Help: Utilize the Laplace transform theorem in order to convert the partial differential equation governing the aforementioned system to an ODE, and then solve the yielded equation by AGM.

6.3.2. Unwanted vibrations originating from various sources are considered as a harmful factor for systems.

Therefore, engineers have been trying to eliminate them by vibrational isolators such as absorbers and dampers, which are the only way to reduce the effect of unwanted vibrations.

As a result, investigate the vibrations of the introduced isolated beam column in the following figure with harmonic load in different modes and force transmission to the foundation of the system.

Also analyze the forced vibration spectrum for choosing the lowest frequency of the nonlinear system, because this frequency may be resonated and damage the system. Furthermore, compare the obtained results by numerical method and AGM in order to assess the scientific validity.

Consider the following beam column with certain physical properties, which are the bending, stiffness *EI*, cross-sectional area *A*, density ρ, and length *L*, which its surrounding is completely fixed and depicted completely as below:

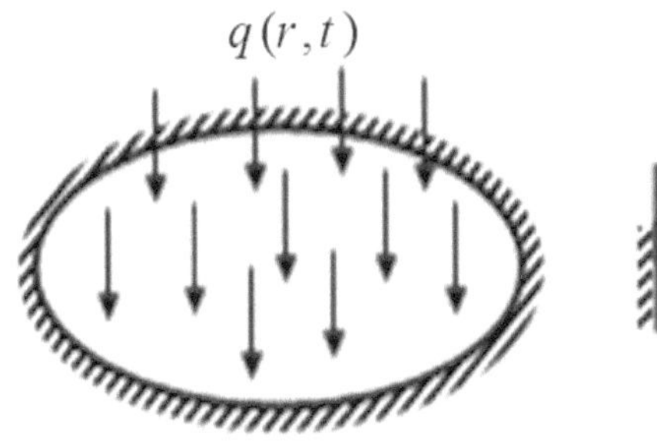

The schematic diagram of the circular plate

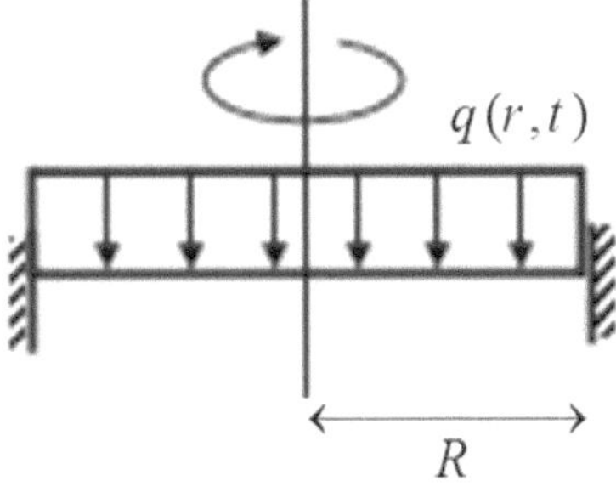

The frontal view of the symmetrical circular plate

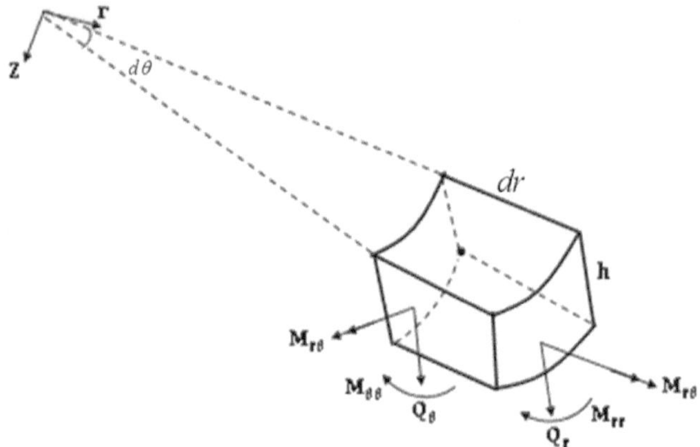

Exerting forces and moments on a differential element of the circular plate

An axial force N and a sinusoidal force $q(r,t)$ are exerted on this vibrational system. Therefore, the nonlinear partial differential equation governing this beam column with external loads is introduced as follows:[22]

$$\rho h \frac{\partial^2 w(r,t)}{\partial t^2} + \frac{1}{r}\frac{\partial}{\partial r}\{r D \frac{\partial}{\partial r}[\frac{1}{r}\frac{\partial}{\partial r}(r \frac{\partial w(r,t)}{\partial r})]\} - \frac{E}{2R}\{\frac{\partial}{\partial r}[A \frac{\partial w(r,t)}{\partial r}]\}\int_0^R (\frac{\partial w(r,t)}{\partial r})^2 dr = q(r,t)$$
.

By applying the method of separation of variables as $w(r,t) = \phi(r).z(t)$, the above equation can be rewritten as

$$\ddot{z}(t) + \frac{D}{\rho h r^3 \phi(r)}\left\{r^3 \phi''''(r) + 2r^2 \phi'''(r) - r\phi''(r) + \phi'(r)\right\} z(t) +$$

$$\frac{E A \phi''(r) \int_0^R [\phi'(r)]^2 dr}{-2R \rho h \phi(r)} . z^3(t) = \frac{q_0 \sin(\Omega t)}{\rho h \phi(r)} \quad , \quad D = \frac{E h^3}{12(1-\upsilon^2)}.$$

Based on the aforementioned equation, the shape function $\phi(r)$ can be obtained according to the modal frequency ω_n; and also, by considering the orthogonality of modes and applying Galerkin method, we will have

$$\ddot{z} + \alpha z + \beta z^3 = \eta \sin(\Omega t).$$

Then in accordance with the orthogonality of the modes, the parameters α, β, γ are defined as follows:

$$\alpha = \frac{D}{\int_0^R \rho h\, r^3 \phi^2(r) dr}\{\int_0^R [r^3 \phi''''(r) + 2r^2 \phi'''(r) - r\,\phi''(r) + \phi'(r)]\phi(r) dr\}$$

$$\beta = \frac{EA \int_0^R \left\{ \phi''(r) \int_0^R (\phi'(r))^2 dr \right\} \phi(r) dr}{-\int_0^L 2\rho h R \phi^2(r) dr} \quad , \quad \eta = \frac{\int_0^R q_0 \phi(r) dr}{\int_0^R \rho h\, \phi^2(r) dr}.$$

In order to calculate $\phi(r)$, we can equal the coefficient of z to the square of modal frequency as

$$\frac{D}{\rho h\, r^3 \phi(r)}\left\{r^3 \phi''''(r) + 2r^2 \phi'''(r) - r\,\phi''(r) + \phi'(r)\right\} = \omega_n^{\ 2}.$$

Therefore, the solution of the above equation can be obtained as follows:

$$\phi(r) = c_1 J_0(\lambda r) + c_2 Y_0(\lambda r) + c_3 J_0(i\lambda r) + c_4 Y_0(i\lambda r).$$

Where $i = \sqrt{-1}$ is the imaginary part of complex number, J_0 , Y_0 are Bessel functions of the first order, and the parameter λ is defined as follows:

$$\lambda = \sqrt[4]{\frac{\rho h}{D}\omega_n^{\ 2}} \quad \text{n} = 1, 2, 3, \ldots$$

On the basis of the given explanations, calculate the mode shape function $\phi(r)$ and modal frequency ω_n by applying boundary conditions of the given beam column as

$$\phi(R) = 0 \quad , \quad \phi'(R) = 0.$$

Then, plot the mode shape function graphs for natural frequency in different modes and then solve the nonlinear characteristic time equation for the second mode of vibration (n=2) while R=1 (m) by AGM.

By considering the following physical quantities for the beam

$$q_0 = 10(\frac{N}{m^2}) \ , \ E = 3\times10^4(\frac{N}{m^2}) \ , \ \rho = 2500(\frac{kg}{m^3}) \ , \ R = 1(m) \, , \upsilon = 0.25$$

$$z_0 = 0.12(m) \ , \ \Omega = 5(\frac{Rad}{\sec}) \ , \ A = 0.00002(m^2) \, , h = 0.01(m) \ , \ I = 0.000001(m^4)$$

and with regard to the given physical values and also the determined domain $t \in [0, 20]$ defined in terms of second (sec), solve the nonlinear characteristic time equation.

By using the separation equation, obtain the displacement of structure elements for the second mode of vibration. Afterward, draw the 3-D graph for the vibrational displacement of structure elements in the second mode.

Since the maximum amount of displacement of circle plate occurs in the first mode $(n = 1)$, the critical load in the first mode of vibration can be obtained by Rayleigh method as follows:

$$N_{cr} = \frac{\int_0^{2\pi}\int_0^R E\,I\,(\phi'')^2\,d\theta dr}{\int_0^{2\pi}\int_0^R (\phi)^2\,d\theta dr}.$$

Therefore, achieve the critical load for $R = 1(m)$.

As regards the Rayleigh method, the bending stiffness of the vibrating beam is yielded as follows:

$$K = \int_0^{2\pi}\int_0^R E\,I\,(\phi'')^2\,d\theta dr.$$

As a result, obtain the bending stiffness of the beam for $R = 1(m)$ in the second mode.

Techniques of preventing and reducing the transmission of system vibrations to its surrounding and vice versa (in earthquake) are defined as “isolation.”

According to the above definition, the amount of transmitted force from the system to the surrounding (foundation of the structure) can be obtained. Therefore, in regard to the obtained bending stiffness and the displacement for any modes of vibration, the aforementioned force can be acquired as follows:

$$F_d(r,t) = KU_n(r,t) = K\phi_n(r).z(t)\sqrt{b^2 - 4ac}.$$

As a result, gain the transmitted force from the vibrational system to its surrounding in the fifth mode of vibration.

In accordance with the introduced beam with the external force $F_0 \sin(\Omega t)$ and the solution of characteristic time function obtained by AGM, the equations of magnification factor and the ability of transmission are presented as follows:

Magnification factor: $TR = \dfrac{K.b}{F_0} = \dfrac{F_d(t)}{F_0}$, $b = \dfrac{\sqrt{\psi}}{\Delta}$

Ability of transmission: $R = 1 - TR$.

Where $F_d(t)$ is the transmitted force from the system to its foundation.

The parameters k, b and F_0 are stiffness, the amplitude of the response to the external force, and the amplitude of the external force, respectively.

At this step, a new parameter named η is defined as the ratio of frequency of harmonic force to the natural frequency as follows:

$$\eta = \frac{\Omega}{\omega} \quad , \quad \omega = \sqrt{\frac{(\alpha + \beta z_0^2)(\alpha + 3\beta z_0^2 - \Omega^2)}{\alpha + \beta z_0^2 - \Omega^2}}.$$

Where Ω is the angular frequency of the external force, and ω is the frequency of the vibrating system. In the isolation procedure, it is very important to calculate the interval of quantity η, which is one of the parameters in the magnification factor. The interval of η in linear vibrational systems (isolation condition) is considered as $\eta > \sqrt{2}$. Here, we are going to calculate the interval of η for nonlinear vibrational systems. As regards the aforementioned equations, the magnification factor (TR) can be obtained as a function of η, initial amplitude of oscillation (Z_0), and vibrational mode (n) as

$TR = f(\eta, n, z_0)$.

In linear vibrational systems, TR is achieved as follows:

$$TR = \frac{1}{\eta^2 - 1} \quad , \quad \eta > \sqrt{2}.$$

To obtain the interval of η in nonlinear systems, it is necessary to draw the magnification factor (TR) with different values of initial amplitude of oscillation (Z_0) or vibration modes (n). Then in accordance with the drawn graph, the interval of (η) can be obtained. Furthermore, vibrational systems must work in the specified interval of η in order to prevent any damage and defect to the system. Eventually, acquire the chart of *TR* with regard to the given physical quantities for different values of initial amplitude of oscillation (Z_0) in the first mode of vibration and for not damaging the vibrational system; distinguish where the system frequency will be greater than the frequency of external force.

Help: It is notable that slope in the center of the circular plate ($r = 0$), with respect to its axial symmetry, is zero. Therefore, we will have

$$\phi'(r) = 0 \quad at \quad r = 0.$$

And we know that the Bessel function $Y_n(r)$ in point $r = 0$ is infinite, so $Y_n(r) = -\infty \quad at \quad r = 0.$

6.4. References

1. T. K. Sherwood and R. L. Pigford, *Absorption and Entraction* (New York: McGraw-Hill, 1952), 256.

2. Richard. H. Pletcher, John. C. Tannehill, Dale Anderson, *Computational Fluid Mechanics and Heat Transfer,* 3rd ed, (Taylor and Francis, 2012).

3. J. D. Achenbach, *Wave Propagation in Elastic Solid* (Elsevier Science B. V., 1984).

4. R. Byron Bird, Warren E. Stewart, and Edwin N. Lightfoot, *Transport Phenomena.*

5. R. E. Treybel, *Mass Transfer Operations* (1990).

6. Robert. H. Perry, Don. W. Green, *Perry's Chemical Engineers' Handbook,* 7th ed, (McGrow-Hill, 1997).

7. Octave Levenspiel, *Chemical Reaction Engineering* (1985).

8. Warren L. McCabe, Julian C. Smith, and Peter Harriott, *Unit Operations* (1975).

9. G. A. Korn and T. M. Korn, *Mathematical Handbook for Scientists and Engineers*, 2nd ed. (McGraw-Hill, 1967).

10. J. Kong et al., "Nanotube Molecular Wires as Chemical Sensors," *Science* 287 (2000): 622–625.

11. B. Vigolo et al., "Macroscopic Fibers and Ribbons of Oriented Carbon Nanotubes," *Science* 290 (2000): 1331–1334.

12. H. W. C. Postma et al., "Carbon Nanotube Single-Electron Transistors at Room Temperature," *Science* 293 (2001): 76–79.

13. R. H. Baughman, A. A. Zakhidov, and W. A. de Heer, "Carbon Nanotubes: The Route towards Applications," *Science* 297 (2002): 787–792.

14. A. H. Nayfeh and D. T. Mook, *Nonlinear Oscillations* (Wiley and Sons, 1995).

15. Zhilum Xu, *Applied Elasticity* (Wiley Eastern).

16. H. Seyed, Hashemi Kachapi, and D. D. Ganji, *Progress in Nonlinear Science*, vol. 3 (2011).

17. K. F. Graff, *Wave Motion in Elastic Solids* (Ohio State University Press).

18. M. P. Morse and H. Feshbach, *Methods of Theoretical Physics* (New York: McGraw-Hill, 1953).

19. J. W. Dettman, *Mathematical Methods in Physics and Engineering* (NewYork: Mcgraw-Hill, 1962).

20. P. Moon and D. E. Spencer, *Field Theory for Engineers* (New Jersey: Van Nostrand, 1961).

21. Clarence Raymond Wylie, *Advanced Engineering Mathematics*, ed. Louis C. Barrett (1982).

22. K. F. Graff, *Wave Motion in Elastic Solids* (Ohio State University Press, 1975).

Appendices

Maple Codes

Eventually, in the last part of this book, the codes of some complicated problems are presented in order to familiarize the students with the entity of AGM.

1. First-Order Nonlinear Differential Equation

restart;

$M := a + b\cdot e^{-c\,x};$

$\boldsymbol{Tt} := \frac{Vo^2}{M^2\cdot \gamma l\cdot g\cdot R\cdot To}\cdot\left(1+\frac{(\gamma l-1)}{2}\cdot M^2\right);$

$$F := \frac{(Tw-Tt)}{4\,Tt}\cdot\frac{\left(1+\gamma l\cdot M^2\right)\cdot f}{d}+\frac{1}{2}\cdot\frac{\gamma l\cdot M^2\cdot f}{d}=\left(\frac{\left(1-\gamma l\cdot M^2\right)}{M}+\frac{(\gamma l-1)\cdot\left(1+\gamma l\cdot M^2\right)\cdot M}{2\cdot\left(1+\frac{(\gamma l-1)}{2}\cdot M^2\right)}\right)\cdot\frac{d}{dx}M;$$

$M := unapply(M, x);$
$F := unapply(F, x);$
$s1 := M(0) = M1;$
$s2 := F(0);$
$s3 := F'(0);$
$S := solve([s1, s2, s3], \{a, b, c\}) :$

$S := convert(S, radical) :$
$M := eval\left(a + b\cdot e^{-c\,x}, S\right) :$

$$\psi_1 = 6\,M1^6\,Tw\,\gamma l^2\,g\,R\,To + 3\,M1^6\,Vo^2\,\gamma l^2 - 3\,M1^6\,\gamma l\,Vo^2 - 5\,Vo^2\,\gamma l^2\,M1^4 - 10\,Tw\,\gamma l^2\,g\,R\,To\,M1^4 + 2\,M1^4\,Tw\,\gamma l\,g\,R\,To + 6\,\gamma l\,Vo^2\,M1^4 + M1^4\,Vo^2 - 6\,Tw\,\gamma l\,g\,R\,To\,M1^2 - 3\,\gamma l\,Vo^2\,M1^2 - Vo^2\,M1^2 + 2\,Vo^2;$$

$$\psi_2 = Vo^2\,\gamma l^2\,M1^4 + 2\,Tw\,\gamma l^2\,g\,R\,To\,M1^4 - \gamma l\,Vo^2\,M1^4 - 2\,Vo^2\,\gamma l^2\,M1^2 - 4\,Tw\,\gamma l^2\,g\,R\,To\,M1^2 + 2\,\gamma l\,Vo^2\,M1^2 - \gamma l\,Vo^2 - 2\,Tw\,\gamma l\,g\,R\,To + Vo^2;$$

$$\psi_3 = \gamma l^2\,M1^6\,Vo^2 + 2\,\gamma l^2\,M1^6\,Tw\,g\,R\,To - \gamma l\,M1^6\,Vo^2 - Vo^2\,\gamma l^2\,M1^4 - 2\,Tw\,\gamma l^2\,g\,R\,To\,M1^4 + 2\,\gamma l\,Vo^2\,M1^4 + 2\,M1^4\,Tw\,\gamma l\,g\,R\,To + M1^4\,Vo^2 - \gamma l\,Vo^2\,M1^2 - 2\,Tw\,\gamma l\,g\,R\,To\,M1^2 - 3\,Vo^2\,M1^2 + 2\,Vo^2;$$

$$'M(x)' = applyrule\Big(\Big[6\,M1^6\,Tw\,\gamma l^2\,g\,R\,To + 3\,M1^6\,Vo^2\,\gamma l^2 - 3\,M1^6\,\gamma l\,Vo^2 - 5\,Vo^2\,\gamma l^2\,M1^4 - 10\,Tw\,\gamma l^2\,g\,R\,To\,M1^4 + 2\,M1^4\,Tw\,\gamma l\,g\,R\,To + 6\,\gamma l\,Vo^2\,M1^4 + M1^4\,Vo^2 - 6\,Tw\,\gamma l\,g\,R\,To\,M1^2 - 3\,\gamma l\,Vo^2\,M1^2 - Vo^2\,M1^2 + 2\,Vo^2 = \psi_1,\ Vo^2\,\gamma l^2\,M1^4 + 2\,Tw\,\gamma l^2\,g\,R\,To\,M1^4 - \gamma l\,Vo^2\,M1^4 - 2\,Vo^2\,\gamma l^2\,M1^2 - 4\,Tw\,\gamma l^2\,g\,R\,To\,M1^2 + 2\,\gamma l\,Vo^2\,M1^2 - \gamma l\,Vo^2 - 2\,Tw\,\gamma l\,g\,R\,To + Vo^2 = \psi_2,\ \gamma l^2\,M1^6\,Vo^2 + 2\,\gamma l^2\,M1^6\,Tw\,g\,R\,To - \gamma l\,M1^6\,Vo^2 - Vo^2\,\gamma l^2\,M1^4 - 2\,Tw\,\gamma l^2\,g\,R\,To\,M1^4 + 2\,\gamma l\,Vo^2\,M1^4 + 2\,M1^4\,Tw\,\gamma l\,g\,R\,To + M1^4\,Vo^2 - \gamma l\,Vo^2\,M1^2 - 2\,Tw\,\gamma l\,g\,R\,To\,M1^2 - 3\,Vo^2\,M1^2 + 2\,Vo^2 = \psi_3\Big], M\Big);$$

$$d := 0.4;\ \gamma l := 1.4;\ f := 0.00045;\ L := 50;\ Tw := 100;\ R := 8;\ g := 9.8;\ Vo := 5;\ To := 25;\ M1 := \frac{Vo}{\sqrt{\gamma l\cdot g\cdot R\cdot (To + 273)}};$$

$M := M;$

fig, AGM;

plot$(M, x = 0\,..L, axes = boxed, labels = [\text{"x"}, \text{"M(x)"}], thickness = 2);$

$$M := a + b\,e^{-c\,x}$$

$$Tt := \frac{Vo^2\left(1 + \frac{1}{2}\,(\gamma l - 1)\,\left(a + b\,e^{-c\,x}\right)^2\right)}{\left(a + b\,e^{-c\,x}\right)^2\,\gamma l\,g\,R\,To}$$

$$F := \frac{1}{4}\left(\left(Tw - \frac{Vo^2\left(1+\frac{1}{2}(\gamma l-1)\left(a+b\,\mathrm{e}^{-cx}\right)^2\right)}{\left(a+b\,\mathrm{e}^{-cx}\right)^2 \gamma l\, g\, R\, To}\right)\left(\left(a+b\,\mathrm{e}^{-cx}\right)^2 \gamma l + 1\right) f \left(a+b\,\mathrm{e}^{-cx}\right)^2 \gamma l\, g\, R\, To\right) \Big/ \left(Vo^2\left(1+\frac{1}{2}(\gamma l-1)\left(a+b\,\mathrm{e}^{-cx}\right)^2\right) d\right) + \frac{1}{2}\,\frac{\gamma l\left(a+b\,\mathrm{e}^{-cx}\right)^2 f}{d} = -\left(\frac{-\left(a+b\,\mathrm{e}^{-cx}\right)^2 \gamma l+1}{a+b\,\mathrm{e}^{-cx}} + \frac{(\gamma l-1)\left(\left(a+b\,\mathrm{e}^{-cx}\right)^2 \gamma l+1\right)\left(a+b\,\mathrm{e}^{-cx}\right)}{2+(\gamma l-1)\left(a+b\,\mathrm{e}^{-cx}\right)^2}\right) b\, c\, \mathrm{e}^{-cx}$$

$$M := x \rightarrow a + b\,\mathrm{e}^{-cx}$$

$$F := x \rightarrow \frac{1}{4}\left(\left(Tw - \frac{Vo^2\left(1+\frac{1}{2}(\gamma l-1)\left(a+b\,\mathrm{e}^{-cx}\right)^2\right)}{\left(a+b\,\mathrm{e}^{-cx}\right)^2 \gamma l\, g\, R\, To}\right)\left(\left(a+b\,\mathrm{e}^{-cx}\right)^2 \gamma l + 1\right) f \left(a+b\,\mathrm{e}^{-cx}\right)^2 \gamma l\, g\, R\, To\right) \Big/ \left(Vo^2\left(1+\frac{1}{2}(\gamma l-1)\left(a+b\,\mathrm{e}^{-cx}\right)^2\right) d\right) + \frac{1}{2}\,\frac{\gamma l\left(a+b\,\mathrm{e}^{-cx}\right)^2 f}{d} = -\left(\frac{-\left(a+b\,\mathrm{e}^{-cx}\right)^2 \gamma l+1}{a+b\,\mathrm{e}^{-cx}} + \frac{(\gamma l-1)\left(\left(a+b\,\mathrm{e}^{-cx}\right)^2 \gamma l+1\right)\left(a+b\,\mathrm{e}^{-cx}\right)}{2+(\gamma l-1)\left(a+b\,\mathrm{e}^{-cx}\right)^2}\right) b\, c\, \mathrm{e}^{-cx}$$

$$s1 := a + b = M1$$

$$s2 := \frac{1}{4}\,\frac{1}{Vo^2\left(1+\frac{1}{2}\,(\gamma l-1)\,(a+b)^2\right)d}\Bigg(\bigg(Tw$$

$$-\frac{Vo^2\left(1+\frac{1}{2}\,(\gamma l-1)\,(a+b)^2\right)}{(a+b)^2\,\gamma l\,g\,R\,To}\bigg)\Big((a+b)^2\,\gamma l+1\Big)f\,(a$$

$$+b)^2\,\gamma l\,g\,R\,To\Bigg)+\frac{1}{2}\,\frac{\gamma l\,(a+b)^2 f}{d}=-\left(\frac{-(a+b)^2\,\gamma l+1}{a+b}\right.$$

$$\left.+\frac{(\gamma l-1)\,\big((a+b)^2\,\gamma l+1\big)\,(a+b)}{2+(\gamma l-1)\,(a+b)^2}\right)b\,c$$

$$s3 := \frac{1}{4}\,\frac{1}{Vo^2\left(1+\frac{1}{2}\,(\gamma l-1)\,(a+b)^2\right)d}\Bigg(\bigg(\frac{Vo^2\,(\gamma l-1)\,b\,c}{(a+b)\,\gamma l\,g\,R\,To}$$

$$-\frac{2\,Vo^2\left(1+\frac{1}{2}\,(\gamma l-1)\,(a+b)^2\right)b\,c}{(a+b)^3\,\gamma l\,g\,R\,To}\bigg)\Big((a+b)^2\,\gamma l+1\Big)f\,(a$$

$$+b)^2\,\gamma l\,g\,R\,To\Bigg)$$

$$-\frac{1}{2}\,\frac{1}{Vo^2\left(1+\frac{1}{2}\,(\gamma l-1)\,(a+b)^2\right)d}\Bigg(\bigg(Tw$$

$$-\frac{Vo^2\left(1+\frac{1}{2}\,(\gamma l-1)\,(a+b)^2\right)}{(a+b)^2\,\gamma l\,g\,R\,To}\bigg)(a+b)^3\,\gamma l^2\,b\,c\,f\,g\,R\,To\Bigg)$$

$$+\frac{1}{4}\frac{1}{Vo^2\left(1+\frac{1}{2}(\gamma l-1)(a+b)^2\right)^2 d}\left(\left(\left(Tw\right.\right.\right.$$

$$\left.-\frac{Vo^2\left(1+\frac{1}{2}(\gamma l-1)(a+b)^2\right)}{(a+b)^2\gamma l\, g\, R\, To}\right)\left((a+b)^2\gamma l+1\right)f(a$$

$$\left.+b)^3\gamma l\, g\, R\, To\,(\gamma l-1)\, b\, c\right)$$

$$-\frac{1}{2}\frac{1}{Vo^2\left(1+\frac{1}{2}(\gamma l-1)(a+b)^2\right) d}\left(\left(\left(Tw\right.\right.\right.$$

$$\left.-\frac{Vo^2\left(1+\frac{1}{2}(\gamma l-1)(a+b)^2\right)}{(a+b)^2\gamma l\, g\, R\, To}\right)\left((a+b)^2\gamma l+1\right)f(a$$

$$\left.+b)\,\gamma l\, g\, R\, To\, b\, c\right)-\frac{\gamma l\,(a+b)\, f\, b\, c}{d}=-\left(2\,\gamma l\, b\, c\right.$$

$$+\frac{\left(-(a+b)^2\gamma l+1\right) b\, c}{(a+b)^2}-\frac{2\,(\gamma l-1)(a+b)^2\gamma l\, b\, c}{2+(\gamma l-1)(a+b)^2}$$

$$-\frac{(\gamma l-1)\left((a+b)^2\gamma l+1\right) b\, c}{2+(\gamma l-1)(a+b)^2}$$

$$\left.+\frac{2\,(\gamma l-1)^2\left((a+b)^2\gamma l+1\right)(a+b)^2\, b\, c}{\left(2+(\gamma l-1)(a+b)^2\right)^2}\right) b\, c$$

$$+\left(\frac{-(a+b)^2\gamma l+1}{a+b}+\frac{(\gamma l-1)\left((a+b)^2\gamma l+1\right)(a+b)}{2+(\gamma l-1)(a+b)^2}\right) b\, c^2$$

$$\psi_1=6\,MI^6\, R\, To\, Tw\, g\,\gamma l^2+3\,MI^6\, Vo^2\,\gamma l^2-10\,MI^4\, R\, To\, Tw\, g\,\gamma l^2$$

$$-3\,MI^6\, Vo^2\,\gamma l+2\,MI^4\, R\, To\, Tw\, g\,\gamma l-5\,MI^4\, Vo^2\,\gamma l^2+6\,MI^4\, Vo^2\,\gamma l$$

$$-6\,MI^2\, R\, To\, Tw\, g\,\gamma l+MI^4\, Vo^2-3\,MI^2\, Vo^2\,\gamma l-MI^2\, Vo^2+2\,Vo^2$$

$$\psi_2 = 2\,M1^4\,R\,To\,Tw\,g\,\gamma l^2 + M1^4\,Vo^2\,\gamma l^2 - 4\,M1^2\,R\,To\,Tw\,g\,\gamma l^2 - M1^4\,Vo^2\,\gamma l - 2\,M1^2\,Vo^2\,\gamma l^2 + 2\,M1^2\,Vo^2\,\gamma l - 2\,R\,To\,Tw\,g\,\gamma l - Vo^2\,\gamma l + Vo^2$$

$$\psi_3 = 2\,M1^6\,R\,To\,Tw\,g\,\gamma l^2 + M1^6\,Vo^2\,\gamma l^2 - 2\,M1^4\,R\,To\,Tw\,g\,\gamma l^2 - M1^6\,Vo^2\,\gamma l + 2\,M1^4\,R\,To\,Tw\,g\,\gamma l - M1^4\,Vo^2\,\gamma l^2 + 2\,M1^4\,Vo^2\,\gamma l - 2\,M1^2\,R\,To\,Tw\,g\,\gamma l + M1^4\,Vo^2 - M1^2\,Vo^2\,\gamma l - 3\,M1^2\,Vo^2 + 2\,Vo^2$$

$$M(x) = \frac{2\,M1^3\,\psi_2}{\psi_1} + \frac{\psi_3\,M1\,e^{-\frac{1}{8}\frac{\psi_1 f x}{(M1^2-1)^2\,Vo^2\,d}}}{\psi_1}$$

$$d := 0.4$$

$$\gamma l := 1.4$$

$$f := 0.00045$$

$$L := 50$$

$$Tw := 100$$

$$R := 8$$

$$g := 9.8$$

$$Vo := 5$$

$$To := 25$$

$$M1 := 0.02764647868$$

$$M := 0.01920289138 + 0.008443587296\,e^{0.006818983152\,x}$$

fig, AGM

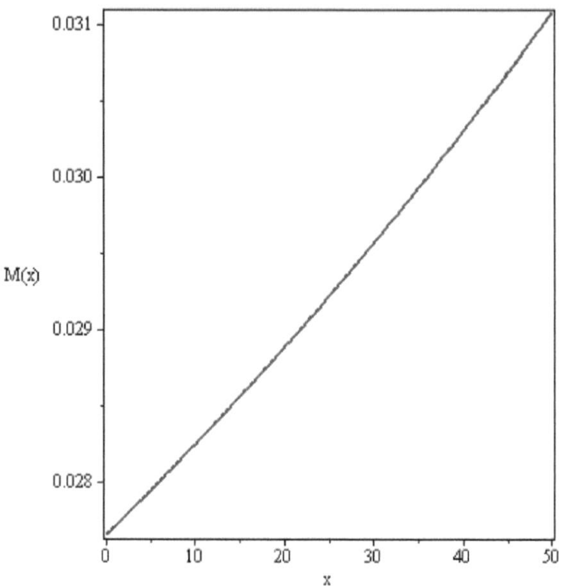

The code of numerical solution is presented as

restart;

$$d := 0.4;\ \gamma l := 1.4; f := 0.00045; L := 50; Tw := 100; R := 8; g := 9.8; Vo := 5; To := 25; M1 := \frac{Vo}{\sqrt{\gamma l \cdot g \cdot R \cdot (To + 273)}};$$

$$\boldsymbol{Tt} := \frac{\boldsymbol{Vo}^2}{M(x)^2 \cdot \gamma l \cdot g \cdot R \cdot To} \cdot \left(1 + \frac{(\gamma l - 1)}{2} \cdot M(x)^2\right);$$

$$F := \frac{(Tw - Tt)}{4\,Tt} \cdot \frac{\left(1 + \gamma l \cdot M(x)^2\right) \cdot f}{d} + \frac{1}{2} \cdot \frac{\gamma l \cdot M(x)^2 \cdot f}{d} = \left(\frac{\left(1 - \gamma l \cdot M(x)^2\right)}{M(x)} + \frac{(\gamma l - 1) \cdot \left(1 + \gamma l \cdot M(x)^2\right) \cdot M(x)}{2 \cdot \left(1 + \frac{(\gamma l - 1)}{2} \cdot M(x)^2\right)} \right) \cdot \frac{\mathrm{d}}{\mathrm{d}x} M(x);$$

```
ic := M(0) = M1;

F := dsolve([F, ic], numeric);
for i from 0  by 10 while  i ≤ 50 do;
([x=i, F(i)[2]]);
end do;

Fig, Numerical;
plots[odeplot](F, [x, M(x) ], x=0 ..L, thickness=2, axes=boxed );
```

$$d := 0.4$$

$$\gamma 1 := 1.4$$

$$f := 0.00045$$

$$L := 50$$

$$Tw := 100$$

$$R := 8$$

$$g := 9.8$$

$$Vo := 5$$

$$To := 25$$

$$M1 := 0.02764647868$$

$$Tt := \frac{0.009110787176\left(1+0.2000000000\,M(x)^2\right)}{M(x)^2}$$

$$F := \frac{1}{1+0.2000000000\,M(x)^2}\left(0.03087000000\left(100 - \frac{0.009110787176\left(1+0.2000000000\,M(x)^2\right)}{M(x)^2}\right)\left(1 + 1.4\,M(x)^2\right)M(x)^2\right) + 0.0007875000000\,M(x)^2 = \left(\frac{1-1.4\,M(x)^2}{M(x)} + \frac{0.4\left(1+1.4\,M(x)^2\right)M(x)}{2+0.4\,M(x)^2}\right)\left(\frac{d}{dx}M(x)\right)$$

$$ic := M(0) = 0.02764647868$$

$F :=$ **proc**(x_rkf45) ... **end proc**

$$[x = 0, M(x) = 0.0276464786800000]$$
$$[x = 10, M(x) = 0.0282426472756346]$$
$$[x = 20, M(x) = 0.0288829259550963]$$
$$[x = 30, M(x) = 0.0295728328464396]$$
$$[x = 40, M(x) = 0.0303191034687777]$$
$$[x = 50, M(x) = 0.0311297646350610]$$

Fig, Numerical

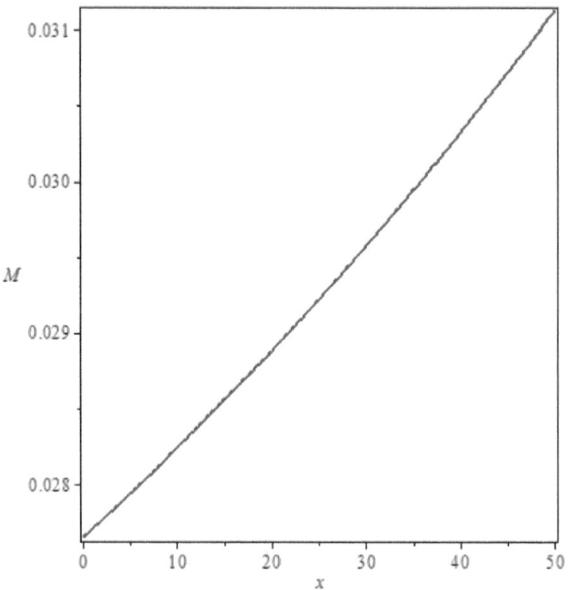

2. Nonvibrational Differential Equation

restart;

$A := 0.2; M := 1; pr := 1; Le := 1; s := 1; Nt := 0.1; Nb := 1;$

$f := \sum_{i=0}^{7} a_i \cdot \eta^i; \quad \theta := \sum_{i=0}^{5} b_i \cdot \eta^i; \quad \phi := \sum_{i=0}^{5} c_i \cdot \eta^i;$

$$F := \frac{d^4}{d\eta^4} f - s \cdot \left(\eta \cdot \frac{d^3}{d\eta^3} f + 3 \cdot \frac{d^2}{d\eta^2} f - 2 f \cdot \frac{d^3}{d\eta^3} f \right) - M^2 \cdot \frac{d^2}{d\eta^2} f = 0;$$

$$G := \frac{d^2}{d\eta^2} \theta + pr \cdot s \cdot (2f - \eta) \cdot \frac{d}{d\eta} \theta + pr \cdot Nb \cdot \left(\frac{d}{d\eta} \theta \right) \cdot \left(\frac{d}{d\eta} \phi \right) + pr \cdot N\!\!\cdot \left(\frac{d}{d\eta} \theta \right)^2 = 0;$$

$$H := \frac{d^2}{d\eta^2} \phi + Le \cdot s \cdot (2f - \eta) \cdot \frac{d}{d\eta} \phi + \frac{Nt}{Nb} \cdot \frac{d^2}{d\eta^2} \theta = 0;$$

$f := unapply(f, \eta);$
$\theta := unapply(\theta, \eta);$
$\phi := unapply(\phi, \eta);$

$F := unapply(F, \eta);$
$G := unapply(G, \eta);$
$H := unapply(H, \eta);$

$s1 := f(0) = A;\quad s2 := f'(0) = 0;$
$s3 := f(1) = \frac{1}{2};\quad s4 := f'(1) = 0;$

$s5 := \theta(0) = 1;\quad s6 := \theta(1) = 0;$
$s7 := \phi(0) = 1;\quad s8 := \phi(1) = 0;$

$s9 := F(0);\ \ s10 := F(1);$

$s11 := G(0);\ \ s12 := G(1);$

$s13 := H(0);\ \ s14 := H(1);$

$s15 := H'(0);\ s16 := H'(1);$

$s17 := F'(0);\ \mathrm{s18} := \mathrm{F}'(1);$

$\mathrm{s19} := \mathrm{G}'(0);\ s20 := G'(1);$

$S := solve\big([s1, s2, s3, s4, s5, s6, s7, s8, s9, s10, s11, s12, s13, s14, s15, s16, s17, s18, s19, s20], \{a_0, a_1, a_2, a_3, a_4, a_5, a_6, a_7, b_0, b_1, b_2, b_3, b_4, b_5, c_0, c_1, c_2, c_3, c_4, c_5\}\big)[6];$

$X := convert(S, radical);$

$\theta := eval\left(\sum_{i=0}^{5} b_i \cdot \eta^i, X\right);$

$\phi := eval\left(\sum_{i=0}^{5} c_i \cdot \eta^i, X\right);$

$f := eval\left(\sum_{i=0}^{7} a_i \cdot \eta^i, X\right);$

AGM;
fig, AGM;

$plot(f, \eta = 0..1, axes = boxed, labels = [\text{"}\eta\text{"}, \text{"f(}\eta\text{)"}], thickness = 2);$
$plot(diff(f, \eta), \eta = 0..1, axes = boxed, labels = [\text{"}\eta\text{"}, \text{"df(}\eta\text{)"}], thickness = 2);$
$plot(\theta, \eta = 0..1, axes = boxed, labels = [\text{"}\eta\text{"}, \text{"}\theta(\eta)\text{"}], thickness = 2);$
$plot(\phi, \eta = 0..1, axes = boxed, labels = [\text{"}\eta\text{"}, \text{"}\phi(\eta)\text{"}], thickness = 2);$

$$A := 0.2$$
$$M := 1$$
$$pr := 1$$
$$Le := 1$$
$$s := 1$$
$$Nt := 0.1$$
$$Nb := 1$$

$$f := a_0 + a_1\eta + a_2\eta^2 + a_3\eta^3 + a_4\eta^4 + a_5\eta^5 + a_6\eta^6 + a_7\eta^7$$
$$\theta := \eta^5 b_5 + \eta^4 b_4 + \eta^3 b_3 + \eta^2 b_2 + \eta b_1 + b_0$$
$$\phi := \eta^5 c_5 + \eta^4 c_4 + \eta^3 c_3 + \eta^2 c_2 + \eta c_1 + c_0$$

$$F := 840\eta^3 a_7 + 360\eta^2 a_6 + 120\eta a_5 + 24a_4 - \eta\left(210\eta^4 a_7 + 120\eta^3 a_6 + 60\eta^2 a_5 + 24\eta a_4 + 6a_3\right) - 168\eta^5 a_7 - 120\eta^4 a_6 - 80\eta^3 a_5 - 48\eta^2 a_4 - 24\eta a_3 - 8a_2 + 2\left(a_0 + a_1\eta + a_2\eta^2 + a_3\eta^3 + a_4\eta^4 + a_5\eta^5 + a_6\eta^6 + a_7\eta^7\right)\left(210\eta^4 a_7 + 120\eta^3 a_6 + 60\eta^2 a_5 + 24\eta a_4 + 6a_3\right) = 0$$

$$G := 20\eta^3 b_5 + 12\eta^2 b_4 + 6\eta b_3 + 2b_2 + \left(2a_0 + 2a_1\eta + 2a_2\eta^2 + 2a_3\eta^3 + 2a_4\eta^4 + 2a_5\eta^5 + 2a_6\eta^6 + 2a_7\eta^7 - \eta\right)\left(5\eta^4 b_5 + 4\eta^3 b_4 + 3\eta^2 b_3 + 2\eta b_2 + b_1\right) + \left(5\eta^4 b_5 + 4\eta^3 b_4 + 3\eta^2 b_3 + 2\eta b_2 + b_1\right)\left(5\eta^4 c_5 + 4\eta^3 c_4 + 3\eta^2 c_3 + 2\eta c_2 + c_1\right) + 0.1\left(5\eta^4 b_5 + 4\eta^3 b_4 + 3\eta^2 b_3 + 2\eta b_2 + b_1\right)^2 = 0$$

$$H := 20\eta^3 c_5 + 12\eta^2 c_4 + 6\eta c_3 + 2c_2 + \left(2a_0 + 2a_1\eta + 2a_2\eta^2 + 2a_3\eta^3 + 2a_4\eta^4 + 2a_5\eta^5 + 2a_6\eta^6 + 2a_7\eta^7 - \eta\right)\left(5\eta^4 c_5 + 4\eta^3 c_4 + 3\eta^2 c_3 + 2\eta c_2 + c_1\right) + 2.0\eta^3 b_5 + 1.2\eta^2 b_4 + 0.6\eta b_3 + 0.2b_2 = 0$$

$$f := \eta \rightarrow a_0 + a_1\eta + a_2\eta^2 + a_3\eta^3 + a_4\eta^4 + a_5\eta^5 + a_6\eta^6 + a_7\eta^7$$
$$\theta := \eta \rightarrow \eta^5 b_5 + \eta^4 b_4 + \eta^3 b_3 + \eta^2 b_2 + \eta b_1 + b_0$$

$$\phi := \eta \to \eta^5 c_5 + \eta^4 c_4 + \eta^3 c_3 + \eta^2 c_2 + \eta c_1 + c_0$$

$$F := \eta \to 840\,\eta^3 a_7 + 360\,\eta^2 a_6 + 120\,\eta a_5 + 24\,a_4 - \eta\left(210\,\eta^4 a_7 + 120\,\eta^3 a_6 + 60\,\eta^2 a_5 + 24\,\eta a_4 + 6\,a_3\right) - 168\,\eta^5 a_7 - 120\,\eta^4 a_6 - 80\,\eta^3 a_5 - 48\,\eta^2 a_4 - 24\,\eta a_3 - 8\,a_2 + 2\left(a_0 + a_1 \eta + a_2 \eta^2 + a_3 \eta^3 + a_4 \eta^4 + a_5 \eta^5 + a_6 \eta^6 + a_7 \eta^7\right)\left(210\,\eta^4 a_7 + 120\,\eta^3 a_6 + 60\,\eta^2 a_5 + 24\,\eta a_4 + 6\,a_3\right) = 0$$

$$G := \eta \to 20\,\eta^3 b_5 + 12\,\eta^2 b_4 + 6\,\eta b_3 + 2\,b_2 + \left(2\,a_0 + 2\,a_1 \eta + 2\,a_2 \eta^2 + 2\,a_3 \eta^3 + 2\,a_4 \eta^4 + 2\,a_5 \eta^5 + 2\,a_6 \eta^6 + 2\,a_7 \eta^7 - \eta\right)\left(5\,\eta^4 b_5 + 4\,\eta^3 b_4 + 3\,\eta^2 b_3 + 2\,\eta b_2 + b_1\right) + \left(5\,\eta^4 b_5 + 4\,\eta^3 b_4 + 3\,\eta^2 b_3 + 2\,\eta b_2 + b_1\right)\left(5\,\eta^4 c_5 + 4\,\eta^3 c_4 + 3\,\eta^2 c_3 + 2\,\eta c_2 + c_1\right) + 0.1\left(5\,\eta^4 b_5 + 4\,\eta^3 b_4 + 3\,\eta^2 b_3 + 2\,\eta b_2 + b_1\right)^2 = 0$$

$$H := \eta \to 20\,\eta^3 c_5 + 12\,\eta^2 c_4 + 6\,\eta c_3 + 2\,c_2 + \left(2\,a_0 + 2\,a_1 \eta + 2\,a_2 \eta^2 + 2\,a_3 \eta^3 + 2\,a_4 \eta^4 + 2\,a_5 \eta^5 + 2\,a_6 \eta^6 + 2\,a_7 \eta^7 - \eta\right)\left(5\,\eta^4 c_5 + 4\,\eta^3 c_4 + 3\,\eta^2 c_3 + 2\,\eta c_2 + c_1\right) + 2.0\,\eta^3 b_5 + 1.2\,\eta^2 b_4 + 0.6\,\eta b_3 + 0.2\,b_2 = 0$$

$$s1 := a_0 = 0.2$$

$$s2 := a_1 = 0$$

$$s3 := a_0 + a_1 + a_2 + a_3 + a_4 + a_5 + a_6 + a_7 = \frac{1}{2}$$

$$s4 := 7\,a_7 + 6\,a_6 + 5\,a_5 + 4\,a_4 + 3\,a_3 + 2\,a_2 + a_1 = 0$$

$$s5 := b_0 = 1$$

$$s6 := b_5 + b_4 + b_3 + b_2 + b_1 + b_0 = 0$$

$$s7 := c_0 = 1$$

$$s8 := c_5 + c_4 + c_3 + c_2 + c_1 + c_0 = 0$$

$$s9 := 12\,a_0 a_3 - 8\,a_2 + 24\,a_4 = 0$$

$$s10 := 462\,a_7 + 120\,a_6 - 20\,a_5 - 48\,a_4 - 30\,a_3 - 8\,a_2 + 2\,(a_0 + a_1 + a_2 + a_3 + a_4 + a_5 + a_6 + a_7)\,(210\,a_7 + 120\,a_6 + 60\,a_5 + 24\,a_4 + 6\,a_3) = 0$$

$$s11 := 2\,b_2 + 2\,a_0\,b_1 + b_1\,c_1 + 0.1\,b_1^2 = 0$$

$$s12 := 20\,b_5 + 12\,b_4 + 6\,b_3 + 2\,b_2 + (2\,a_0 + 2\,a_1 + 2\,a_2 + 2\,a_3 + 2\,a_4 + 2\,a_5 + 2\,a_6 + 2\,a_7 - 1)\,(5\,b_5 + 4\,b_4 + 3\,b_3 + 2\,b_2 + b_1) + (5\,b_5 + 4\,b_4 + 3\,b_3 + 2\,b_2 + b_1)\,(5\,c_5 + 4\,c_4 + 3\,c_3 + 2\,c_2 + c_1) + 0.1\,(5\,b_5 + 4\,b_4 + 3\,b_3 + 2\,b_2 + b_1)^2 = 0$$

$$s13 := 2\,c_2 + 2\,a_0\,c_1 + 0.2\,b_2 = 0$$

$$s14 := 20\,c_5 + 12\,c_4 + 6\,c_3 + 2\,c_2 + (2\,a_0 + 2\,a_1 + 2\,a_2 + 2\,a_3 + 2\,a_4 + 2\,a_5 + 2\,a_6 + 2\,a_7 - 1)\,(5\,c_5 + 4\,c_4 + 3\,c_3 + 2\,c_2 + c_1) + 2.0\,b_5 + 1.2\,b_4 + 0.6\,b_3 + 0.2\,b_2 = 0$$

$$s15 := 6\,c_3 + (-1 + 2\,a_1)\,c_1 + 4\,a_0\,c_2 + 0.6\,b_3 = 0$$

$$s16 := 60\,c_5 + 24\,c_4 + 6\,c_3 + (14\,a_7 + 12\,a_6 + 10\,a_5 + 8\,a_4 + 6\,a_3 + 4\,a_2 + 2\,a_1 - 1)\,(5\,c_5 + 4\,c_4 + 3\,c_3 + 2\,c_2 + c_1) + (2\,a_0 + 2\,a_1 + 2\,a_2 + 2\,a_3 + 2\,a_4 + 2\,a_5 + 2\,a_6 + 2\,a_7 - 1)\,(20\,c_5 + 12\,c_4 + 6\,c_3 + 2\,c_2) + 6.0\,b_5 + 2.4\,b_4 + 0.6\,b_3 = 0$$

$$s17 := 48\,a_0\,a_4 + 12\,a_1\,a_3 - 30\,a_3 + 120\,a_5 = 0$$

$$s18 := 630\,a_7 - 240\,a_6 - 300\,a_5 - 144\,a_4 - 30\,a_3 + 2\,(7\,a_7 + 6\,a_6 + 5\,a_5 + 4\,a_4 + 3\,a_3 + 2\,a_2 + a_1)\,(210\,a_7 + 120\,a_6 + 60\,a_5 + 24\,a_4 + 6\,a_3) + 2\,(a_0 + a_1 + a_2 + a_3 + a_4 + a_5 + a_6 + a_7)\,(840\,a_7 + 360\,a_6 + 120\,a_5 + 24\,a_4) = 0$$

$$s19 := 6\,b_3 + (-1 + 2\,a_1)\,b_1 + 4\,a_0\,b_2 + 2\,b_2\,c_1 + 2\,b_1\,c_2 + 0.4\,b_1\,b_2 = 0$$

$$s20 := 60\,b_5 + 24\,b_4 + 6\,b_3 + \left(14\,a_7 + 12\,a_6 + 10\,a_5 + 8\,a_4 + 6\,a_3 + 4\,a_2 + 2\,a_1 - 1\right)\left(5\,b_5 + 4\,b_4 + 3\,b_3 + 2\,b_2 + b_1\right) + \left(2\,a_0 + 2\,a_1 + 2\,a_2 + 2\,a_3 + 2\,a_4 + 2\,a_5 + 2\,a_6 + 2\,a_7 - 1\right)\left(20\,b_5 + 12\,b_4 + 6\,b_3 + 2\,b_2\right) + \left(20\,b_5 + 12\,b_4 + 6\,b_3 + 2\,b_2\right)\left(5\,c_5 + 4\,c_4 + 3\,c_3 + 2\,c_2 + c_1\right) + \left(5\,b_5 + 4\,b_4 + 3\,b_3 + 2\,b_2 + b_1\right)\left(20\,c_5 + 12\,c_4 + 6\,c_3 + 2\,c_2\right) + 0.2\left(5\,b_5 + 4\,b_4 + 3\,b_3 + 2\,b_2 + b_1\right)\left(20\,b_5 + 12\,b_4 + 6\,b_3 + 2\,b_2\right) = 0$$

$$S := \{a_0 = 0.2000000000,\ a_1 = 0.,\ a_2 = 0.9836441577,\ a_3 = -0.9117907018,\ a_4 = 0.4190604561,\ a_5 = -0.2614725119,\ a_6 = 0.09470567439,\ a_7 = -0.02414707442,\ b_0 = 1.,\ b_1 = -0.6463728195,\ b_2 = -0.2661050519,\ b_3 = -0.1308319847,\ b_4 = 0.1358560524,\ b_5 = -0.09254619626,\ c_0 = 1.,\ c_1 = -1.158742179,\ c_2 = 0.2583589410,\ c_3 = -0.2144883569,\ c_4 = 0.1716792890,\ c_5 = -0.05680769387\}$$

$$X := \{a_0 = 0.2000000000,\ a_1 = 0.,\ a_2 = 0.9836441577,\ a_3 = -0.9117907018,\ a_4 = 0.4190604561,\ a_5 = -0.2614725119,\ a_6 = 0.09470567439,\ a_7 = -0.02414707442,\ b_0 = 1.,\ b_1 = -0.6463728195,\ b_2 = -0.2661050519,\ b_3 = -0.1308319847,\ b_4 = 0.1358560524,\ b_5 = -0.09254619626,\ c_0 = 1.,\ c_1 = -1.158742179,\ c_2 = 0.2583589410,\ c_3 = -0.2144883569,\ c_4 = 0.1716792890,\ c_5 = -0.05680769387\}$$

$$\theta := -0.09254619626\,\eta^5 + 0.1358560524\,\eta^4 - 0.1308319847\,\eta^3 - 0.2661050519\,\eta^2 - 0.6463728195\,\eta + 1.$$

$$\phi := -0.05680769387\,\eta^5 + 0.1716792890\,\eta^4 - 0.2144883569\,\eta^3 + 0.2583589410\,\eta^2 - 1.158742179\,\eta + 1.$$

$$f := 0.2000000000 + 0.9836441577\,\eta^2 - 0.9117907018\,\eta^3 + 0.4190604561\,\eta^4 - 0.2614725119\,\eta^5 + 0.09470567439\,\eta^6 - 0.02414707442\,\eta^7$$

fig, AGM

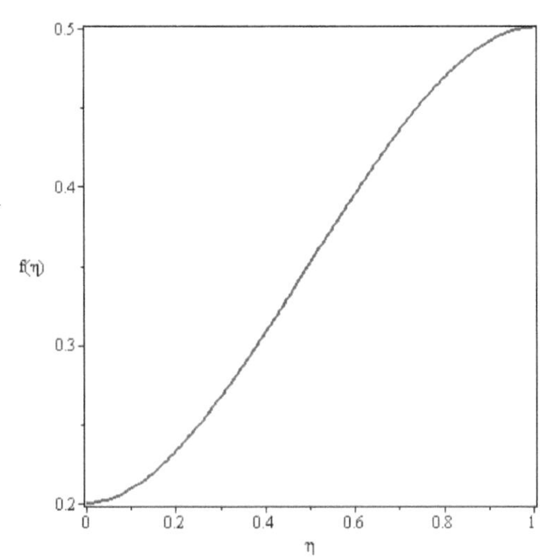

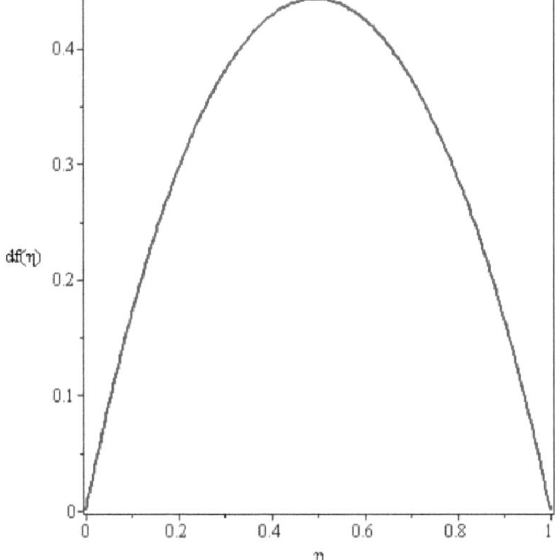

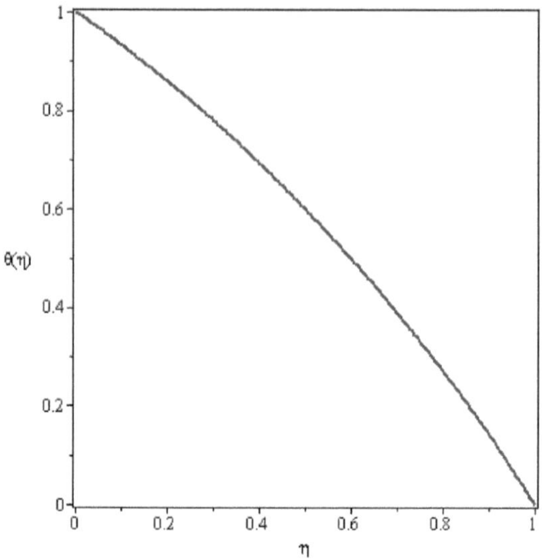

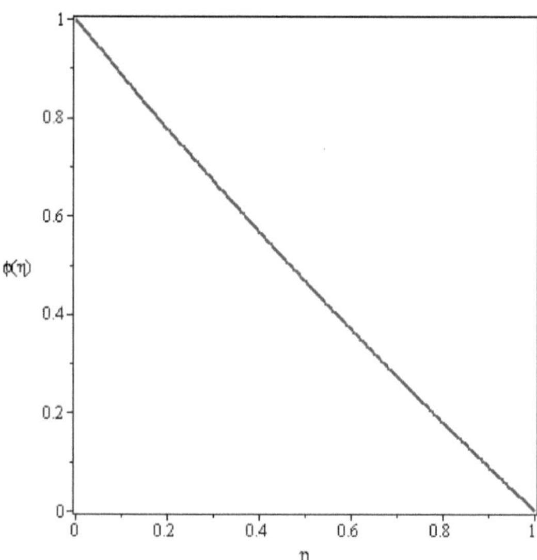

And the code of numerical solution of the presented problem is written as

restart;
$A := 0.2; M := 1; pr := 1; Le := 1; s := 1; Nt := 0.1; Nb := 1;$

$$f := \frac{d^4}{dy^4} v(y) - s \cdot \left(y \cdot \frac{d^3}{dy^3} v(y) + 3 \cdot \frac{d^2}{dy^2} v(y) - 2\, v(y) \cdot \frac{d^3}{dy^3} v(y) \right) - M^2 \cdot \frac{d^2}{dy^2} v(y) = 0, \frac{d^2}{dy^2} u(y) + pr \cdot s \cdot \left(2 \cdot v(y) \cdot \frac{d}{dy} u(y) - y \cdot \frac{d}{dy} u(y) \right) + pr \cdot Nb \cdot \left(\frac{d}{dy} u(y) \right) \cdot \frac{d}{dy} w(y) + pr \cdot Nt \cdot \left(\frac{d}{dy} u(y) \right)^2 = 0, \frac{d^2}{dy^2} w(y) + Le \cdot s \cdot (2 \cdot v(y) - y) \cdot \frac{d}{dy} w(y) + \frac{Nt}{Nb} \cdot \frac{d^2}{dy^2} u(y) = 0;$$

$$ic := v(0) = A, v(1) = \frac{1}{2}, \mathrm{D}(v)(0) = 0, \mathrm{D}(v)(1) = 0, u(0) = 1, u(1) = 0, w(0) = 1, w(1) = 0;$$

```
f := dsolve([f, ic], numeric);
for i from 0 by 0.2 while i ≤ 1 do;
([x=i, f(i)[2], f(i)[4], f(i)[8]]);
end do;

Fig, Numerical;
plots[odeplot](f, [y, v(y)], y=0..1, axes=boxed, labels=["η", "f(y)"], thickness=2);
plots[odeplot](f, [y, diff(v(y), y)], y=0..1, axes=boxed, labels=["η", "df(η)"], thickness=2);
plots[odeplot](f, [y, u(y)], y=0..1, axes=boxed, labels=["η", "θ(η)"], thickness=2);
plots[odeplot](f, [y, w(y)], y=0..1, axes=boxed, labels=["η", "φ(η)"], thickness=2);
```

$$A := 0.2$$
$$M := 1$$
$$pr := 1$$
$$Le := 1$$
$$s := 1$$
$$Nt := 0.1$$
$$Nb := 1$$

$$f := \frac{d^4}{dy^4} v(y) - y\left(\frac{d^3}{dy^3} v(y)\right) - 4\left(\frac{d^2}{dy^2} v(y)\right) + 2\, v(y)\left(\frac{d^3}{dy^3} v(y)\right)$$
$$= 0, \frac{d^2}{dy^2} u(y) + 2\, v(y)\left(\frac{d}{dy} u(y)\right) - y\left(\frac{d}{dy} u(y)\right)$$
$$+ \left(\frac{d}{dy} u(y)\right)\left(\frac{d}{dy} w(y)\right) + 0.1\left(\frac{d}{dy} u(y)\right)^2 = 0, \frac{d^2}{dy^2} w(y)$$
$$+ (2\, v(y) - y)\left(\frac{d}{dy} w(y)\right) + 0.1\left(\frac{d^2}{dy^2} u(y)\right) = 0$$

$$ic := v(0) = 0.2, v(1) = \frac{1}{2}, D(v)(0) = 0, D(v)(1) = 0, u(0) = 1, u(1) = 0,$$
$$w(0) = 1, w(1) = 0$$

f := **proc**(x_bvp) ... **end proc**

[x=0, $u(y)$ = 1.00000000000000, $v(y)$ = 0.200000000000000, $w(y)$ = 1.00000000000000]

[x=0.2, $u(y)$ = 0.864352514964776, $v(y)$ = 0.232846609716063, $w(y)$ = 0.774823428037221]

[x=0.4, $u(y)$ = 0.702284466988763, $v(y)$ = 0.307848802177926, $w(y)$ = 0.564194824118406]

[x=0.6, $u(y)$ = 0.508517836879244, $v(y)$ = 0.395253747353458, $w(y)$ = 0.365653464110325]

[x=0.8, $u(y)$ = 0.277348236192227, $v(y)$ = 0.468563921400436, $w(y)$ = 0.178126101030995]

[x= 1.0, $u(y)$ = 0., $v(y)$ = 0.500000000000000, $w(y)$ = 0.]

Fig, Numerical

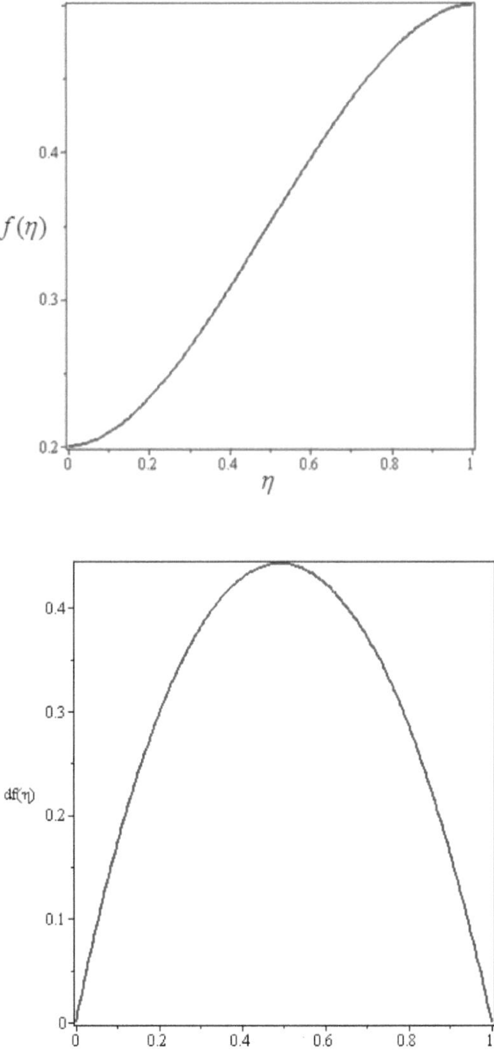
$f(\eta)$
0.2
0.3
0.4
0
0.2
0.4
0.6
0.8
1
η
$df(\eta)$
0
0.1
0.2
0.3
0.4
0
0.2
0.4
0.6
0.8
1
η

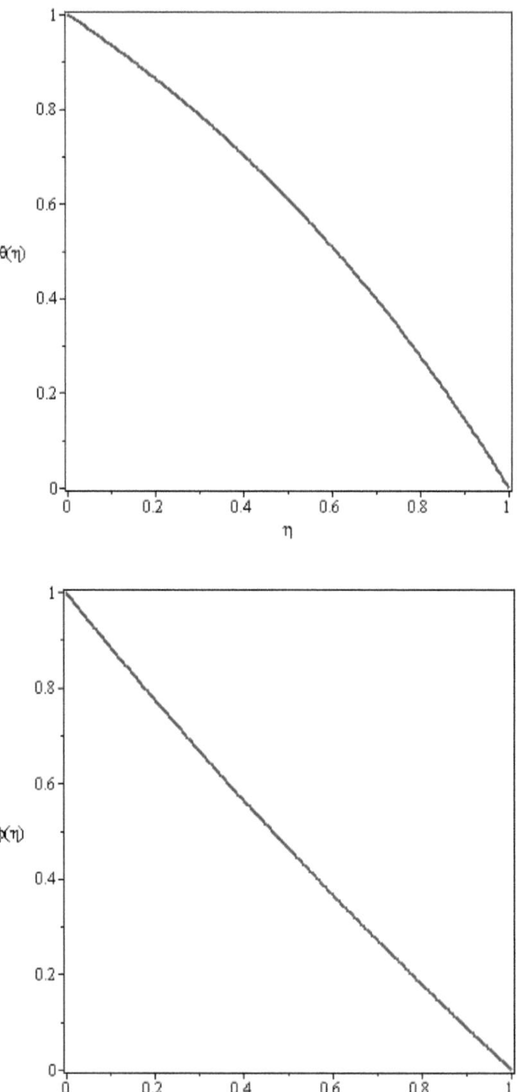
θ(η)
η
ϕ(η)
η

3. A Vibrational Differential Equation

restart;

$\theta := e^{-a\,t} \cdot (b \cdot \cos(\omega\, t + \varphi));$

$f := \frac{d^2}{dt^2}\theta + \alpha \cdot \frac{d}{dt}\theta + \beta\,\theta + \gamma 1 \cdot \theta^3 + \eta = 0;$

$\theta := unapply(\theta, t);$
$f := unapply(f, t);$

$s1 := \theta(0) = uo;$
$s2 := \theta'(0) = 0;$
$s3 := f(0);$
$s4 := f'(0);$

$S := solve([s1, s2, s3, s4], \{a, b, \omega, \varphi\}):$
$S := convert(S, radical):$

$\omega := rhs(S[3]);$
$\theta := eval\left(e^{-a\,t} \cdot (b \cdot \cos(\omega\, t + \varphi)), S\right);$

$\psi = 4\,\beta\, uo + 4\,\gamma 1\, uo^3 + 4\,\eta;$
$'\theta' = applyrule\left(\left[-4\,\beta\, uo - 4\,\gamma 1\, uo^3 - 4\,\eta = -\psi, 4\,\beta\, uo + 4\,\gamma 1\, uo^3 + 4\,\eta = \psi\right], \theta\right);$

$\alpha := \frac{c}{\rho\, A}; \beta := \frac{\left(E\, Ie\, \pi^4 - N\,\pi^2 \cdot L^2\right)}{\rho\, A\, L^4}; \gamma 1 := \frac{2E\,\pi^4}{\rho\, L^4}; \eta := \frac{q\,\pi}{4\,\rho\, A\, L^4};$
$uo := 0.06; L := 4; N := 2; E := 200000; Ie := 0.0005; \rho := 1800; A := 0.001; q := 3; c := 0.5;$

$AGM;\ \omega := evalf(\omega);\ \theta := evalf(\theta);\ d\theta := diff(\theta, t);$
fig, AGM;
$plot(\theta, t = 0\,..20, axes = boxed, labels = [\text{"t"}, \text{"θ(t)"}], thickness = 2);$
deraviton;
$plot(d\theta, t = 0\,..20, axes = boxed, labels = [\text{"t"}, \text{"dθ(t)"}], thickness = 2);$

Phase Plan;
$plot([\theta, d\theta, t = 0\,..5], scaling = constrained, thickness = 2, axes = boxed, label = [\text{"θ"}, \text{"dθ"}]);$

%%%%%%%%%%%%%%%%%%%%%%%%%%%%%%%%%%;

$$ul := -\theta\cdot\left(2\sin\left(\frac{\pi x}{L}\right) - \sin\left(\frac{2\pi x}{L}\right)\right);$$

$$plot\left(\frac{ul}{\theta}, x=0..L, axes=boxed, labels=["x", "u(x)"], thickness=2\right);$$

$$U := \theta\cdot ul;$$

$$plot3d(U, x=0..L, t=0..5, axes=box);$$

$$with(plots):$$

$$contourplot(U, t=0..5, x=0..L, contours=7, axes=boxed);$$

%%%%%%%%%%%%%%%%%%%%%%%%%%%%%%%%%%;

$$a := rhs(S[1]);$$

$$b := rhs(S[2]);$$

Lost energy **in** *each cycle*;

$$\boldsymbol{k} := \mathbf{1};$$

$$Fd1 := evalf\left(\alpha\cdot\left(a\,x+\omega\cdot\sqrt{b^2\cdot e^{-\frac{4\cdot k\cdot a\cdot\pi}{\omega}} - x^2}\right)\right);$$

$$Fd2 := evalf\left(\alpha\cdot\left(a\,x-\omega\cdot\sqrt{b^2\cdot e^{-\frac{4\cdot k\cdot a\cdot\pi}{\omega}} - x^2}\right)\right);$$

$$Wd := \int_0^{\frac{2k\pi}{\omega}} \alpha\cdot d\theta^2\,dt;$$

$$plot([Fd1, Fd2], x=-0.1..0.1, labels=["x", "Fd"], thickness=2);$$

```
k := 2;
```

$$Fd1 := evalf\left(\alpha\cdot\left(a\,x+\omega\cdot\sqrt{b^2\cdot e^{-\frac{4\cdot k\cdot a\cdot\pi}{\omega}}-x^2}\right)\right);$$

$$Fd2 := evalf\left(\alpha\cdot\left(a\,x-\omega\cdot\sqrt{b^2\cdot e^{-\frac{4\cdot k\cdot a\cdot\pi}{\omega}}-x^2}\right)\right);$$

$$Wd := \int_0^{\frac{2k\pi}{\omega}} \alpha\cdot d\theta^2\,dt;$$

```
plot([Fd1, Fd2], x=-0.1..0.1, labels=["x", "Fd"], thickness=2);
k := 3;
```

$$Fd1 := evalf\left(\alpha\cdot\left(a\,x+\omega\cdot\sqrt{b^2\cdot e^{-\frac{4\cdot k\cdot a\cdot\pi}{\omega}}-x^2}\right)\right);$$

$$Fd2 := evalf\left(\alpha\cdot\left(a\,x-\omega\cdot\sqrt{b^2\cdot e^{-\frac{4\cdot k\cdot a\cdot\pi}{\omega}}-x^2}\right)\right);$$

$$Wd := \int_0^{\frac{2k\pi}{\omega}} \alpha\cdot d\theta^2\,dt;$$

```
plot([Fd1, Fd2], x=-0.1..0.1, labels=["x", "Fd"], thickness=2);
FD := c U;
plot3d(FD, x=0..L, t=0..5, axes=box);
```

$$\theta := e^{-a\,t}\,b\cos(t\,\omega+\varphi)$$

$$f := a^2 e^{-a\,t} b\cos(t\,\omega+\varphi)+2\,a\,e^{-a\,t} b\sin(t\,\omega+\varphi)\,\omega-e^{-a\,t} b\cos(t\,\omega+\varphi)\,\omega^2+\alpha\left(-a\,e^{-a\,t} b\cos(t\,\omega+\varphi)-e^{-a\,t} b\sin(t\,\omega+\varphi)\,\omega\right)+\beta\,e^{-a\,t} b\cos(t\,\omega+\varphi)+\gamma 1\left(e^{-a\,t}\right)^3 b^3\cos(t\,\omega+\varphi)^3+\eta=0$$

$$\theta := t\rightarrow e^{-a\,t}\,b\cos(t\,\omega+\varphi)$$

$$f := t \to a^2 e^{-at} b\cos(t\omega+\varphi) + 2a e^{-at} b\sin(t\omega+\varphi)\,\omega - e^{-at} b\cos(t\omega+\varphi)\,\omega^2 + \alpha\left(-a e^{-at} b\cos(t\omega+\varphi) - e^{-at} b\sin(t\omega+\varphi)\,\omega\right) + \beta e^{-at} b\cos(t\omega+\varphi) + \gamma 1\left(e^{-at}\right)^3 b^3\cos(t\omega+\varphi)^3 + \eta = 0$$

$$s1 := b\cos(\varphi) = uo$$

$$s2 := -a\,b\cos(\varphi) - b\sin(\varphi)\,\omega = 0$$

$$s3 := a^2 b\cos(\varphi) + 2\,a\,b\sin(\varphi)\,\omega - b\cos(\varphi)\,\omega^2 + \alpha\left(-a\,b\cos(\varphi) - b\sin(\varphi)\,\omega\right) + \beta b\cos(\varphi) + \gamma 1\, b^3\cos(\varphi)^3 + \eta = 0$$

$$s4 := -a^3 b\cos(\varphi) - 3a^2 b\sin(\varphi)\,\omega + 3\,a\,b\cos(\varphi)\,\omega^2 + b\sin(\varphi)\,\omega^3 + \alpha\left(a^2 b\cos(\varphi) + 2\,a\,b\sin(\varphi)\,\omega - b\cos(\varphi)\,\omega^2\right) - \beta\,a\,b\cos(\varphi) - \beta b\sin(\varphi)\,\omega - 3\gamma 1\, b^3\cos(\varphi)^3 a - 3\gamma 1\, b^3\cos(\varphi)^2\sin(\varphi)\,\omega = 0$$

$$\omega := \sqrt{-\frac{1}{4}\,\frac{-4uo^3\gamma 1 + uo\,\alpha^2 - 4uo\,\beta - 4\eta}{uo}}$$

$$\theta := \left(e^{-\frac{1}{2}\alpha t} uo\cos\left(t\sqrt{-\frac{1}{4}\,\frac{-4uo^3\gamma 1 + uo\,\alpha^2 - 4uo\,\beta - 4\eta}{uo}} + \arctan\left(-\frac{1}{2}\,\frac{\alpha\sqrt{-\frac{-4uo^3\gamma 1 + uo\,\alpha^2 - 4uo\,\beta - 4\eta}{4uo^3\gamma 1 + 4uo\,\beta + 4\eta}}}{\sqrt{-\frac{1}{4}\,\frac{-4uo^3\gamma 1 + uo\,\alpha^2 - 4uo\,\beta - 4\eta}{uo}}},\ \sqrt{-\frac{-4uo^3\gamma 1 + uo\,\alpha^2 - 4uo\,\beta - 4\eta}{4uo^3\gamma 1 + 4uo\,\beta + 4\eta}}\right)\right)\right) \Big/ \sqrt{-\frac{-4uo^3\gamma 1 + uo\,\alpha^2 - 4uo\,\beta - 4\eta}{4uo^3\gamma 1 + 4uo\,\beta + 4\eta}}$$

$$\psi = 4uo^3\gamma 1 + 4uo\,\beta + 4\eta$$

$$\theta = \frac{1}{\sqrt{-\frac{uo\,\alpha^2 - \psi}{\psi}}}\left(e^{-\frac{1}{2}\alpha t}\,uo\cos\left(t\sqrt{-\frac{1}{4}\frac{uo\,\alpha^2 - \psi}{uo}} + \arctan\left(-\frac{1}{2}\frac{\alpha\sqrt{-\frac{uo\,\alpha^2 - \psi}{\psi}}}{\sqrt{-\frac{1}{4}\frac{uo\,\alpha^2 - \psi}{uo}}}, \sqrt{-\frac{uo\,\alpha^2 - \psi}{\psi}}\right)\right)\right)$$

$$\alpha := \frac{c}{\rho A}$$

$$\beta := \frac{E\,Ie\,\pi^4 - L^2 N \pi^2}{\rho A L^4}$$

$$\gamma 1 := \frac{2 E \pi^4}{\rho L^4}$$

$$\eta := \frac{1}{4}\frac{q\pi}{\rho A L^4}$$

$$uo := 0.06$$

$$L := 4$$

$$N := 2$$

$$E := 200000$$

$$Ie := 0.0005$$

$$\rho := 1800$$

$$A := 0.001$$

$$q := 3$$

$$c := 0.5$$

AGM

$$\omega := 4.563339885$$

$$\theta := 0.06002778372\,e^{-0.1388888889\,t}\cos(4.563339885\,t - 0.03042640484)$$

$$d\theta := -0.008337192184\,e^{-0.1388888889\,t}\cos(4.563339885\,t - 0.03042640484) - 0.2739271797\,e^{-0.1388888889\,t}\sin(4.563339885\,t - 0.03042640484)$$

fig, AGM

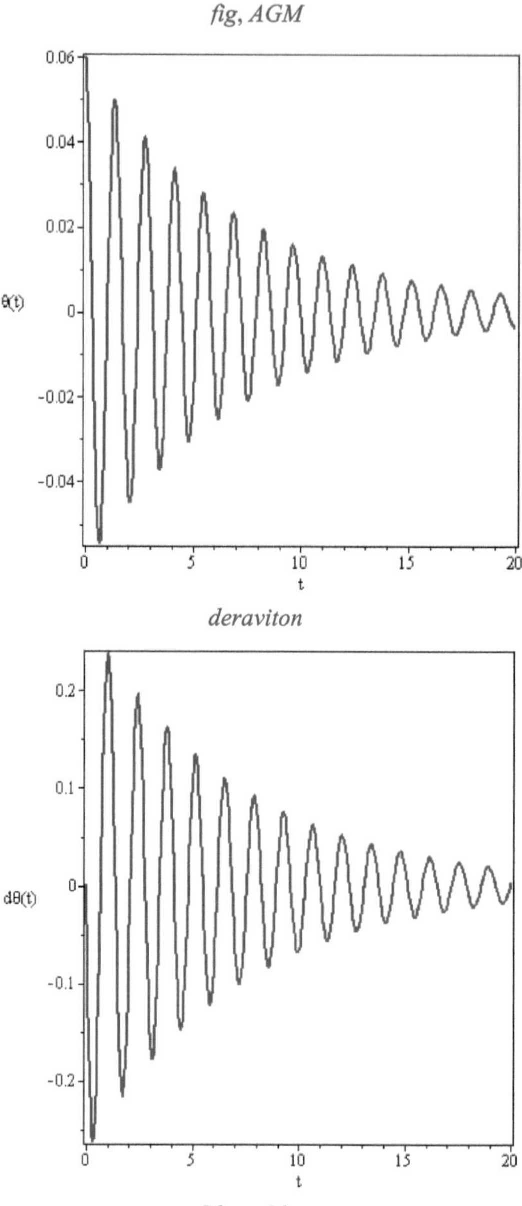

Phase Plan

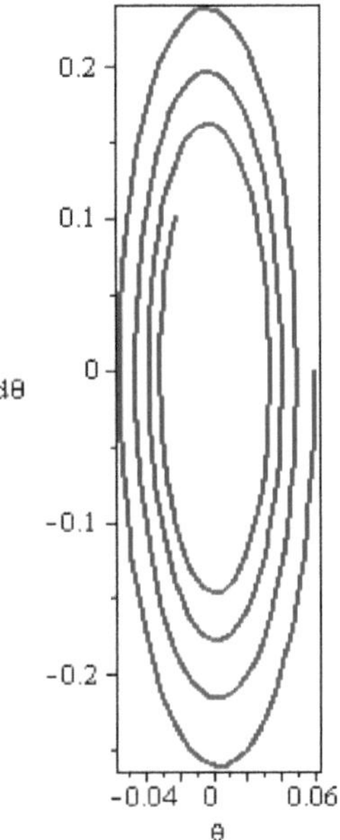

%%%%%%%%%%%%%%%%%%%%%%%%%%%%%%%%%%%%%%%
%

$$ul := -0.06002778372\, e^{-0.1388888889\, t} \cos(4.563339885\, t - 0.03042640484) \left(2 \sin\left(\frac{1}{4} \pi x\right) - \sin\left(\frac{1}{2} \pi x\right)\right)$$

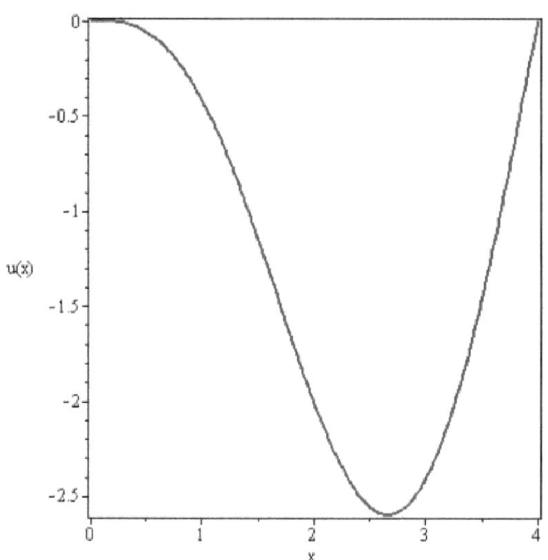

$$U := -0.003603334818\left(e^{-0.1388888889\,t}\right)^2 \cos(4.563339885\,t - 0.03042640484)^2 \left(2\sin\left(\frac{1}{4}\pi x\right) - \sin\left(\frac{1}{2}\pi x\right)\right)$$

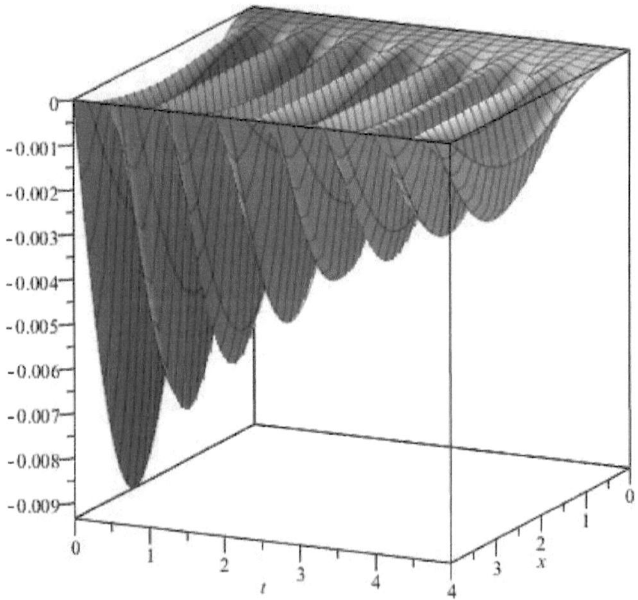

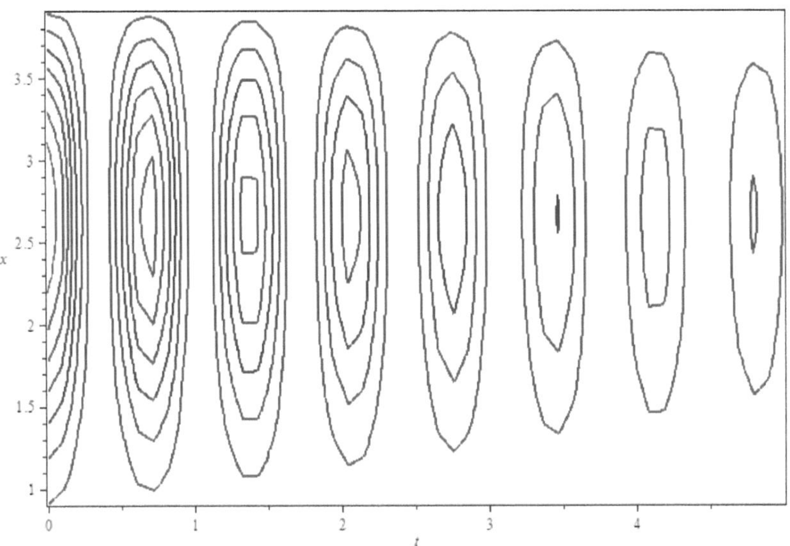

%%%

$$a := 0.1388888889$$

$$b := 0.06$$

$$\left(-\left(-0.05283333332\,\pi^4 + 0.004629629630 + 0.01666666667\,\pi^2 - 0.006510416668\,\pi\right) \Big/ \left(0.05283333332\,\pi^4 - 0.01666666667\,\pi^2 + 0.006510416668\,\pi\right)\right)^{1/2}$$

Lost energy **in** *each cycle*

$$k := 1$$

$$Fd1 := 0.03858024692\,x + 1.267594413\,\sqrt{0.002458108724 - 1.\,x^2}$$

$$Fd2 := 0.03858024692\,x - 1.267594413\,\sqrt{0.002458108724 - 1.\,x^2}$$

$$Wd := 0.01192413469$$

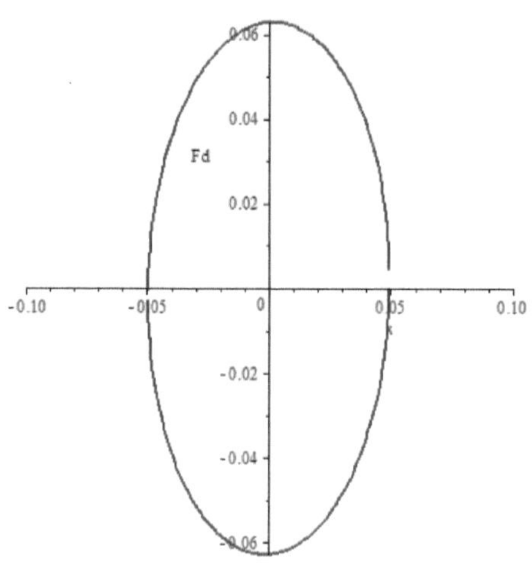

$$k := 2$$

$$Fd1 := 0.03858024692\,x + 1.267594413\,\sqrt{0.001676862907 - 1.\,x^2}$$

$$Fd2 := 0.03858024692\,x - 1.267594413\,\sqrt{0.001676862907 - 1.\,x^2}$$

$$Wd := 0.02005849384$$

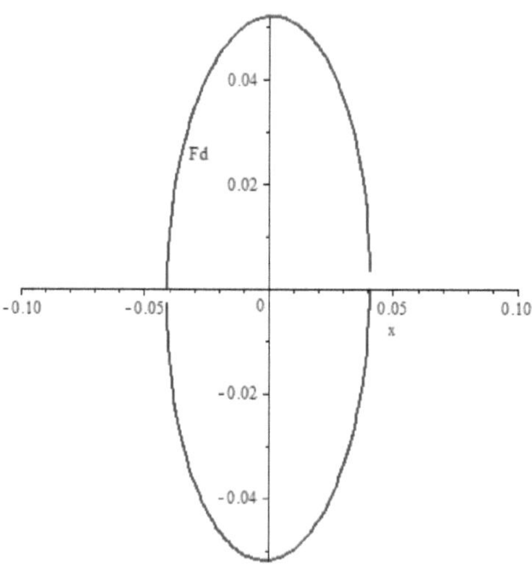

$$k := 3$$

$$Fd1 := 0.03858024692\,x + 1.267594413\sqrt{0.001143915720 - 1.\,x^2}$$

$$Fd2 := 0.03858024692\,x - 1.267594413\sqrt{0.001143915720 - 1.\,x^2}$$

$$Wd := 0.02560755885$$

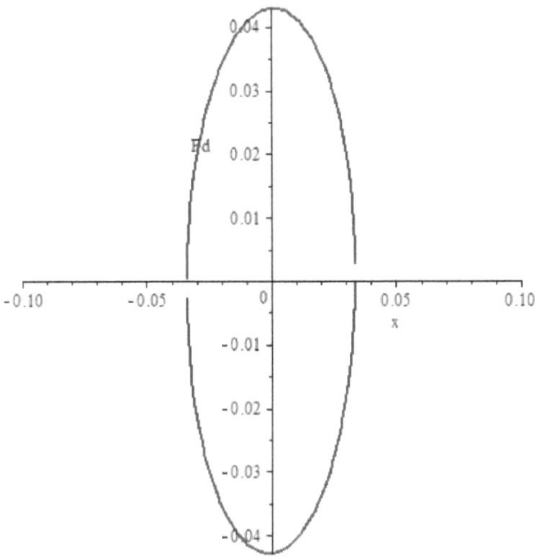

$$FD := -0.001801667409\left(e^{-0.1388888889\,t}\right)^2 \cos(4.563339885\,t - 0.03042640484)^2 \left(2\sin\left(\frac{1}{4}\pi x\right) - \sin\left(\frac{1}{2}\pi x\right)\right)$$

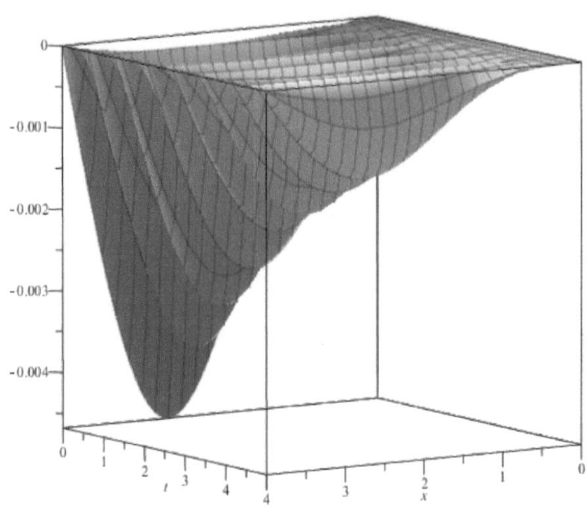

Afterward, the numerical code of the aforementioned problem is defined as follows:

Numerical;
with(*plots*) :
with(*Student*[*Calculus1*]) :
$\theta := u(x); t := x;$

$$\alpha := \frac{c}{\rho A}; \beta := \frac{\left(E\,Ie\,\pi^4 - N\pi^2 \cdot L^2\right)}{\rho A L^4}; \gamma 1 := \frac{2E\,\pi^4}{\rho\, L^4}; \eta := \frac{q\,\pi}{4\rho A L^4};$$

$uo := 0.06; L := 4; N := 2; E := 200000; Ie := 0.0005; \rho := 1800; A := 0.001; q := 3; c := 0.5;$

$$f := \frac{d^2}{dt^2}\theta + \alpha \cdot \frac{d}{dt}\theta + \beta\theta + \gamma 1 \cdot \theta^3 + \eta = 0; ic := u(0) = uo, D(u)(0) = 0;$$

f1 := *dsolve*({*f*, *ic*}, *numeric*, *output* = *listprocedure*);
for *i* **from** 0 **by** 4 **while** $i \le 20$ **do**;
([*x*=*i*,*f1*(*i*)[2],*f1*(*i*)[3]]);
end do;

```
Figures;
plots[odeplot](f1, [x, u(x) ], x=0..20, thickness=2, axes=boxed);
plots[odeplot](f1, [x, diff(u(x), x) ], x=0..20, thickness=2, axes
    =boxed);
plots[odeplot](f1, [u(x), diff(u(x), x) ], x=0..5, scaling=constrained, axes
    =boxed, labels=["du(t)", "u(t)"], thickness=2);

%%%%%%%%%%%%%%%%%%%%%%%%%%%%%%%%%%%%%%%%%%%%%%%%%%%%%%%%%%%%;

u := subs(f1, u(x)) :
du := subs(f1, d/dx u(x)) :
```

$\theta := 0.05002323785\,\mathrm{e}^{-0.1388888889\,t}\cos(4.555008562\,t - 0.03048202170);$

$d\theta := -0.006947671924\,\mathrm{e}^{-0.1388888889\,t}\cos(4.555008562\,t - 0.03048202170) - 0.2278562767\,\mathrm{e}^{-0.1388888889\,t}\sin(4.555008562\,t - 0.03048202170);$

```
plot((abs(θ - u(x))), x=0..15, thickness=2, labels=["t",
    "Er,AGM,Num,u(t)"]);
plot(abs(dθ - du(x)), x=0..15, thickness=2, labels=["t",
    "Er,AGM,Num,du(t)"]);
```

Numerical

$$\theta := u(x)$$

$$t := x$$

$$\alpha := 0.2777777778$$

$$\beta := 0.2170138889\,\pi^4 - 0.06944444445\,\pi^2$$

$$\gamma 1 := \frac{125}{144}\pi^4$$

$$\eta := 0.001627604167\,\pi$$

$$uo := 0.06$$

$$L := 4$$

$$N := 2$$

$$E := 200000$$

$$Ie := 0.0005$$

$$\rho := 1800$$

$$A := 0.001$$

$$q := 3$$

$$c := 0.5$$

$$f := \frac{d^2}{dx^2}u(x) + 0.2777777778\left(\frac{d}{dx}u(x)\right) + \left(0.2170138889\,\pi^4 - 0.06944444445\,\pi^2\right)u(x) + \frac{125}{144}\pi^4 u(x)^3 + 0.001627604167\,\pi = 0$$

$$ic := 0.0600000000000000 = 0.06,\ D(u)(0) = 0$$

$$\left[x = 0, u(x)(0) = 0.0600000000000000, \frac{d}{dx}u(x)(0) = 0.\right]$$

$$\left[x = 4, u(x)(4) = 0.0254030274540773, \frac{d}{dx}u(x)(4) = 0.101850274348737\right]$$

$$\left[x = 8, u(x)(8) = 0.00147398363655162, \frac{d}{dx}u(x)(8) = 0.0893408935375857\right]$$

$$\left[x = 12, u(x)(12) = -0.00737333241788741, \frac{d}{dx}u(x)(12) = 0.0412900544678873\right]$$

$$\left[x = 16, u(x)(16) = -0.00674107641356286, \frac{d}{dx}u(x)(16) = 0.00476676434847314\right]$$

$$\left[x = 20, u(x)(20) = -0.00327196261766695, \frac{d}{dx}u(x)(20) = -0.00966860531468078\right]$$

Figures

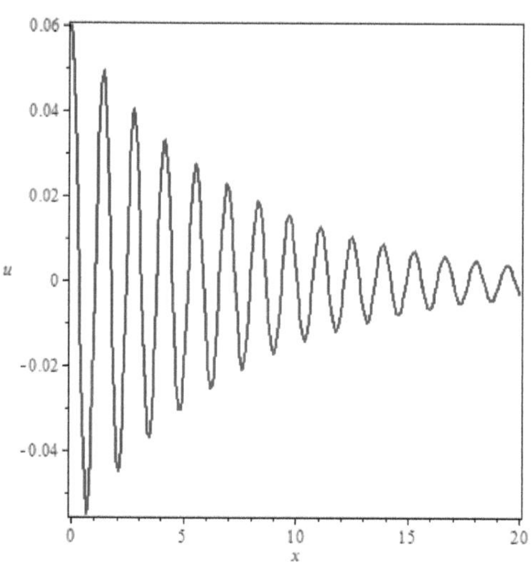

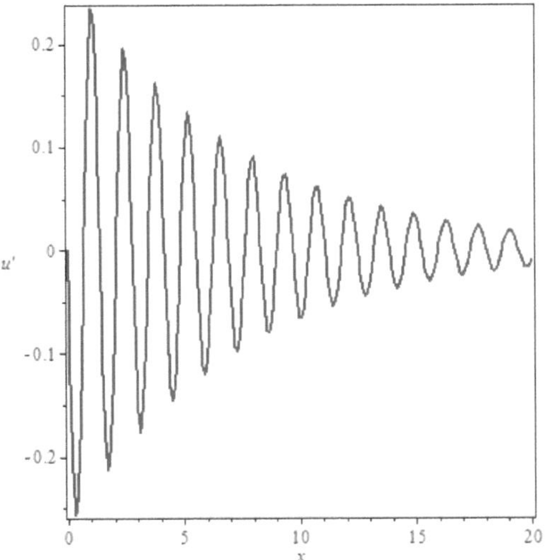

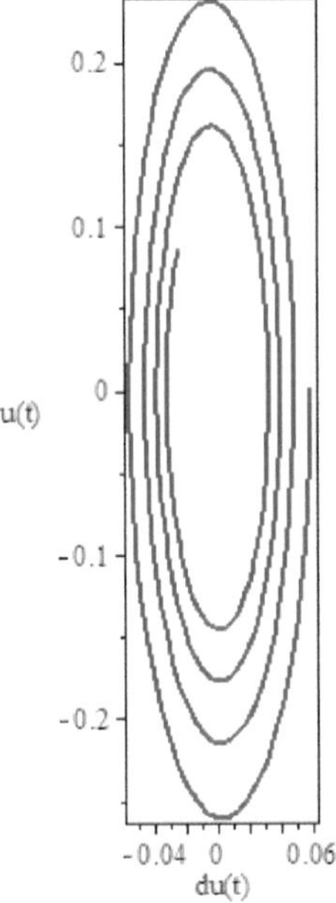

%%%%%%%%%%%%%%%%%%%%%%%%%%%%%%%%%%%%%

$$\theta := 0.05002323785\, e^{-0.1388888889\, x} \cos(4.555008562\, x - 0.03048202170)$$

$$d\theta := -0.006947671924\, e^{-0.1388888889\, x} \cos(4.555008562\, x - 0.03048202170) - 0.2278562767\, e^{-0.1388888889\, x} \sin(4.555008562\, x - 0.03048202170)$$

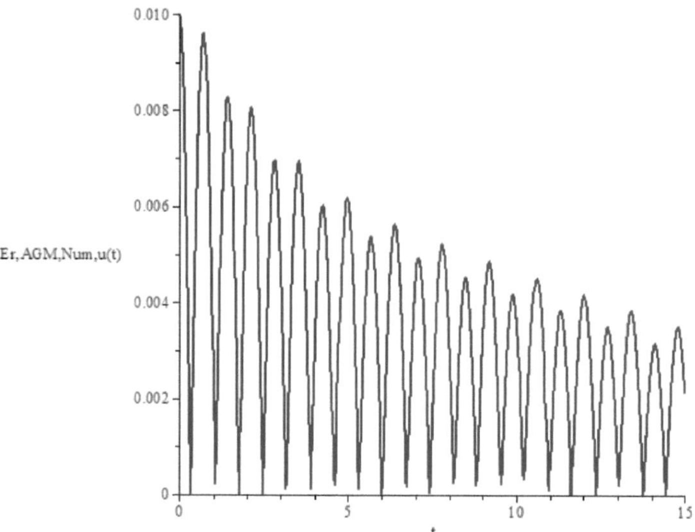

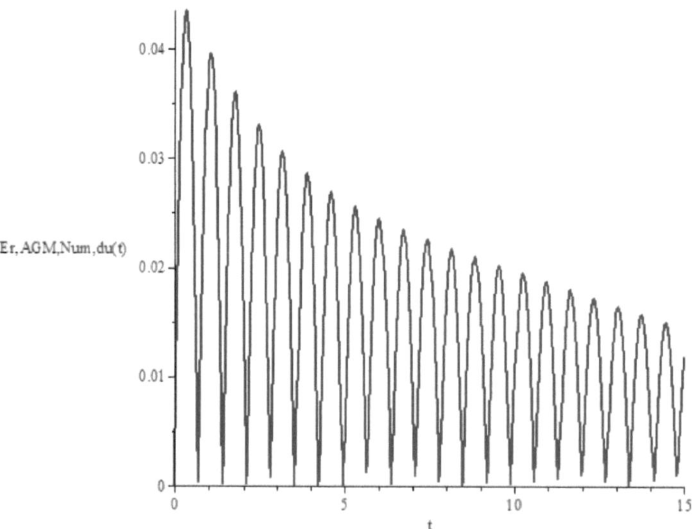

4. Nonlinear Integrodifferential Equation

restart;

Nonlinear Integrodifferential Equation;

$$\frac{d^2}{dx^2}u(x)+\left(\frac{d}{dx}u(x)\right)=\int_0^x\left(u(t)\cdot\sinh^2(x-t)-u(t)^2+u(t)\right)dt;$$

$$ut:=\sum_{i=0}^{7}a_i\cdot t^i;$$

$$ux:=\sum_{i=0}^{7}a_i\cdot x^i;$$

$$f:=\frac{d^2}{dx^2}ux+\left(\frac{d}{dx}ux\right)=\int_0^x\left(ut\cdot\sinh^2(x-t)-ut^2+ut\right)dt:$$

$ux := unapply(ux, x);$

$f := unapply(f, x):$

$s1 := ux(1) = 0;$

$s2 := ux(0) = 1;$

$s3 := f(0);$

$s4 := f(1);$

$s5 := f'(0);$

$s6 := f'(1);$

$s7 := f''(0);$

$s8 := f''(1);$

$S := solve([s1, s2, s3, s4, s5, s6, s7, s8], \{a_0, a_1, a_2, a_3, a_4, a_5, a_6, a_7\}):$

$S := evalf(convert(S, radical));$

$$u:=\left(eval\left(\sum_{i=0}^{7}a_i\cdot x^i, S\right)\right);$$

$du := diff(u, x);$

fig, AGM;

plot(*u*, *x*=0 ..1, *axes*=*boxed*, *labels*= ["x", "u(x)"], *thickness*=2);

plot(*du*, *x*=0 ..1, *axes*=*boxed*, *labels*= ["x", "du(x)"], *thickness*=2);

Nonlinear Integrodifferential Equation

$$\frac{d^2}{dx^2}u(x)+\frac{d}{dx}u(x)=\int_0^x\left(u(t)\sinh(-x+t)^2-u(t)^2+u(t)\right)dt$$

$$ut := a_0 + a_1 t + a_2 t^2 + a_3 t^3 + a_4 t^4 + a_5 t^5 + a_6 t^6 + a_7 t^7$$

$$ux := a_0 + a_1 x + a_2 x^2 + a_3 x^3 + a_4 x^4 + a_5 x^5 + a_6 x^6 + a_7 x^7$$

$$ux := x \to a_0 + a_1 x + a_2 x^2 + a_3 x^3 + a_4 x^4 + a_5 x^5 + a_6 x^6 + a_7 x^7$$

$$s1 := a_0 + a_1 + a_2 + a_3 + a_4 + a_5 + a_6 + a_7 = 0$$

$$s2 := a_0 = 1$$

$$s3 := a_1 + 2a_2 = 0$$

$$\begin{aligned}
s4 := {} & 49a_7 + 36a_6 + 25a_5 + 16a_4 + 9a_3 + 4a_2 + a_1 = \frac{15}{8}a_5\cosh(1)^2 \\
& + \frac{3}{8}a_3\cosh(1)^2 + \frac{315}{16}a_7\cosh(1)^2 + \frac{1}{4}a_1\cosh(1)^2 \\
& + \frac{45}{8}a_6\cosh(1)\sinh(1) + \frac{3}{4}a_4\cosh(1)\sinh(1) \\
& + \frac{1}{4}a_2\cosh(1)\sinh(1) + \frac{1}{2}a_0\cosh(1)\sinh(1) - \frac{2}{3}a_0a_2 - \frac{1}{2}a_1a_2 \\
& - \frac{1}{3}a_2a_3 - \frac{2}{7}a_2a_4 - \frac{1}{2}a_3a_0 - \frac{2}{5}a_3a_1 - \frac{2}{9}a_3a_5 - \frac{2}{5}a_4a_0 \\
& - \frac{1}{3}a_4a_1 - \frac{1}{4}a_4a_3 - \frac{1}{5}a_4a_5 - \frac{1}{3}a_5a_0 - \frac{2}{7}a_5a_1 - \frac{1}{4}a_5a_2 \\
& - \frac{1}{6}a_5a_6 - \frac{2}{13}a_5a_7 - \frac{2}{7}a_6a_0 - \frac{1}{4}a_6a_1 - \frac{2}{9}a_6a_2 - \frac{1}{5}a_6a_3 \\
& - \frac{2}{11}a_6a_4 - \frac{1}{7}a_6a_7 - \frac{1}{4}a_7a_0 - \frac{2}{9}a_7a_1 - \frac{1}{5}a_7a_2 - \frac{2}{11}a_7a_3 \\
& - \frac{1}{6}a_7a_4 - \frac{1}{3}a_1^2 - \frac{1}{5}a_2^2 - \frac{1}{7}a_3^2 - \frac{1}{9}a_4^2 - \frac{1}{11}a_5^2 - \frac{1}{13}a_6^2 - \frac{1}{15}a_7^2 \\
& + \frac{1}{2}a_0 - \frac{1}{12}a_2 - \frac{5}{8}a_3 - \frac{23}{20}a_4 - \frac{103}{24}a_5 - \frac{563}{56}a_6 - \frac{187}{4}a_7 \\
& - a_0a_1 - a_0^2
\end{aligned}$$

$$s5 := 2a_2 + 6a_3 = -a_0^2 + a_0$$

$$s6 := 252\, a_7 + 150\, a_6 + 80\, a_5 + 36\, a_4 + 12\, a_3 + 2\, a_2 = \frac{3}{4}\, a_4 \sinh(1)^2 + \frac{3}{4}\, a_4 \cosh(1)^2 + \frac{1}{4}\, a_2 \sinh(1)^2 + \frac{1}{4}\, a_2 \cosh(1)^2 + \frac{1}{2}\, a_0 \sinh(1)^2 + \frac{1}{2}\, a_0 \cosh(1)^2 + \frac{45}{8}\, a_6 \sinh(1)^2 + \frac{45}{8}\, a_6 \cosh(1)^2 + \frac{1}{2}\, a_1 \cosh(1)\sinh(1) + \frac{3}{4}\, a_3 \cosh(1)\sinh(1) + \frac{315}{8}\, a_7 \cosh(1)\sinh(1) + \frac{15}{4}\, a_5 \cosh(1)\sinh(1) - 2\, a_0\, a_2 - 2\, a_1\, a_2 - 2\, a_2\, a_3 - 2\, a_2\, a_4 - 2\, a_3\, a_0 - 2\, a_3\, a_1 - 2\, a_3\, a_5 - 2\, a_4\, a_0 - 2\, a_4\, a_1 - 2\, a_4\, a_3 - 2\, a_4\, a_5 - 2\, a_5\, a_0 - 2\, a_5\, a_1 - 2\, a_5\, a_2 - 2\, a_5\, a_6 - 2\, a_5\, a_7 - 2\, a_6\, a_0 - 2\, a_6\, a_1 - 2\, a_6\, a_2 - 2\, a_6\, a_3 - 2\, a_6\, a_4 - 2\, a_6\, a_7 - 2\, a_7\, a_0 - 2\, a_7\, a_1 - 2\, a_7\, a_2 - 2\, a_7\, a_3 - 2\, a_7\, a_4 - a_1^2 - a_2^2 - a_3^2 - a_4^2 - a_5^2 - a_6^2 - a_7^2 + \frac{1}{2}\, a_0 + \frac{1}{2}\, a_1 + \frac{1}{4}\, a_2 - \frac{1}{4}\, a_3 - \frac{7}{4}\, a_4 - \frac{23}{4}\, a_5 - \frac{161}{8}\, a_6 - \frac{563}{8}\, a_7 - 2\, a_0\, a_1 - a_0^2$$

$$s7 := 6\, a_3 + 24\, a_4 = -2\, a_0\, a_1 + a_1$$

$$s8 := 1050\,a_7 + 480\,a_6 + 180\,a_5 + 48\,a_4 + 6\,a_3 = \frac{1}{2}\,a_1 \sinh(1)^2 + \frac{3}{4}\,a_3 \sinh(1)^2 + \frac{15}{4}\,a_5 \sinh(1)^2 + \frac{315}{8}\,a_7 \sinh(1)^2 + \frac{15}{4}\,a_5 \cosh(1)^2 + \frac{3}{4}\,a_3 \cosh(1)^2 + \frac{315}{8}\,a_7 \cosh(1)^2 + \frac{1}{2}\,a_1 \cosh(1)^2 + \frac{45}{2}\,a_6 \cosh(1)\sinh(1) + 3\,a_4 \cosh(1)\sinh(1) + a_2 \cosh(1)\sinh(1) + 2\,a_0 \cosh(1)\sinh(1) - 4\,a_0 a_2 - 6\,a_1 a_2 - 10\,a_2 a_3 - 12\,a_2 a_4 - 6\,a_3 a_0 - 8\,a_3 a_1 - 16\,a_3 a_5 - 8\,a_4 a_0 - 10\,a_4 a_1 - 14\,a_4 a_3 - 18\,a_4 a_5 - 10\,a_5 a_0 - 12\,a_5 a_1 - 14\,a_5 a_2 - 22\,a_5 a_6 - 24\,a_5 a_7 - 12\,a_6 a_0 - 14\,a_6 a_1 - 16\,a_6 a_2 - 18\,a_6 a_3 - 20\,a_6 a_4 - 26\,a_6 a_7 - 14\,a_7 a_0 - 16\,a_7 a_1 - 18\,a_7 a_2 - 20\,a_7 a_3 - 22\,a_7 a_4 - 2\,a_1^2 - 4\,a_2^2 - 6\,a_3^2 - 8\,a_4^2 - 10\,a_5^2 - 12\,a_6^2 - 14\,a_7^2 + \frac{1}{2}\,a_1 + a_2 + \frac{3}{4}\,a_3 - a_4 - \frac{35}{4}\,a_5 - \frac{69}{2}\,a_6 - \frac{1127}{8}\,a_7 - 2\,a_0 a_1$$

$$S := \{a_0 = 1.,\ a_1 = -1.637459677,\ a_2 = 0.8187298387,\ a_3 = -0.2729099462,\ a_4 = 0.1364549731,\ a_5 = -0.06544906120,\ a_6 = 0.02343499126,\ a_7 = -0.002801118184\}$$

$$u := 1. - 1.637459677\,x + 0.8187298387\,x^2 - 0.2729099462\,x^3 + 0.1364549731\,x^4 - 0.06544906120\,x^5 + 0.02343499126\,x^6 - 0.002801118184\,x^7$$

$$du := -1.637459677 + 1.637459677\,x - 0.8187298386\,x^2 + 0.5458198924\,x^3 - 0.3272453060\,x^4 + 0.1406099476\,x^5 - 0.01960782729\,x^6$$

fig, AGM

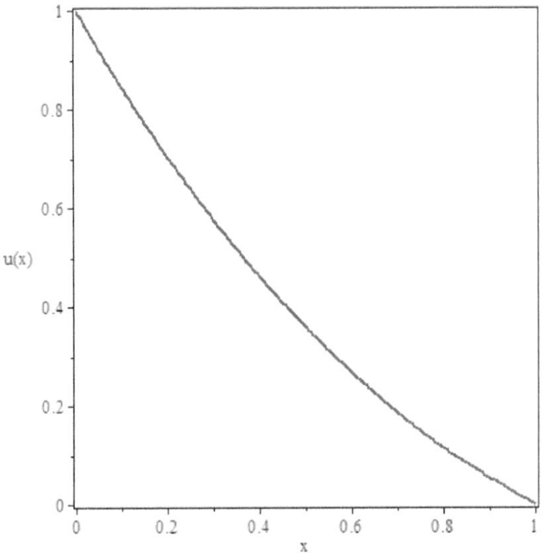
u(x)
x
0
0.2
0.4
0.6
0.8
1

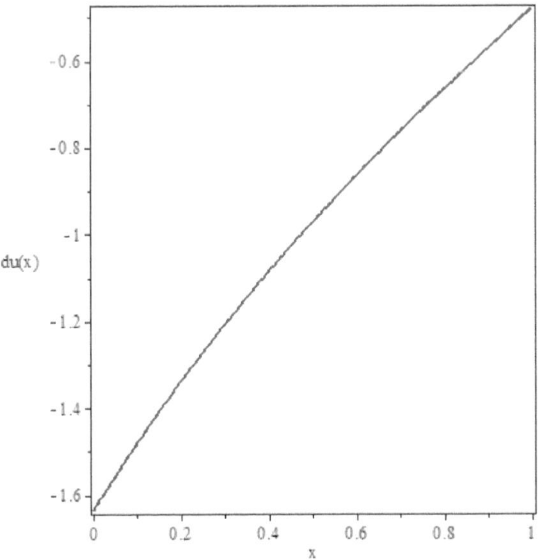
du(x)
x
-0.6
-0.8
-1
-1.2
-1.4
-1.6
0
0.2
0.4
0.6
0.8
1

After that, the numerical solution of the above problem can be written in the following form:

restart;

$$f := \frac{d^2}{dx^2}u(x) + \left(\frac{d}{dx}u(x)\right) = \int_0^x \left(u(x)\cdot \sinh^2(x-t) - u(x)^2 + u(x)\right)$$

d*t*;

$ic := u(1)=0,\ (u)(0)=1;$

f1 := *dsolve*([*f*, *ic*], *numeric*, *output* = *listprocedure*);
for *i* **from** 0 **by** 0.4 **while** $i \le 1$ **do**;
([*x*=*i*,*f1*(*i*)[2],*f1*(*i*)[3]]);
end do;

Figures;
plots[*odeplot*](*f1*, [*x*, *u*(*x*)], *x*=0 ..1, *thickness*=2, *axes*=*boxed*);
plots[*odeplot*](*f1*, [*x*, *diff*(*u*(*x*), *x*)], *x*=0 ..1, *thickness*=2, *axes*=*boxed*);

$$f := \frac{d^2}{dx^2}u(x) + \frac{d}{dx}u(x) = -\frac{1}{8}u(x)\left(8u(x)\,x\,e^{2x} - e^{4x} - 4x\,e^{2x} + 1\right)e^{-2x}$$

$$ic := u(1)=0,\ u(0)=1$$

$$f1 := \left[x = \mathbf{proc}(x)\ \ldots\ \mathbf{end\ proc},\ u(x) = \mathbf{proc}(x)\ \ldots\ \mathbf{end\ proc},\ \frac{d}{dx}u(x) = \mathbf{proc}(x)\ \ldots\ \mathbf{end\ proc}\right.$$

$$\left[x = 0,\ u(x)(0) = 1.00000000000000,\ \frac{d}{dx}u(x)(0) = -1.62618983709882\right]$$

$$\left[x = 0.4,\ u(x)(0.4) = 0.465851742868689,\ \frac{d}{dx}u(x)(0.4) = -1.07325409516433\right]$$

$$\left[x = 0.8,\ u(x)(0.8) = 0.121165695727478,\ \frac{d}{dx}u(x)(0.8) = -0.676582939675142\right]$$

Figures

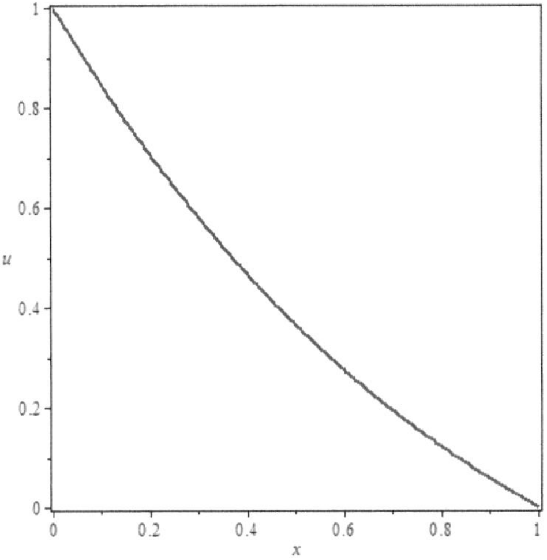
1
0.8
0.6
u
0.4
0.2
0
0
0.2
0.4
0.6
0.8
1
x

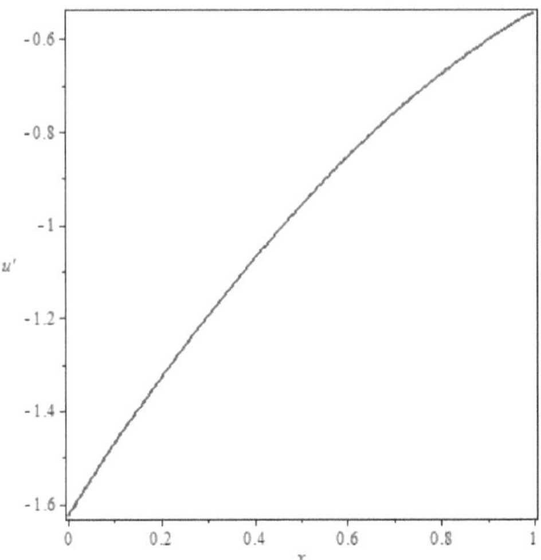
-0.6
-0.8
-1
u'
-1.2
-1.4
-1.6
0
0.2
0.4
0.6
0.8
1
x

5. Separation

restart;

$$U := \phi(x)\cdot z(t);$$

$$f := \rho\cdot A\cdot\frac{\partial^2}{\partial t^2}U + \frac{\partial^2}{\partial x^2}\left(EIe\cdot\frac{\partial^2}{\partial x^2}U\right) + N\cdot\frac{\partial^2}{\partial x^2}U - \frac{EA}{2L}\cdot\frac{\partial^2}{\partial x^2}U\cdot\int_0^L\left(\frac{\partial}{\partial x}U\right)^2 dx = Fo\cdot\sin(\Omega);$$

$$expand\left(\frac{f}{\rho\cdot A\cdot\phi(x)}\right);$$

%%%%%%%%%%%%%%%%%%%%%%%%%%%%%%%;

restart;

$$\phi := \frac{\left(\cos\left(\frac{1}{2}\alpha L\right)\sinh\left(\frac{1}{2}\alpha x\right) - \sin\left(\frac{1}{2}\alpha x\right)\cosh\left(\frac{1}{2}\alpha L\right)\right)}{\cosh\left(\frac{1}{2}\alpha L\right)};$$

$$\alpha := \frac{\mathbf{1}}{\mathbf{2}}\,\frac{(\mathbf{4}\,\boldsymbol{n}+\mathbf{1})\,\boldsymbol{\pi}}{\boldsymbol{L}};$$

$$n := 1;$$

$$Ncr := evalf\left(\frac{\left(E\cdot Ie\cdot\int_0^L(\phi'')^2\,dx\right)}{\int_0^L(\phi)^2\,dx}\right);$$

$$n := 2;$$

$$K := evalf\left(E\cdot Ie\cdot\int_0^L(\phi'')^2\,dx\right) + \frac{EA}{L};$$

$$L := 2;$$

$$\phi := \phi;$$

$$plot(\phi, x = 0..L);$$

$$\alpha := expand\left(evalf\left(\frac{\left(\int_0^L (EIe\cdot\phi'''' + N\cdot\phi'')\cdot\phi \, dx\right)}{\rho\cdot S\cdot\int_0^L (\phi)^2 \, dx}\right)\right);$$

$$\beta := -evalf\left(\frac{\left(\int_0^L \left(E\cdot\phi''\cdot\int_0^L (\phi')^2 \, dx\right)\cdot\phi \, dx\right)}{2\,\rho\cdot S\cdot\int_0^L (\phi)^2 \, dx}\right);$$

$$\gamma 1 := evalf\left(\frac{\left(\int_0^L Fo\cdot\phi \, dx\right)}{\rho\cdot S\cdot\int_0^L (\phi)^2 \, dx}\right);$$

%%%%%%%%%%%%%%%%%%%%%%%%%%%%%%%%%%%;

restart;

$\eta := I\lambda;$

$$\phi := c1\sinh\left(\frac{1}{2}\eta x\right) + c2\sinh\left(\frac{1}{2}\lambda x\right) + c3\cosh\left(\frac{1}{2}\eta x\right) + c4\cosh\left(\frac{1}{2}\lambda x\right);$$

%%%%%%%%%%%%%%%%%%%%%%%%%%%%%;

♠ *Boundary Conditions Hinged* **and** *Fixed Ends* ;

$s1 := eval(\phi, x=0) = 0;$
$s2 := eval(\phi, x=L) = 0;$
$s3 := eval(diff(\phi, x, x), x=0) = 0;$
$s4 := (eval(diff(\phi, x), x=L)) = 0;$
with(LinearAlgebra) :
$q1 := [s1, s2, s3, s4] :$
$q2 := [c1, c2, c3, c4] :$
Coefficients Matrix;
Mat := [GenerateMatrix(q1, q2)][1];
Determinant;
det := factor(Determinant(Mat)) = 0;

%%%%%%%%%%%%%%%%%%%%%%%%%%%%%;

```
restart;
```
$$\boldsymbol{det} := \cosh\left(\frac{1}{2}\mu\right)\sin\left(\frac{1}{2}\mu\right) - \sinh\left(\frac{1}{2}\mu\right)\cos\left(\frac{1}{2}\mu\right);$$

plot(*det*, μ = 0 .. ∞, *labels* = ["μ=λ L", "det(μ)"], *thickness* = 1);

%%%%%%%%%%%%%%%%%%%%%%%%%%;

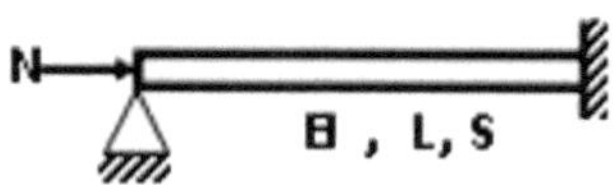

restart;

Determinent of Bending Vibration **and** *The Solution of Modal Frequency*;

$$det := \cosh\left(\frac{1}{2}\alpha L\right)\sin\left(\frac{1}{2}\alpha L\right) - \sinh\left(\frac{1}{2}\alpha L\right)\cos\left(\frac{1}{2}\alpha L\right);$$

$\beta = \alpha L$;

Shape Function;

$$yx_n = \frac{\left(\cos\left(\frac{1}{2}\alpha_n L\right)\sinh\left(\frac{1}{2}\alpha_n x\right) - \sin\left(\frac{1}{2}\alpha_n x\right)\cosh\left(\frac{1}{2}\alpha_n L\right)\right)}{\cosh\left(\frac{1}{2}\alpha_n L\right)};$$

det := applyrule(α L = β, det) = 0;

Angular Frequency;

for i by 2 to 1200 do

g_i := fsolve(det, β = 0.1 ..i) ;

end do:

for *i* to 9 do

$$\alpha_i := \frac{\sqrt{-2\, E\, Ie\left(N + \sqrt{N^2 + 4\, E\, Ie\, \omega_i^2\ S\rho}\right)}}{E\, Ie};$$

β_i := *sort*({ *seq*(g_i, *i* = 1 ..600) }) [*i*] = $\alpha_i \cdot L$;

ω_i := *solve*(β_i, ω_i) [1];

end do:

%%%%%%%%%%% %%%%%%%%%%;

for n to 8 do
$\beta_n := lhs\left(\beta_n\right);$
$\omega_n := \omega_n;$
end do;

%%%%%%%%%%%%%%%%%%%%%;

with(Statistics) :
X := Vector([1, 2, 3, 4, 5, 6, 7, 8]) :
e1 := $\left[\mathrm{seq}\left(\beta_i, i=1..8\right)\right]$:
Y1 := Vector(e1) :

$$ff := \frac{(\,a1\cdot k + a2)\cdot\pi}{2}:$$

$$\alpha_k = \frac{\sqrt{-2\,E\,Ie\left(N+\sqrt{N^2+4\,E\,Ie\,\omega_k^2\;S\,\rho}\right)}}{E\,Ie};$$

$$\alpha_k := \frac{(\mathrm{NonlinearFit}(\,ff, X, Y1, k)\,)}{L};$$

$$zz := \;\alpha_k = \frac{\sqrt{-2\,E\,Ie\left(N+\sqrt{N^2+4\,E\,Ie\,\omega^2\;S\,\rho}\right)}}{E\,Ie}:$$

$\omega_k := simplify(\,[\,\mathrm{solve}(zz, \omega)\,][\,1\,]);$
$L := 4;$

%%%%%%%%%%%%%%%%%%%%%%%%%%%;

for n to 6 do

$$\beta_n := \frac{\beta_n}{L}:$$

$$yx := \frac{\cos\left(\frac{1}{2}\,\beta_n\,L\right)\sinh\left(\frac{1}{2}\,\beta_n\,x\right) - \sin\left(\frac{1}{2}\,\beta_n\,x\right)\cosh\left(\frac{1}{2}\,\beta_n\,L\right)}{\cosh\left(\frac{1}{2}\,\beta_n\,L\right)};$$

plot(*yx*, *x* = 0 ..*L*, *labels* = ["x", "ϕ(x)"], *thickness* = 2);
end **do**;

$$U := \phi(x)\;z(t)$$

$$f := \rho A \phi(x) \left(\frac{d^2}{dt^2} z(t) \right) + E\,Ie \left(\frac{d^4}{dx^4} \phi(x) \right) z(t) + N \left(\frac{d^2}{dx^2} \phi(x) \right) z(t) - \frac{1}{2} \frac{E A \left(\frac{d^2}{dx^2} \phi(x) \right) z(t) \left(\int_0^L \left(\frac{d}{dx} \phi(x) \right)^2 z(t)^2 \, dx \right)}{L} = Fo \sin(\Omega)$$

$$\frac{d^2}{dt^2} z(t) + \frac{E\,Ie \left(\frac{d^4}{dx^4} \phi(x) \right) z(t)}{\rho A \phi(x)} + \frac{N \left(\frac{d^2}{dx^2} \phi(x) \right) z(t)}{\rho A \phi(x)} - \frac{1}{2} \frac{E \left(\frac{d^2}{dx^2} \phi(x) \right) z(t)^3 \left(\int_0^L \left(\frac{d}{dx} \phi(x) \right)^2 dx \right)}{\rho \phi(x) L} = \frac{Fo \sin(\Omega)}{\rho A \phi(x)}$$

%%

$$\phi := \frac{\cos\left(\frac{1}{2} \alpha L \right) \sinh\left(\frac{1}{2} \alpha x \right) - \sin\left(\frac{1}{2} \alpha x \right) \cosh\left(\frac{1}{2} \alpha L \right)}{\cosh\left(\frac{1}{2} \alpha L \right)}$$

$$\alpha := \frac{1}{2} \frac{(4n+1)\pi}{L}$$

$$n := 1$$

$$Ncr := \frac{1}{L} \left(2.002146810\, E\,Ie \left(\int_{0.}^{L} 0.001551607900 \left(-0.7071067810 \left(\frac{d^2}{dx^2} \sinh\left(\frac{5}{4} \frac{\pi x}{L} \right)(x) \right) - 25.38686122 \left(\frac{d^2}{dx^2} \sin\left(\frac{5}{4} \frac{\pi x}{L} \right)(x) \right) \right)^2 dx \right) \right)$$

$$n := 2$$

$$K := E\,Ie\left(\left(\int_{0.}^{L} 0.000002899784793\left(0.7071067810\left(\frac{d^2}{dx^2}\sinh\left(\frac{9}{4}\frac{\pi x}{L}\right)(x)\right)\right.\right.\right.$$
$$\left.\left.\left.- 587.2420093\left(\frac{d^2}{dx^2}\sin\left(\frac{9}{4}\frac{\pi x}{L}\right)(x)\right)\right)^2 dx\right) + \frac{EA}{L}$$

$$L := 2$$

$$\phi := \frac{\frac{1}{2}\sqrt{2}\,\sinh\left(\frac{9}{8}\pi x\right) - \sin\left(\frac{9}{8}\pi x\right)\cosh\left(\frac{9}{4}\pi\right)}{\cosh\left(\frac{9}{4}\pi\right)}$$

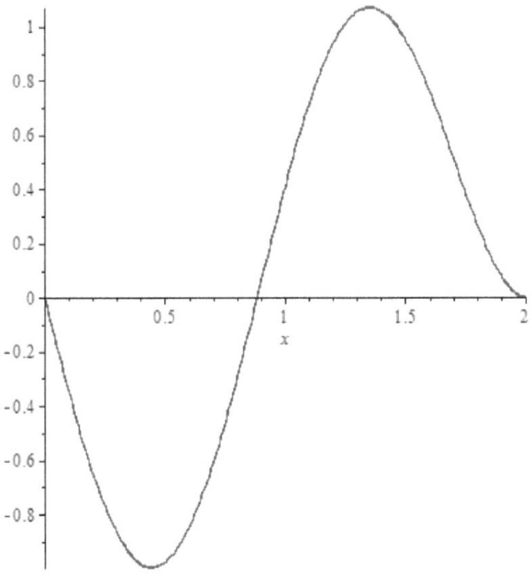

$$\alpha := \frac{1}{\rho S}\left(0.001702878443\left(\int_{0.}^{2.}\left(0.0008514377253\, Ie\, E\left(\frac{d^4}{dx^4}\sinh\left(\frac{9}{8}\pi x\right)(x)\right)\sinh(3.534291736\, x)\right.\right.\right.$$
$$- 0.7071067812\, Ie\, E\left(\frac{d^4}{dx^4}\sinh\left(\frac{9}{8}\pi x\right)(x)\right)\sin(3.534291736\, x)$$
$$- 0.7071067809\, Ie\, E\left(\frac{d^4}{dx^4}\sin\left(\frac{9}{8}\pi x\right)(x)\right)\sinh(3.534291736\, x)$$
$$+ 587.2420083\, Ie\, E\left(\frac{d^4}{dx^4}\sin\left(\frac{9}{8}\pi x\right)(x)\right)\sin(3.534291736\, x)$$
$$+ 0.0008514377253\, N\left(\frac{d^2}{dx^2}\sinh\left(\frac{9}{8}\pi x\right)(x)\right)\sinh(3.534291736\, x)$$
$$- 0.7071067812\, N\left(\frac{d^2}{dx^2}\sinh\left(\frac{9}{8}\pi x\right)(x)\right)\sin(3.534291736\, x)$$
$$- 0.7071067809\, N\left(\frac{d^2}{dx^2}\sin\left(\frac{9}{8}\pi x\right)(x)\right)\sinh(3.534291736\, x)$$
$$\left.\left.\left.+ 587.2420083\, N\left(\frac{d^2}{dx^2}\sin\left(\frac{9}{8}\pi x\right)(x)\right)\sin(3.534291736\, x)\right)dx\right)\right)$$

$$\beta := -\frac{1}{\rho S}\left(0.5000008789\left(\int_{0.}^{2.} 0.00003109755771\, E\left(0.7071067810\left(\frac{d^2}{dx^2}\sinh\left(\frac{9}{8}\pi x\right)(x)\right)\right.\right.\right.$$
$$\left.- 587.2420093\left(\frac{d^2}{dx^2}\sin\left(\frac{9}{8}\pi x\right)(x)\right)\right)$$
$$(0.7071067810\sinh(3.534291736\, x)$$
$$- 587.2420093\sin(3.534291736\, x))\, dx))$$

$$\gamma 1 := \frac{0.1168579734\, Fo}{\rho S}$$

%%

$$\eta := \mathrm{I}\lambda$$

$$\phi := \mathrm{I}\, c1 \sin\left(\frac{1}{2}\lambda x\right) + c2 \sinh\left(\frac{1}{2}\lambda x\right) + c3 \cos\left(\frac{1}{2}\lambda x\right) + c4 \cosh\left(\frac{1}{2}\lambda x\right)$$

%%%%%%%%%%%%%%%%%%%%%%%%%%%%%%

♠ *Boundary Conditions Hinged* **and** *Fixed Ends*

$$s1 := c3 + c4 = 0$$

$$s2 := \mathrm{I}\, c1 \sin\left(\frac{1}{2}\lambda L\right) + c2 \sinh\left(\frac{1}{2}\lambda L\right) + c3 \cos\left(\frac{1}{2}\lambda L\right) + c4 \cosh\left(\frac{1}{2}\lambda L\right) = 0$$

$$s3 := -\frac{1}{4} c3 \lambda^2 + \frac{1}{4} c4 \lambda^2 = 0$$

$$s4 := \frac{1}{2}\mathrm{I}\, c1 \cos\left(\frac{1}{2}\lambda L\right)\lambda + \frac{1}{2} c2 \cosh\left(\frac{1}{2}\lambda L\right)\lambda - \frac{1}{2} c3 \sin\left(\frac{1}{2}\lambda L\right)\lambda + \frac{1}{2} c4 \sinh\left(\frac{1}{2}\lambda L\right)\lambda = 0$$

Coefficients Matrix

$$Mat := \begin{bmatrix} 0 & 0 & 1 & 1 \\ \mathrm{I}\sin\left(\frac{1}{2}\lambda L\right) & \sinh\left(\frac{1}{2}\lambda L\right) & \cos\left(\frac{1}{2}\lambda L\right) & \cosh\left(\frac{1}{2}\lambda L\right) \\ 0 & 0 & -\frac{1}{4}\lambda^2 & \frac{1}{4}\lambda^2 \\ \frac{1}{2}\mathrm{I}\cos\left(\frac{1}{2}\lambda L\right)\lambda & \frac{1}{2}\cosh\left(\frac{1}{2}\lambda L\right)\lambda & -\frac{1}{2}\sin\left(\frac{1}{2}\lambda L\right)\lambda & \frac{1}{2}\sinh\left(\frac{1}{2}\lambda L\right)\lambda \end{bmatrix}$$

LinearAlgebra:-Determinant

$$det := -\frac{1}{4}\mathrm{I}\lambda^3 \left(\sin\left(\frac{1}{2}\lambda L\right)\cosh\left(\frac{1}{2}\lambda L\right) - \cos\left(\frac{1}{2}\lambda L\right)\sinh\left(\frac{1}{2}\lambda L\right)\right) = 0$$

%%%%%%%%%%%%%%%%%%%%%%%%%

$$det := \cosh\left(\frac{1}{2}\mu\right)\sin\left(\frac{1}{2}\mu\right) - \sinh\left(\frac{1}{2}\mu\right)\cos\left(\frac{1}{2}\mu\right)$$

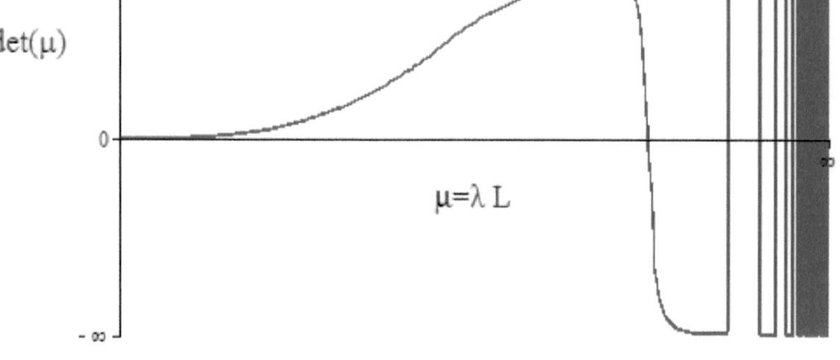

%%%%%%%%%%%%%%%%%%%%%%%%%

Bending Vibration Determinent **and** *The Solution of Modal Frequency*

$$det := \cosh\left(\frac{1}{2}\alpha L\right)\sin\left(\frac{1}{2}\alpha L\right) - \sinh\left(\frac{1}{2}\alpha L\right)\cos\left(\frac{1}{2}\alpha L\right)$$

$$\beta = \alpha L$$

Shape Function

$$yx_n = \frac{\cos\left(\frac{1}{2}\alpha_n L\right)\sinh\left(\frac{1}{2}\alpha_n x\right) - \sin\left(\frac{1}{2}\alpha_n x\right)\cosh\left(\frac{1}{2}\alpha_n L\right)}{\cosh\left(\frac{1}{2}\alpha_n L\right)}$$

$$det := \cosh\left(\frac{1}{2}\beta\right)\sin\left(\frac{1}{2}\beta\right) - \sinh\left(\frac{1}{2}\beta\right)\cos\left(\frac{1}{2}\beta\right) = 0$$

Angular Frequency

%%%%%%%%%%%

$$\beta_1 := 7.853204624$$

$$\omega_1 := \frac{1}{S\rho L^2}\left(3.141281850\ 10^{-8}\sqrt{S\rho\left(1.562500000\ 10^{16} N L^2 + 2.409094643\ 10^{17} E\,Ie\right)}\right)$$

$$\beta_2 := 14.13716549$$

$$\omega_2 := \frac{1}{S\rho L^2}\left(3.534291372\ 10^{-8}\sqrt{S\rho\left(4.000000000\ 10^{16} N L^2 + 1.998594481\ 10^{18} E\,Ie\right)}\right)$$

$$\beta_3 := 20.42035225$$

$$\omega_3 := \frac{1}{S\rho L^2}\left(0.000001276272016\sqrt{S\rho\left(6.400000000\ 10^{13} N L^2 + 6.671852576\ 10^{15} E\,Ie\right)}\right)$$

$$\beta_4 := 26.70353756$$

$$\omega_4 := \frac{1}{S\rho L^2}\left(2.670353756\ 10^{-7}\sqrt{S\rho\left(2.500000000\ 10^{15} N L^2 + 4.456743239\ 10^{17} E\,Ie\right)}\right)$$

$$\beta_5 := 32.98672286$$

$$\omega_5 := \frac{1}{S\rho L^2}\left(1.649336143\ 10^{-7}\sqrt{S\rho\left(1.000000000\ 10^{16} N L^2 + 2.720309713\ 10^{18} E\,Ie\right)}\right)$$

$$\beta_6 := 39.26990817$$

$$\omega_6 := \frac{1}{S\rho L^2}\left(9.817477042\ 10^{-8}\sqrt{S\rho\left(4.000000000\ 10^{16}\ N L^2 + 1.542125688\ 10^{19}\ E\, Ie\right)}\right)$$

$$\beta_7 := 45.55309348$$

$$\omega_7 := \frac{1}{S\rho L^2}\left(4.555309348\ 10^{-7}\sqrt{S\rho\left(2.500000000\ 10^{15}\ N L^2 + 1.296927703\ 10^{18}\ E\, Ie\right)}\right)$$

$$\beta_8 := 51.83627878$$

$$\omega_8 := \frac{1}{S\rho L^2}\left(2.591813939\ 10^{-7}\sqrt{S\rho\left(1.000000000\ 10^{16}\ N L^2 + 6.717499494\ 10^{18}\ E\, Ie\right)}\right)$$

%%%%%%%%%%%%%%%

$$\alpha_k = \frac{\sqrt{-2\, E\, Ie\left(N + \sqrt{4\, E\, Ie\, S\rho\, \omega_k^2 + N^2}\right)}}{E\, Ie}$$

$$\alpha_k := \frac{1}{2}\,\frac{(4.00004127645260\, k + 0.999752308172145)\,\pi}{L}$$

$$\omega_k := \frac{1}{S\rho L^2}\Big(2.796017462\ 10^{-25}\left(1.1236071\ 10^{7}\, k + 2.808293\ 10^{6}\right)\left(S\rho\left(9.869808093\ 10^{36}\ E\, Ie\, k^2 + 4.933630801\ 10^{36}\ E\, Ie\, k + 1.000000000 + 6.165447354\ 10^{35}\ E\, Ie\right)\right)^{1/2}\Big)$$

$$L := 4$$

%%%%%%%%%%%%%%%%%%%%%%%%%%%

$$\beta_1 := 1.963301156$$

$$yx := -0.02787489553 \sinh(0.9816505780\, x) - 1.000000000 \sin(0.9816505780\, x)$$

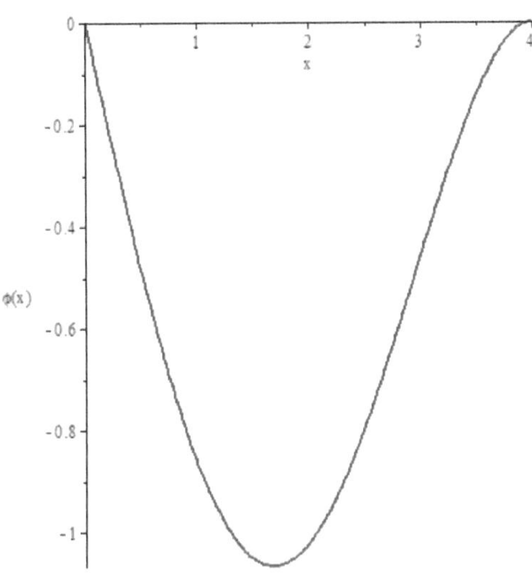

$$\beta_2 := 3.534291372$$

$$yx := 0.001204116527 \sinh(1.767145686\, x) - 1.000000000 \sin(1.767145686\, x)$$

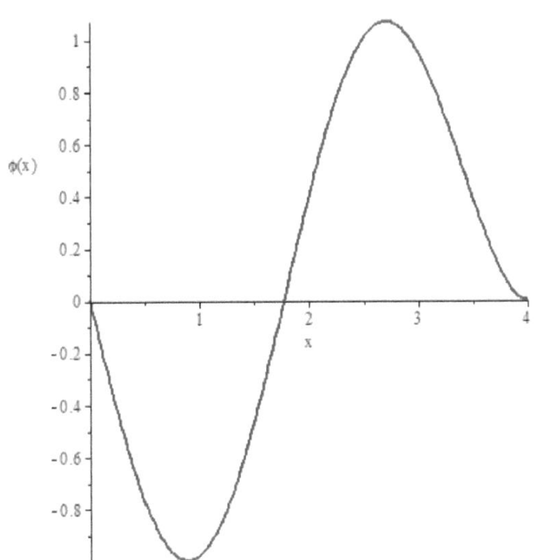

$$\beta_3 := 5.105088062$$

$$yx := -0.00005203455571 \sinh(2.552544031\,x) - 1.000000000 \sin(2.552544031\,x)$$

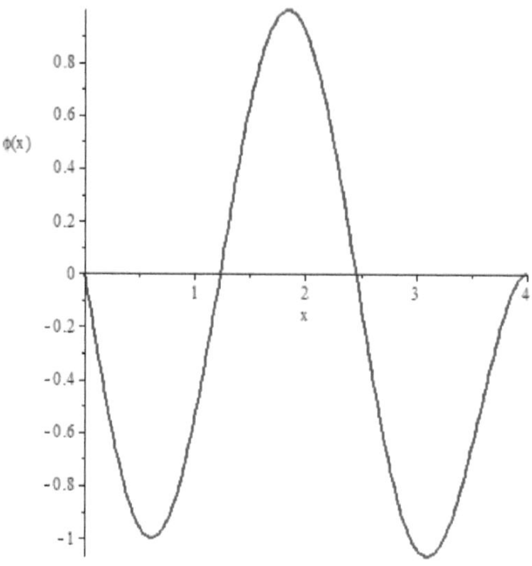

$$\beta_4 := 6.675884390$$

$$yx := 0.000002248617013 \sinh(3.337942195\,x) - 1.000000000 \sin(3.337942195\,x)$$

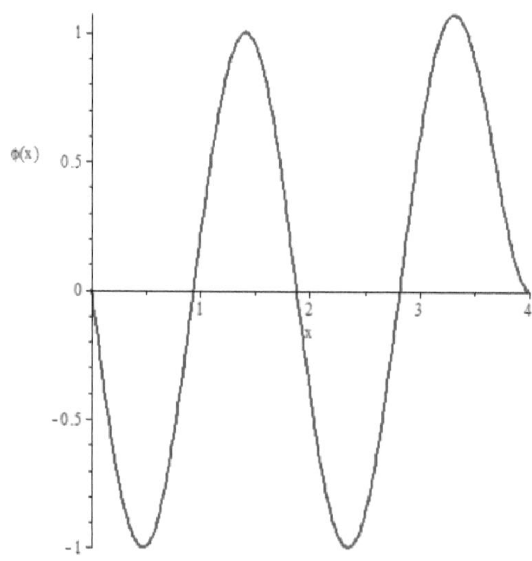

$$\beta_5 := 8.246680715$$

$$yx := -9.717155247\ 10^{-8} \sinh(4.123340358\, x) - 1.000000000 \sin(4.123340358\, x)$$

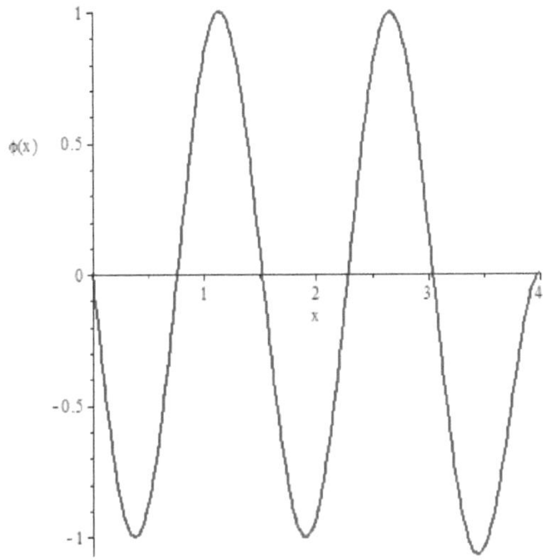

$$\beta_6 := 9.817477042$$

$$yx := 4.199163557\ 10^{-9} \sinh(4.908738521\, x) - 1.000000000 \sin(4.908738521\, x)$$

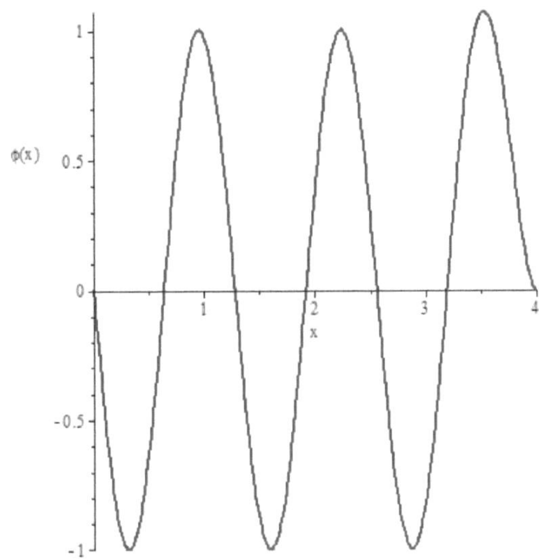

Index

N

O

P

R

S

T

V

W

www.ingramcontent.com/pod-product-compliance
Ingram Content Group UK Ltd.
Pitfield, Milton Keynes, MK11 3LW, UK
UKHW041431210726
13854UKWH00010B/1872